# Omics Technologies for Sustainable Agriculture and Global Food Security Volume 1

Anirudh Kumar • Rakesh Kumar •
Pawan Shukla • Manish K. Pandey
Editors

# Omics Technologies for Sustainable Agriculture and Global Food Security Volume 1

Springer

*Editors*
Anirudh Kumar
Department of Botany
Indira Gandhi National Tribal University (IGNTU)
Amarkantak, Madhya Pradesh, India

Rakesh Kumar
Department of Life Science
Central University of Karnataka
Kalaburagi, Karnataka, India

Pawan Shukla
Biotechnology Division
Seribiotech Research Laboratory (SBRL)
Bangalore
Karnataka, India

Manish K. Pandey
Center of Excellence in Genomics & Systems Biology (CEGSB)
International Crops Research Institute for the Semi-Arid Tropics (ICRISAT)
Hyderabad, Telangana, India

ISBN 978-981-16-0830-8 ISBN 978-981-16-0831-5 (eBook)
https://doi.org/10.1007/978-981-16-0831-5



This Springer imprint is published by the registered company Springer Nature Singapore Pte Ltd.
The registered company address is: 152 Beach Road, #21-01/04 Gateway East, Singapore 189721, Singapore

# Preface

Metabolic plasticity to adopt changing environmental conditions played a significant role during evolution and colonization of terrestrial plant from aquatic life. The extensive urbanization and industrialization in addition to several other environmentally non-friendly human activities in the last half-century have brought significant change in the climatic conditions, making the living conditions challenging for plants and animals. According to Intergovernmental Panel on Climate Change (IPCC) report, agricultural crops will face extreme climate changes such as increase in temperature, drought, flood, and increase in sea level that leads to intrusion of salt on arable land.

Climate change and its adverse impact on food production and quality are worrisome for the food and nutritional security of human civilization. Furthermore, in view of exponentially increasing global population in excess of 9.7 billion by 2050 and sharp diminishing natural resources, there is an untiring importunity to augment food production. We need almost 70% more food production than is consumed today. Achieving humongous goal of global food security will be possible only by improving the qualitative and quantitative traits of crops through exploitation and modification of metabolic pathways involving advanced analytical tools and technologies. We have emphasized the challenges ahead in the food and agriculture sectors in the face of climate change and global megatrend like increasing agricultural productivity with minimum inputs. Collaborative and transdisciplinary approaches for global food security with a focus on technological intervention from producer to consumer have been realized. In the last two decades, unprecedented progress has been made in the implementation of high-throughput omics technology in plant biology, which provides better understanding and support in developing climate-resilient technologies.

Omics technology provides a platform of different modern tools to ponder upon the integrative approach of omics to deal with complex biological problems. It touches multiple branches of biology. In the last two decades, omics platform has added many new tools and application and modernized itself. For instance, it started with genomics followed by transcriptomics, proteomics, and metabolomics. In addition, different high-throughput next generation sequencing (NGS) platforms have evolved different sequencing purposes, which are quick and cost-effective as well. With the passage of time, omics has made a significant contribution in the field of agriculture from crop breeding to gene editing to address the food security issues for sustainable crop production.

Genomics, transcriptomics, proteomics, and metabolomics are the four most important branches of omics technologies which imply genome, transcriptome, proteome, and metabolome quantification and characterization with extremely automated methods. These advancements in technologies disclose the genomic regions, expression of genes, proteins, and metabolites, which provide molecular insights into growth, development, resistance, and yield in changing environments. Genomics has opened new avenues and opportunities for crop development. The discovery of sequencing technology helped in unraveling the secrets of genome information in many plants like *Arabidopsis thaliana,* maize, rice, wheat, soybean, tobacco, canola, etc. which have changed the fate of agricultural science immensely. NGS has provided huge information about the genome of crop plants and led to the discovery of SNPs, microRNA, and QTLs which were earlier a time-consuming task.

Gene discovery and expression profiling of major crop species have given outre opportunities to plant breeders in terms of enhancing diversity in germplasm collection, introgression of useful traits from novel sources, and identification of key traits controlling genes. The genomics technologies have been found very beneficial in deciphering the mutagenicity in biotic and abiotic stress tolerance through genome sequence, stress-specific gene and transcript collection, protein and metabolite profiling, their dynamic changes, protein interactions, and mutant screening. As a result of high-throughput omics technologies and growing data, bioinformatics would help in analyzing in functionally relevant context.

Further, proteome study provides valuable information about the total proteins profile getting expressed under stress or any developmental changes of a cell/tissue/organ. Proteins profile helps in identifying the protein expression, posttranslational modification, and interaction with other proteins that regulate several cellular and biological processes of a living system. Proteomics study involves gel-based proteomics like 2D PAGE and gel-free proteomic for the detection of low complexity and low abundance proteins, respectively. Furthermore, quantitative proteomics techniques like iTRAQ have gained popularity in the recent past. These advancements in proteomics have broaden our understanding about the biological pathways and help in improving the traits of the plant. In fact, integration of proteome with genome sequencing and RNA-seq is being used for improvisation of genome annotation.

Further, metabolomics research has opened up a new avenue in identifying specific metabolites involved in plant development, fruit ripening, shelf life of fruits, nutrient contents of food crop, and aromatic/phenolics compound in the plants. One should not forget that secondary metabolites played a vital role in evolution of plant especially on colonization of terrestrial life. They help the plants to survive under different climatic conditions and pathogen attacks. Therefore, metabolomics research has widened our knowledge to pinpoint the metabolite involved in survival of plants under the adverse climatic conditions and required for sustainable production of crop. Metabolomics utilized advanced techniques like GC-MS, MALTI-

TOF-MS, ESI-MS, LC-MS, and MS-MS for metabolite profiling. These platforms have been used for the exploration of several new intermediate metabolites for rediscovery of existing pathways and the genes which encode them.

Furthermore, meta-omics techniques have also been utilized for understanding microbial system and their ecological function. Different meta-omics techniques like meta-genomics have been employed for identification of microbial community involved in plant disease tolerance, bioremediation, soil improvement, plant growth-promoting rhizobacteria (PGPR), plant–microbe interaction, virulence mechanism, and biofuel.

In addition, genome editing has emerged as a promising tool for introducing precise mutation in a gene to improve the plant performance. Genome editing techniques such as CRISPR/Cas have been tried in many agriculturally important crops like rice, wheat, maize, and tomato. Researchers and plant breeders are preferring CRISPR/Cas technique over other genome editing techniques due to its precision and easy to use. One of the attractive features of genome editing is generation of desire plants in an environmentally safe manner. Furthermore, few countries like U.S.A. and Canada have approved genome edited canola and mushroom. It is likely that genome editing might be instrumental in meeting the challenge of feeding 12 billion people by the end of the twenty-first century.

Precisely, this book comprehensively and coherently reviews the application of various aspects of rapidly growing omics technology including genomics, transcriptomics, proteomics, and metabolomics for crop development. It provides a detailed examination of how omics can help crop science and introduces the benefits of using these technologies to enhance crop production, genetic resilience, and other values to feed the increasing world population. It also provides platform to ponder upon the integrative approach of omics to deal with complex biological problems. The book highlights crop improvements such as yield enhancement, biotic and abiotic resistance, genetic modification, bioremediation, food security, etc. It explores how different omics technologies independently and collectively would be used to improve the quantitative and qualitative traits of crop plants. This book is also an asset to modern plant breeders and agriculture biotechnologists.

Amarkantak, Madhya Pradesh, India Anirudh Kumar

# Contents

# About the Editors

**Anirudh Kumar** PhD, is a research faculty member in the Department of Botany, Indira Gandhi National Tribal University (IGNTU), Amarkantak, MP, India. Over a 10 plus year career as a researcher in the area of plant molecular biology and plant pathology. He has received M.Sc. and Ph.D. degree from University of Hyderabad and post doctorate from Centre for Cellular and Molecular Biology (CCMB), Hyderabad, India and ARO Israel. His current research interests span from antioxidants studies of medicinal plants to plant pathology. He is author and co-author of several papers on different aspect of plant biology. He also teaches courses for B.Sc., M. Sc. and Ph.D. degree. For the past few years, his research group is trying to studies the phytochemical properties of native plants traditionally used by tribal healers of Amarkantak, MP, India. This book explores how the diverse omics technology autonomously and collectively would be used to expand the quantitative and qualitative traits of crop plants.

**Rakesh Kumar** is a plant biologist, Assistant Professor at Department of Life Sciences, Central University of Karnataka, India. He has received doctoral degree from University of Hyderabad-India, and an alumnus of International Crops Research Institute for the Semi-Arid Tropics, India. He has made significant contribution in the area of plant biotechnology, and published several articles in high impact journals, and received research grants from national and international funding agencies. His expertise includes functional genomics approaches such as TILLING, sequencing, transcriptomics, QTL-seq, proteomics, and LC- and GC-MS. This book provides in-depth information related to OMICS based crop improvement.

**Pawan Shukla** completed his postgraduate and doctorate degree from University of Hyderabad. He was also a Dr. D. S. Kothari post-doctoral fellow at University of Hyderabad. Currently, he is associated as Scientist with Central Silk Board and working at the biotechnology division of Seribiotech Research Laboratory (SBRL), Kodathi, Carmelram Post, Bangalore, Karnataka. He also worked at Central Sericultural Research and Training Institute (CSR&TI), Central Silk Board, Pampore, Jammu and Kashmir (UT), India. His laboratory is working on the development of cold tolerant mulberry variety using modern omics tools. He has published several research papers in reputed International and National Journals and editor in 03 books published by CSR&TI, Central Silk Board Pampore.

**Manish K. Pandey** completed his PhD from Osmania University, India and post-doctoral from University of Georgia/USDA-ARS, USA. He is currently leading the Groundnut Genomics and Trait discovery group at International Crops Research Institute for the Semi-Arid Tropics (ICRISAT), Hyderabad, India. His research interest includes reference genome, cost-effective high throughput genotyping assays, sequencing-based gene(s) discovery, diagnostic markers, genomics-assisted breeding and genomic selection. He is also Adjunct Associate Professor in University of Southern Queensland, Australia. He has also been selected as Associate Fellow/Member in two Indian national (NAAS and NASI) and two state (Telangana and Andhra Pradesh) science academies of India.

# Recent Advances in OMICS Technologies 1

Vikas Gupta, Satish Kumar, Disha Kamboj, Chandra Nath Mishra, Charan Singh, Gyanendra Singh, and Gyanendra Pratap Singh

**Abstract**

Omics is a collaborative, wide-ranging science primarily concerned with the study of the relationships of biological knowledge resulting from genome profiling, transcriptomes, proteomes, metabolomes, and a variety of other related omes. These high-throughput approaches generate huge information related to differentially expressed genes/proteins/metabolites. The interrelationship between different omics studies and high-dimensional biological data produce from these omics are analysed through statistical methods. These studies have helped to create new resistant varieties by elucidating metabolic responses of crop plants to different biotic and abiotic stresses. Interpretation of omics research has enhanced biological study resolution, which contributed to analyses of metabolic pathways and biological system research. However, it is important for the creation of strategy to work on problem-oriented and process-oriented objectives leading to crop improvement to proceed to organised disciplines such as structural genomics, transcriptomics, proteomics, and metabolomics with plant physiology, biochemistry, and plant breeding.

**Keywords**

Omics · Genomics · Proteomics · Metabolomics · Transcriptomics

V. Gupta · S. Kumar (✉) · D. Kamboj · C. N. Mishra · C. Singh · G. Singh · G. P. Singh
ICAR—Indian Institute of Wheat and Barley Research, Karnal, India



A. Kumar et al. (eds.), *Omics Technologies for Sustainable Agriculture and Global Food Security Volume 1*, https://doi.org/10.1007/978-981-16-0831-5_1

## 1.1 Introduction

Food supply is a requirement for food security, and is an important factor to reduce hunger. The agriculture industry has shown significant improvements over the past 50 years with respect to production and productivity enhancement, dominated by the cereals which form the major staples across countries. The twentieth century has seen tremendous increase in food production with the introduction of high yielding crop varieties, especially since the first 'green revolution' thus saving millions of people from starvation (Rajaram 2005). This has been achieved mainly because of development and adoption of high yielding cultivars, increased area under irrigation, and widespread use of fertilisers. However, the present crop yield is not sufficient enough to feed the 9 billion of global population by 2050 (Ray et al. 2013). In addition, global climate change has further worsened this scenario with altered rainfall patterns, extreme weather events, and changing patterns of pathogens and pests in terms of severity and distribution (Abberton et al. 2016). With population rising gradually and natural capital shrinking, the production of food must be improved that can be accomplished with the use of modern analytical methods and technology to increase the quality and quantitative improvement of crop plants (Agrawal et al. 2013; Nelson et al. 2014; Haddeland et al. 2014).

Cultural change relies on current or created genetic variation, and the genetic benefit rate in breeding systems will increase either by expanding the quantity of variation available for selection or by speeding up the selection process so that varieties can be developed more quickly. During the last few decades, the development and use of molecular markers in crop genetics led to the identification and mapping of genes and QTLs controlling key traits. The marker system initially started with the use of restriction fragment length polymorphisms (RFLPs) followed by RAPDs, SSRs, AFLPs and culminated with the SNP markers based on genome sequencing (Tanksley et al. 1989; Varshney et al. 2009). The SSR markers have been used extensively for linkage map construction, tagging of new genes/QTLs due to their abundance in the crop genomes as well as because of their high polymorphism. (Somers et al. 2004; McCouch et al. 2002; Varshney et al. 2007; Smith et al. 1997; Sharma et al. 2013). Use of SSR markers has few limitations like SSRs are not uniformly distributed in genomes, gel based assay and cannot be integrated across platforms, which limit its potential to be exploited in plant breeding programmes.

With the development of low-cost sequencing technologies and genotypic platforms, there is an exponential increase of plant sequences in databases and enhanced our understanding of molecular and physiological roles of genes. Plant genomics includes both structural and functional genomics, which covers both basic and applied aspects. Given the rapid evolution of novel technologies, especially the advent of bioinformatics, the understanding from gene to genome level has increased. The new developments in the 'omics' technologies have brought revolution in plant science research due to the next-generation sequencing (NGS) 454, Solexa Illumina, Applied Biosystems (ABI) SOLiD, and high-throughput marker genotyping technologies. Consequently, the availability of complete genome sequences for different crop plants holds great promise for research targeted towards crop improvement.

Many different functional genomics approaches have been used to identify genes that linked to the key biological processes. The crop traits are generally complex quantitative traits, which are controlled by multiple genes involving complex genic interactions (Core et al. 2008). Omics is a collaborative, broad-based field that focuses on the study of bio informative interactions obtained from genome profiling, transcriptomes, proteomes, metabolomes, and a variety of other related omes. The three main omic technologies for the analyses of genomics, proteomics, and metabolomics are the genetic, proteome and metabolism analyses of organisms' cells and tissues (Varshney et al. 2013). In this chapter we will discuss these technologies in detail and how these can be integrated in breeding programmes for better understanding of the functioning of key genes.

## 1.2 Genome and Genomics

The term genome refers to the entire genome compliment contained in a haploid chromosome set of any organism and the term was coined by German botanist Hans Winkler in 1920 at the University of Hamburg (Vukmirovic and Tilghman 2000). Genomics is the study of complete DNA sequence of organisms. It has been proven as one of the most transformative impacts on biological studies. Genomics study involves recombinant DNA technology, DNA sequencing, assembly, and annotation of genome to decipher the structure, function, and evolution of genes.

The understanding of genome sequences of organisms is crucial for unravelling the roles and networks of genes, for defining evolutionary relationships and processes, and to discover previously unaware control pathways that coordinate gene operation. Genomics has become one of the most transformative forces in biological sciences, the study of the whole organism's DNA sequence. Genomics is the genetic characterisation and cloning technique of entire genomes to explain gene structure, function, and evolution. The genetic material in eukaryotic organism is deoxyribosenucleic acid (DNA), which is a double stranded molecule composed of nucleotide bases, viz., adenine (A), guanine (G), cytosine (C), and thymine (T). The sequence of nucleotides in a gene on a DNA strand determines the sequence of amino acids that make up a protein.

For protein synthesis, the DNA in a gene is first transcribed to messenger RNA, followed by its translation into a sequence of amino acids through ribosomes in the cytoplasm of the cell. The proteins and products are fundamentally responsible for all cellular mechanisms. Finding out the pattern of arrangement of nucleotide bases in the entire genome is called genome sequencing. Time line for key discoveries in molecular biology has been depicted in Table 1.1 that led to the enormous data generation and formed a basis for better understanding of the key functions of the genomes. In 1972, gene sequence for bacteriophage MS2 coat protein was first determined by Walter Fiers and his team at the Laboratory of Molecular Biology of the University of Ghent, Belgium. The first DNA-based genome of bacteriophage

**Table 1.1** Timeline showing key discoveries in DNA Sequencing

| S. no. | Year | Discovery |
|---|---|---|
| 1 | 1977 | Sanger's chain termination method<br>Chemical cleavage Maxam Gilbert's method |
| 2 | 1980 | Non-radioactive sequencing |
| 3 | 1986 | First semi-automated DNA sequencing machine |
| 4 | 1987 | ABI 370 fully automated sequencing machine |
| 5 | 1990 | Base by base sequencing |
| 6 | 1995 | ABI prism 310 genetic analyzer |
| 7 | 1996 | Pyrosequencing |
| 8 | 2000 | Massively parallel signature sequencing (MPSS) |
| 9 | 2003 | SMRT [single molecule real-time sequencing] |
| 10 | 2004 | ABI 3130 Genetic Analyzer |
| 11 | 2005 | First NGS machine Roche 45-4GS-20 |
| 12 | 2006 | Solexa/Illumina sequence analyzer |
| 13 | 2007 | Second commercial NGS platform ABI SOLID sequencer |
| 14 | 2008 | Third commercial NGS platform 454 titanium GAII |
| 15 | 2009 | Illumina GAIIx SOLID 3.0 |
| 16 | 2010 | Illumina Hi Seq2000<br>Ion Torrent |
| 17 | 2011 | Pac Bio RS sequencer<br>CMOS non-optical sequencing MI Seq |
| 18 | 2012 | Ion proton |
| 19 | 2014 | Nanopore sequencer |
| 20 | 2015 | Sequel system (Pac bio) |
| 21 | 2016 | Oxford nanopore mini ion sequencer |

Φ-X174; (5368 bp) was sequenced by Frederick Sanger (Sanger 1977). Since then genomes are being sequenced at a rapid pace ranging from viruses, human, bacteria, fungus, and many crop plants. Scientists, through a variety of functional genomics approaches, are characterising the genes that control key processes. The crop genome sequences are valuable for the discovery and insight into genetic diversity by resequencing of various genomes. The positional cloning and seed breeding promote sequence dependent markers linked to the rare elite alleles.

Affymetrix (Santa Clara, CA, USA; www.affimetrix.com) and the Illumina (San Diego, CA, USA; www.illumina.com) BeadArrayTM technology are commonly used for microarrays from GeneChipTM (GeneChipTM) and chip related technologies. Some recently built commercial platforms such as Affymetrix ® EurekaTM, and Illumina's Infinium also rely on SNP markers of high density. Furthermore, plant breeding and agricultural biotechnology have been revolutionised by the micro-array-based characterisation of plant genomes.

Ganal et al. (2011) reported a uniform large-scale SNP genotype array of over 8,00,000 SNPs, which were spread evenly throughout genome of maize. In addition, competitive PCR (KASP), TaqMan real-time PCR, and STARP experiments use low-cost PCRs to detect SNPs and provide versatility with regard to the number of samples to be analysed (He et al. 2014; Semagn et al. 2014; Long et al. 2017). Diagnostic markers may be used without further confirmation, developed on target gene sequences, to improve application of marker-assisted selection (MAS) in breeding. The production of diagnostic markers is rapidly assisted by genome sequencing/resequencing. Such markers allow plant breeders to precisely produce new crops with targeted characteristics in future to satisfy different food requirements.

Most significant agronomic and economic characteristics in culture write traits in cultural breeding methods are several QTL regulated quantitative characteristics. Therefore, it is of considerable significance for marker-assisted breeding to explore QTL or even underlying causal genes/alleles. Linkage mapping is a standard approach for the genetic dissection of the quantitative trait loci genetic foundation (Emebiri et al. 2017; Liu et al. 2017; Zhang et al. 2017). To date, this approach has detected a significant number of QTLs. Moreover, the meta-QTL study was developed by combining QTL published to classify reliable QTL for crop enhancement with the achievement of biological informatics and large genetic knowledge (Van and McHale 2017). The extensive mapping of genomes in large populations makes it easier to map and clone target genes in crops.

Due to the available high-density SNP markers, the GWAS (Genome-Wide Association Studies) analysis, which includes natural populations, opened a path. The GWAS analysis will solve a range of traditional mapping shortcomings and deliver a powerful alternate technique for the dissection of complex features. By bringing together high-performance phenotypic and genotypic results, GWAS provides inspections of maize's complex characteristic genetic architecture, particularly in view of the rapid loss of corn imbalance (Yan et al. 2011).

Genome selection (GS) estimates the genomic estimated breeding values (GEBVs) of lines by a study of traits and high-density marker values in an artificially generated whole-genome population (Meuwissen et al. 2001; Crossa et al. 2017). GS has shown as important breeding method to develop dynamic characteristics easily. Closer relationships, including polygenic features, between the genomic estimate and true breeding values have been found (Jia and Jannink 2012). Even if also expensive, the number of QTLs that regulate a trait has proven superior to the recurring collection of markings for changes in complex characteristics in crops, so it can efficiently prevent problems.

## 1.3 Transcriptomics

Transcriptome is the study of total RNA, which gets edited and becomes mRNA, and finally translated into proteins. It involve the analysis of expression profiling of mRNAs in a given cell population.

The transcriptome is complex, since it is mainly a representation of the genes actively expressed at any given time under different circumstances. It assesses the changes in gene expression pattern due to inner and outer factors such as biotic and abiotic stress. It is an important method for understanding responses of biological system under different circumstances. Transcriptome research offers a crucial forum for the review of the connexions between genotype and phenotype, which provides a deeper understanding of the fundamental pathways and mechanisms involved in the regulation of the cell fate (Ruan et al. 2004).

Transcriptomic approaches such as next-generation sequencing (NGS) facilitate the comprehension of functional components of genome. From past two decades, a number of agricultural researchers have made use of advances NGS techniques that enabling them to study high-resolution linkage between gene variants and traits of interest resulting in an upsurge in the scope of transcriptomic studies, not only in number but also in the range of agricultural systems that are being studied (Imadi et al. 2015).

Microarrays, serial analysis of gene expression (SAGE), massive parallel signature sequence (MPSS), and next-generation sequencing platforms (NGSPs) are typical technologies applied for genome wide or high-throughput genetic expression analysis (Costa et al. 2010; Harbers et al. 2005; Mardis et al. 2008). High-performance, quantitative PCR is used to confirm the expression of a broad variety of genes that are elucidated by tissue omics and choice care (Czechowski et al. 2004).

### 1.3.1 Differential Display

Differential Display technology is a PCR based technique for analysing gene expression using an Oligo-dT (3′) and an arbitrary 5′-end oligonucleotide primers (Liang and Pardee 1992). In order to achieve good visualisation, all RNA fingerprint variants need a low annealing temperature during PCR amplification. Therefore, not only the initial concentration of the cDNA depends on the quantity of the actual amplification components, but also the consistency of a precise match between primer and prototype (McClelland et al. 1995).

Throughout the course of PCR amplification, plentiful cDNAs with bad matches to the primers used would possibly outperform rare species with perfect matching. Despite this drawback, the classification of genes expressed distinctly by plants is recorded in many ways. This include the cloning of the last (1-aminocyclopropane-1-carboxylate-oxidase) enzyme involved in the synthesis of ethylene, a hormone which is involved in tomato ripening (Barry et al. 1996); and senescence, particular genes involved in the synthesis of flavonoids (Saito et al. 1999; Yamazaki et al. 1999); genes expressed differentially during carpel growth (Yung et al. 1999) or floral transformation (Yu and Goh 2000); module such as plant/environmental interaction based nitrate conveyors (Filleur, and Daniel-Vedele 1999); light photoreceptor-regulated genes, phytochrome (Kuno et al. 2000), and abiotic stress-response proteins (Brosche and Strid 1999; Baldi et al. 1999; Kim et al. 2000). This method is also used to differentiate within a species between individuals (Lapopin et al. 1999; Ni et al. 2000).

### 1.3.2 cDNA-AFLP

This fingerprinting approach depends on the sub-set of DNA molecules from a diverse pool being selectively amplified. The creation of an amplified polymorphic fragment length (AFLP) gives a consistent means with which two similar individuals can distinguish the few differences (Vos et al. 1995). The method is based on highly stringent conditions that are supported by the insertion of dual strand adapters at the ends of limit fragments that act as primary sites during amplification. Selective fragment enhancement is obtained by adding one or two foundations to the PCR primers and is only successfully improved by a fragment flanked by the limiting site, minimising the amount of visualised band.

More than 260 studies presented the advantages of this technique for exposing DNA fingerprinting. As cDNA was used to amplify transcripts in plants due to the benefit of this approach in producing a good individual polymorphism. As a tool for producing differently expressed products, the use of any pair of four and six simple restricting enzymes was suggested and successfully used as a combination of Pst I/Mse I and Asen I/Taq I (Bachem et al. 1996; Money et al. 1996). This technique seems more effective than that to recognise the individual RNA fingerprint and discern it; it is possible to quickly clone and subsequently characterise a certain band, particularly if silver stain is used to establish the fingerprint pattern. There have been accounts of definitions of many ESTs (Suarez et al. 2000; Durrant et al. 2000; Bachem et al. 2000).

### 1.3.3 DNA Microarray

A high efficiency, most frequently used technique for gene expression analysis, and a central factor in today's practical genomics research (Aharoni and Vorst 2001) is the DNA microarray (also referred to as a DNA chip or microarray). In short, this procedure involves the extraction, first of all by means of a laser-capture micro-dissection (LCM) technology of individual cells from two or more biological samples, and then the extraction from cells captured of complete RNA. The copies of each RNA that has been expressed are multiplied. Thus generated RNAs (and hybridised into a DNA chip under machine control be radioactively labelled or labelled with fluorescent colours) are engineered to visualise which genes are transcribed into the RNA in the cell when the sample is taken.

For any colouring position of the chip, scanners are used to read signals and fluorescence measurements are made. For data retrieval and interpretation (http://www.parisdevelopment.com), advanced applications and data collection systems are then used. A variety of plant procedures including the circulation clock, plant protection, environmentally stressful responses, fruit maturing, phytochrome signalling, seed growth, or nitrate assimilation have been investigated using microarray technology (Aharoni and Vorst 2001).

### 1.3.4 Serial Analysis of Gene Expression (SAGE)

The key difference between DNA hybridisation and SAGE approaches is that the latter does not need previous sequence knowledge to be studied, as SAGE is a gene expression profiling technique focused on sequencing (Velculescu et al. 2000). SAGE can be used to achieve full transcriptional profiles of expressed genes, although unknown, for species with poorly defined and expressed sequences. A new modification of SAGE known as Long SAGE makes it possible to use the derived transcriptome in the annotation of genetic expression in the genome (Saha et al. 2002). SAGE is a genuinely global and neutral technique for gene expression. The SAGE procedure uses several enzyme steps, amplifying and cleaning PCR and cloning. The SAGE protocol begins with mRNA cleaning connected with solid-phase magnetic oligo(dT) beads. The cDNA is synthesised directly on the oligo (dT) bead and then digested with the anchoring enzyme NlaIII (AE) to reveal the 3′ most restriction site on the oligo(dT) bead. The 4-bp site-detection enzyme NlaIII is expected to take place per 256 bp and thus to be present on most mRNA organisms in most SAGE experiments. However, a second SAGE library can be used for the identification of transcripts without a NlaIII site with a separate anchoring enzyme and also for the reconfirmation of a transcript identity in both sites. This could decrease the work relevant to research substantially, but the marginal importance of this method remains to be seen. Next, the sample is also divided into two separate tubes and attached to two separate connectors, A and B. The BsmFI, a type IIS restrictive enzyme that reduces 10 bp 3′ from the position of anchoring enzyme recognition, is recognised by both linked instruments. A special oligonucleotide known as the SAGE tag, hence known as the Tagging Enzyme (TE) is produced by BsmFI. The SAGE tags released from the beads of oligo (dT) are then removed and binding to each other for the purpose of making the ditag was amplified by the help of an electronic sequencer (Patino et al. 2002) and released from the ligators. The gel is purified, serially ligated, clone and sequenced.

## 1.4 Proteomics

Proteomics includes the recognition and characterisation of the full collection of genome proteins (Wilkins et al. 1996) sufficient for the preservation of the structure and maintenance of important regulatory functions (Whitelegge 2002). Proteins are huge amino acid molecules, each with a special genetically-defined series that specify the basic shape and role of the proteins in various ways, to form tens of thousands of proteins. Proteomics analyses the sequence of amino acids and various post-translation modifications to determine their relative abundance (Brygos and Joyards 2004). A cell, tissue, or organism's protein is the complete complement of proteins expressed by its genome at any given time (Renaut et al. 2006).

It includes systematic study of proteins expressed by a genome, from their amino acid sequence to relative quantities assessment, post-translocation transition state and the interplay with other proteins or molecules of various forms to protein function and structure classification (Brygoo and Joyard 2004; Rhee et al. 2006). It starts with functionally modified proteins and the corresponding gene sequences, as opposed to genomics.

Proteomics is simply a method for plant functional genomics. The proteomics research leads to understanding diverse biological mechanisms and cellular response to environmental stress. It has emerged as an effective method to understand the role of proteins that establish homeostasis in cells, participate in cell signals, and are needed to maintain the structure. Protein-level studies are thus important to demonstrate molecular mechanisms that underlie plant growth, production, and environmental interactions (Chen and Harmon 2006). These are particularly compatible with crops because they can provide insights, not only in terms of nutritional worth, but also in terms of yield and the impact of adverse conditions on these factors (Solekdeh and Komatsu 2007).

Many emerging proteome approaches are available, each seeking to increase isolation, resolution, and automation in conjunction with various experimental objectives and with the chemical and physical properties of the target proteins. The significant technology used during research is the identifying and characterisation of 2-D polyacrylamide gel electrophoresis of separated proteins (Klose 1975; Fenn et al. 1989; Kara et al. 1989) and Mass Spectroscopy (Fenn et al. 1989; Kara et al. 1989) for the identification and characterisation of separated proteins. 2-DE is based on isoelectric concentration, where the proteins are segregated in polyacrylamide gels pH gradient (first dimension) and in the SDS (sulphate sodium dodecyl) PAGE according to their pI (second dimension) (Klose 1975; Klose and Kobalz 1995) and are separated according to their molecular weight.

The visualisation is accomplished with the use of staining methods of isolated protein spots. Electron-spray-ionisation (ESI) is carried out to transform peptides into ions by transmitting them through high-voltage columns. Proteins within the points of interest are first detected by digesting them into peptides usually with trypsin and then analysed by the mass spectrometry (MS). Flight Time (TOF) is an operation that analyses the mass of peptide ions in mass spectrometry. The MALDI TOF (Matrix-assisted laser desorption and ionisation time-of-flight) technique is the most widely used MS technique. These methods are used to trace peptides to classify protein and its complex interaction (Kersten et al. 2002; Henzel et al. 1993).

An alternative approach for analysis of protein products directly by MS, without isolation of gel, has been developed that enables the automatic analyses of peptide mixtures that are formed from complex protein samples, to be carried out using a capillary, high-performance liquid chromatography pair (HPLC), MS or MS/MS (Appella et al. 1995; Washburn et al. 2001, in French). In addition, the ground breaking reagent known as isotope encoded affinity tag (ICAT) in the LC-MS/MS method is made possible for quantitative proteomics. Yeast two-hybrid (Y2H) assay helps in detection of the protein–protein interactions (Fields and Song 1989). The large-scale protein-interaction map was developed using Y2H systems, which were used in the study of signalling complexes (Chen and Harmon 2006). These maps are also available. This genetic technique enables the speedup detection and separation of the relationship between in vivo protein–protein interactions of the related nucleic acid sequences. In holding the intracellular balance, the relationship between protein and protein is important (Ozbabacan et al. 2011). The study of proteomics is therefore important for understanding the dynamics of complex biological processes. Proteomic analytics are becoming a good method for tracking growth and for obtaining insight into the molecular workings of plants. Attempts were made to improve the photosynthesis and abiotic stress resistance in crops. C4 plants were found to possess two forms of chloroplasts, making them more energy-efficient. A comparative proteomic analysis of C3 and C4 plant chloroplast was carried out to determine the proteins that cause more effective light fastening (Zhao et al. 2013).

In Arabidopsis the application of proteome analysis resulted in better understanding of the complex cellular events while analysing the role of gibberellic acid (GA) during the initial stages of seed germination and the effect of scarification on seed germination. During the production of food, after proteome review, a number of previously unknown, novel enzyme encoding genes were found in maize. The genetic diversity at the level of expressed proteins has been extensively tested with proteomics (Canovas et al. 2004). Genetically heterogeneity at the level of the expressed proteins has been commonly applied to proteomics (Canovas et al. 2004). The male sterility proteins fundamentally required for the selection of hybrids have been identified by Wang et al. (2013). The conditions of heat stress influence crop plant growth and yield. The organ-specific proteomic trial was conducted in various crop plants (Hossain and Komatsu 2013) to determine proteins directly acquired during abiotic stress. The proteomic information has been extended to the development of desired transgenic plants (Gong and Wong 2013).

## 1.5 Metabolomics

The quality and derivative products of crop plants depend directly on their metabolites. The taste, fragrance, colour and texture of crops, their storage properties, and their field efficiency are determined by metabolite concentration (Memelink 2005). A plant's metabolite material is its metabolome. It is a technology for comprehensive analysis which identifies and quantifies all the metabolites of an organism at a particular time (Fiehn 2002). Building on their existence at the developmental stage, metabolites are categorised as primary and secondary.

The metabolome is very diverse: lipid soluble chemical usually present in membranes, acidic and simple ions, acidic and basic chemicals for aqueous cell elements, stable oxidisation systems and structures (Maloney 2004). This metabolome is very diverse. The plant kingdom contains an estimated 100,000–200,000 metabolites (Oksman-Caldentey and Inze 2004). A broad view of the biochemical status, which can then be used to track and determine gene expression (Fiehn et al. 2000), is provided by quantitative and qualitative measurements of this vast number of cellular metabolites. In several plant resistance and stress reactions the metabolites are stated to be working (Bino et al. 2004).

The organism's biochemical reaction to conditional disruption is characterised by the impact it has on differential aggregation of each metabolite (Raamsdonk et al. 2001). As metabolites reflect catabolic and anabolic behaviour by proteins at a given time (Maloney 2004), macro-molecular processes such as feedback inhibition and signalising molecules (Dixon et al. 2006) are increasingly understood as modulating (metabolite). It originated as a practical genome approach to help us understand the diverse molecular dynamics in biological processes (Hall et al. 2002).

In contrast with genes and proteins, metabolites are much more variable in the order for atoms and subgroups. The invention of new methods such as mass spectrometry (MS), nuclear resonance magnetism (NMR), and four-transform infrared spectroscopy (FTIR) makes the study of metabolites much simpler. The study of metabolomics relies on the sample being extracted at a certain point in time (Hall 2006). The techniques of metabolomics give us an insight into the diverse metabolism and diversity of the organic reactions involved in plant growth, differentiation, stress tolerance, and pathogens defence mechanisms. The performance of the metabolome analysis depends on a few important factors, such as biological material/sample production and the processing of samples/metabolite detection (Hall 2006).

Bino et al. (2004) proposed a metabolomics research (MIAMET) protocol that includes experimental design; sampling; planning, extraction, and deriving of metabolites; design and design of metabolite profiles; and calculation of metabolites and parameters of metabolites. Because of complicated metabolic data for complex analytics, various analytical methods have been developed. These methods include target analysis, profiling of metabolites, metabolomics, and metabolism. Metabolite target analysis requires a variety of approaches for the preparation and analysis of samples for one or a few compounds from diverse mixtures. Samples based on genotype, phenotype, or biologic significance are tested by metabolic fingerprinting on a variety of metabolites (Shanks 2005).

Metabolomics in the strict sense is the calculation of all metabolites in a given organism. It is not yet theoretically feasible, and would possibly require platform of complementary technologies, since no single methodology is systematic, selective, and efficient enough to quantify them all (Weckwerth 2003). Technology for the study of target compounds and metabolic profiling is such as gas chromatography (GC), high-performance liquid chromatography (HPLC), and NMR. These methods focus on chromatographic isolation, often paired with sophisticated analytical calibrations. Samples are studied using NMR, direct injection mass spectrometry (MS), or FTIR spectroscopy for metabolic fingerprinting as crude extracts without separation. Often paired with multivariate analyses, these fingerprinting methods get the best out of the results.

In order to produce metabolic profile for study of different stress, economically significant characteristics, and heredity results, studies in metabolomics were carried out in the agricultural sector. Seasonal variations, geographical area, and natural variation have all been successfully used to determine the effect of metabolic profiles. The metabolic profiles of transgenic plants have been studied in order to characterize them. Metabolomics allows researchers to classify and distinguish genotypes and phenotypes based on metabolic levels in a non-biased way. This information will be important for tracking crop quality characteristics (Hall et al. 2005). The detection and enhancement of trait production in farm and biological refining products (Reid et al. 2004) will identify possible biological markers for commodity contaminations and adulteration (Dixon et al. 2006).

Plant properties can be enhanced in different ways, for example, through raising metabolic fluxes by metabolic engineering to useful biochemical routes (e.g. increasing food's nutritional value, reducing pesticidal or fertiliser requirements, etc.) or through utilising the required routes for pharmaceutical development in plants (Giddings et al. 2000). In the same way, the incorporation of a foreign group of enzymes that contribute to the creation of desired finished products from other or more upstream precursors can lead to metabolistic shortcuts, and the synthesis of new metabolites may also generate foreign enzymes (Bino et al. 2004; Memelink 2005). Metabolic fluid analyses helped to define the similar catalytic properties of orthologous enzymes. Metabolistic experiments to classify the growth profile, production, and chemical taxis research were also conducted. Biochemical network construction has been carried out by creating relative metabolite profiles (Mendes 2002). The incorporation of metabolome and transcriptome evidence indicates that regulatory networks are being recognised and genetic material is correlated with phenotypic characteristics (Urano et al. 2009).

## 1.6 Bioinformatics

Over the past 15 years, bioinformatics has steadily developed and has become a modern major field in biology. A vast range of genetically and genome dependent results have been created using next-generation sequence technology to produce massive genetic and genome data, too detailed to be processed by the human brain. Bioinformatics includes one discipline: genetics, computer science, and IT (Rhee et al. 2006; Vassilev et al. 2005). It uses specialised tools and computers to search and interpret datasets from gene sequences and other sources. The National Science Foundation (NSF) began a project following the genome sequence of *Arabidopsis thaliana* to determine the function of 25,000 genes of Arabidopsis (Ruperao et al. 2014). Rice was the first genome-sequenced crop (International Rice Genome Sequencing Project 2005), and subsequently a variety of large crop genomes were sequenced. A vast number of data were published from all these sequencing initiatives. A series of bioinformatics instruments have been developed to organise and interpret these data that have significantly contributed to the drawing of significant biological conclusions, gene function predictions, etc. In addition, it allows researchers to define quantitative trait loci characteristics (QTLs) in the production of unorthodox mapping communities and online marker tools.

For the DNA sequencing research, transcriptomic, proteomics and metabolomics, multiple bioinformatics instruments and databases were developed and different researchers have given a detailed list of the same for separate purposes (Bino et al. 2004; Varshney et al. 2005; Rhee et al. 2006). These databases formed the foundations for generating theories, developing tests, and inferring knowledge of a single organism. Furthermore, various species have used datasets and omic tools to evaluate the 'omic' properties of species which allow for further analysis of preserved genes and evolutionary relationships. The role of bioinformatics in gluing fundamental science to applied research will continue to be important, and biotechnology is crucial in solving the pressing issues of society such as sustainable energy production, reduction of world malnutrition, hunger and protection of the environment (Rhee et al. 2006).

## 1.7 Conclusion

The genomics, proteomics, and metabolomics technology platform is extremely successful. They significantly increase the number of proteins/genes which can be concurrently identified and are able to connect complex mixtures in gene/protein expression profiles with complex results. Constitute infrastructure for the implementation of genomics information and techniques (and other 'omics') was considerably advanced to allow alterations in crop plant characteristics to boost real and potential production, increase resource use efficiency, and improve health in the crop system. The need for developing new crop varieties with higher yield, heat resistance, and lower use of pesticides is to meet the demands of the growing population, improvements in climate trends, and environmental stress.

High-dimensional biological data have been produced by advanced technologies, analysed using statistical methods to detect interrelationships within omics studies. In order to grow new resistant varieties, the study of metabolic responses of cultivars against different biotic and abiotic stresses will be helpful. However, over time, stage, and environmental factors the transcriptome, proteome, and metabolomic data have become extremely variable and thus the processing and interpretation of samples needs to be conducted with great caution. The understanding of omics also enhanced the resolution of biological experiments, which led to analyses of metabolic pathways and biological system research. Disciplines including structural genomics, transcriptomics, proteomics, and metabolomics are also required for the production of strategy for problem-oriented and process-oriented processes in plant physiology, bio-methods, and plant breeding.

## References

Abberton M, Batley J, Bentley A, Bryant J, Cai H et al (2016) Global agricultural intensification during climate change: a role for genomics. Plant Biotechnol J 14:1095–1098

Agrawal GK, Sarkar A, Righetti PG, Pedreschi R, Carpentier S et al (2013) A decade of plant proteomics and mass spectrometry: translation of technical advancements to food security and safety issues. Mass Spectrom Rev 32:335–365

Aharoni A, Vorst O (2001) DNA microarrays for functional plant genomics. Plant Mol Biol 48:99–118

Appella E, Padlan EA, Hunt DF (1995) Analysis of the structure of naturally processed peptides bound by class I and class II major histocompatibility complex molecules. EXS 73:105–119
Bachem CW, Van Der Hoeven RS, De Bruijn SM, Vreugdenhil D, Zabeau M et al (1996) Visualization of differential gene expression using a novel method of RNA fingerprinting based on AFLP: analysis of gene expression during potato tuber development. Plant J 9:745–753
Bachem CW, Oomen RJF, Kuyt S, Horvath BM, Claassens MM et al (2000) Antisense suppression of a potato alpha-SNAP homologue leads to alterations in cellular development and assimilate distribution. Plant Mol Biol 43:473–482
Baldi P, Grossi M, Pecchioni N, Vale G, Cattivelli L (1999) High expression level of a gene coding for a chloroplastic amino acid selective channel protein is correlated to cold acclimation in cereals. Plant Mol Biol 41:233–243
Barry CS, Blume B, Bouzayen M, Cooper W, Hamilton AJ et al (1996) Differential expression of the 1-aminocyclopropane-1-carboxylate oxidase gene family of tomato. Plant J 9:525–535
Bino RJ, Hall RD, Fiehn O et al (2004) Potential of metabolomics as a functional genomics tool. Trends Pl Sci 9:418–425
Brosche M, Strid A (1999) Cloning, expression, and molecular characterization of a small pea gene family regulated by low levels of ultraviolet B radiation and other stresses. Plant Physiol 121:479–487
Brygoo H, Joyard J (2004) Introduction – focus on plant proteomics. Plant Physiol Biochem 42:913–917
Canovas FM, Dumas-Gaudot E, Recorbet G et al (2004) Plant proteome analysis. Proteomics 4:285–298
Chen S, Harmon AC (2006) Advances in plant proteomics. Proteomics 6:5504–5516
Core LJ, Waterfall JJ, Lis JT (2008) Nascent RNA sequencing reveals wide spread pausing and divergent initiation at human promoters. Science 322:1845–1848
Costa V, Gallo MA, Letizia F et al (2010) The molecular basis of PPAR function. PPAR Res 2010:409168. https://doi.org/10.1155/2010/409168
Crossa J, Pérez-Rodríguez P, Cuevas J, Montesinos-López O, Jarquín D et al (2017) Genomic selection in plant breeding: methods, models, and perspectives. Trends Plant Sci 22:961–975
Czechowski T, Bari RP, Stitt M et al (2004) Real-time RT-PCR profiling of over 1400 Arabidopsis transcription factors: unprecedented sensitivity reveals novel root- and shoot-specific genes. Plant J 38:366–379
Dixon RA, Gang DR, Charlton AD et al (2006) Applications of metabolomics in agriculture. J Agric Food Chem 54:8984–8994
Durrant WE, Rowland O, Piedras P, Hammond-Kosack KE, Jones JD (2000) cDNA-AFLP reveals a striking overlap in race specific resistance and wound response gene expression profiles. Plant Cell 12:963–977
Emebiri LC, Tan MK, El-Bouhssini M, Wildman O, Jighly A et al (2017) QTL mapping identifies a major locus for resistance in wheat to Sunn pest (Eurygaster integriceps) feeding at the vegetative growth stage. Theor Appl Genet 130:309–318
Fenn JB, Mann M, Meng CK et al (1989) Electrospray ionization for mass spectrometry of large biomolecules. Science 246:64–71
Fiehn O (2002) Metabolomics – the link between genotypes and phenotypes. Plant Mol Biol 48:151–171
Fiehn O, Kopka DP et al (2000) Metabolic profiling for plant functional genomics. Nat Biotechnol 18:1157–1161
Fields S, Song O (1989) A novel genetic system to detect protein-protein interactions. Nature 340:245–246
Filleur S, Daniel-Vedele F (1999) Expression analysis of a high affinity nitrate transporter isolated from *Arabidopsis thaliana* by differential display. Planta 207:461–469

Ganal MW, Durstewitz G, Polley A, Berard A, Buckler ES et al (2011) A large maize (*Zea mays* L.) SNP genotyping array: development and germplasm genotyping, and genetic mapping to compare with the B73 reference genome. PLoS One 6:e28334

Giddings G, Allison G, Brooks D, Carter A (2000) Transgenic plants as factories for pharmaceuticals. Nat Biotechnol 18:1151–1155

Gong CY, Wang T (2013) Proteomic evaluation of genetically modified crops current status and challenges. Front Plant Sci 4:41

Haddeland I, Heinke J, Biemans H, Eisner S, Flörke M et al (2014) Global water resources affected by human interventions and climate change. Proc Natl Acad Sci U S A 111:3251–3256

Hall RD (2006) Plant metabolomics: from holistic hope to hype, to hot topic. New Phytol 169:453–468

Hall R, Beale M, Fiehn O et al (2002) Plant metabolomics—the missing link in functional genomics strategies. Plant Cell 14:1437–1440

Hall RD, Vos CHR, Varhoeven HA, Bino RJ (2005) Metabolomics for the assessment of functional diversity and quality traits in plants. In: Vaidyanathan S, Harrigan GG, Goodacre R (eds) Metabolome analyses: strategies for systems biology. Springer, New York

Harbers M, Carninci P (2005) Tag-based approaches for transcriptome research and genome annotation. Nat Methods 2:495–502

He C, Holme J, Anthony J (2014) SNP genotyping: the KASP assay. Crop Breed: Meth Protocol 1145:75–86

Henzel WJ, Billeci TM, Stults JT et al (1993) Identifying proteins from two-dimensional gels by molecular mass searching of peptide fragments in protein sequence databases. Proc Natl Acad Sci U S A 90:5011–50115

Hossain Z, Komatsu S (2013) Contribution of proteomic studies towards understanding plant heavy metal stress responses front. Plant Sci 3:310

Imadi SR, Kazi AG, Ahanger MA, Gucel S, Ahmad P (2015) Plant transcriptomics and responses to environmental stress: an overview. J Genet 94:525–537

International Rice Genome Sequencing Project., Sasaki, T. The map-based sequence of the rice genome. Nature 436, 793–800 (2005). https://doi.org/10.1038/nature03895

Jia Y, Jannink JL (2012) Multiple-trait genomic selection methods increase genetic value prediction accuracy. Genetics 192:1513–1522

Kara M, Bahr U, Ingendoh AI, Hillenkamp F (1989) Laser description-ionization mass spectrometry of proteins with masses 100,000 to 250,000 Dalton. Angew Chem Int 28:760–761

Kersten B, Bürkle L, Kuhn EJ et al (2002) Large-scale plant proteomics. Plant Mol Biol 48:133–141

Kim C, Lee S, Park H, Bae C, Cheong Y et al (2000) Identification of rice blast fungal elicitor responsive genes by differential display analysis. Mol Plant Microbe Interact 13:470–474

Klose J (1975) Protein mapping by combined isoelectric focusing and electrophoresis of mouse tissues. A novel approach to testing for induced point mutations in mammals. Humangenetik 26:231–243

Klose J, Kobalz U (1995) Two-dimensional electrophoresis of proteins: an updated protocol and implications for a functional analysis of the genome. Electrophoresis 16:1034–1059

Kuno N, Muramatsu T, Hamazato F, Furuya M (2000) Identification by large scale screening of phytochrome regulated genes in etiolated seedlings of *Arabidopsis* using a fluorescent differential display technique. Plant Physiol 122:15–24

Lapopin L, Gianinazzi-Pearson V, Franken P (1999) Comparative differential RNA display analysis of arbuscular mycorrhiza in *Pisum sativum* wild type and a mutant defective in late stage development. Plant Mol Biol 41:669–677

Liang P, Pardee AB (1992) Differential display of eukaryotic messenger RNA by means of the polymerase chain reaction. Science 257:967–971

Liu M, Lei L, Miao F, Powers C, Zhang X et al (2017) The STENOFOLIA gene from Medicago alters leaf width, flowering time and chlorophyll content in transgenic wheat. Plant Biotechnol J 16:186–196. https://doi.org/10.1111/pbi.12759

Long YM, Chao WS, Ma GJ, Xu SS, Qi LL (2017) An innovative SNP genotyping method adapting to multiple platforms and throughputs. Theor Appl Genet 130:597–607

Maloney V (2004) Plant metabolomics. Biotechnol J 2:92–99

Mardis ER (2008) The impact of next-generation sequencing technology on genetics. Trends Genet 24:133–141

Mcclelland M, Mathieu-Daude F, Welsh J (1995) RNA fingerprinting and differential display using arbitrary primed PCR. Trends Genet 11:242–246

McCouch SR, Teytelman L, Xu Y, Lobos KB, Clare K et al (2002) Development and mapping of 2240 new SSR markers for rice (Oryza sativa L.). DNA Res 9:199–207

Memelink J (2005) Tailoring the plant metabolome without a loose stitch. Trends Pl Sci 10:305–307

Mendes P (2002) Emerging bioinformatics for the metabolome. Brief Bioinform 3:34–45

Meuwissen THE, Hayes BJ, Goddard ME (2001) Prediction of total genetic value using genome-wide dense marker maps. Genetics 157:1819–1829

Money T, Reader S, Qu L, Dunford R, Moore G (1996) AFLP-based mRNA fingerprinting. Nucleic Acids Res 24:2616–2617

Nelson GC, Valin H, Sands RD, Havlík P, Ahammad H et al (2014) Climate change effects on agriculture: economic responses to biophysical shocks. Proc Natl Acad Sci U S A 111:3274–3279

Ni Z, Sun Q, Liu Z, Wu L, Wang X (2000) Identification of a hybrid specific expressed gene encoding novel RNA binding protein in wheat seedling leaves using differential display of mRNA. Mol Gen Genet 263:934–938

Oksman-Caldentey KM, Inz D (2004) Plant cell factories in the post genomic era: new ways to produce designer secondary metabolites. Trends Pl Sci 9:433–440

Ozbabacan SE, Engin HB, Gursoy A, Keskin O (2011) Transient protein protein interactions. Protein Eng Des Sel 24:635–648

Patino WD, Mian OY, Hwang PM (2002) Serial analysis of gene expression. Circ Res 91:565–569

Raamsdonk LM et al (2001) A functional genomics strategy that uses metabolome data to reveal the phenotype of silent mutations. Nat Biotechnol 19:45–50

Rajaram S (2005) Role of conventional plant breeding and biotechnology in future wheat production. Turk J Agric For 29:105–111

Ray DK, Mueller ND, West PC, Foley JA (2013) Yield trends are insufficient to double global crop production by 2050. PLoS One 8:e66428

Reid LM, Donell CPO, Downey G (2004) Potential of SPME-GC and chemometrics to detect adulteration of soft fruit purees. J Agric Food Chem 52:421–427

Renaut J, Hausman JF, Wisniewski ME (2006) Proteomics and low temperature studies: bridging the gap between gene expression and metabolism. Physiol Plant 126:97–109

Rhee SY, Dickerson J, Xu D (2006) Bioinformatics and its applications in plant biology. Annu Rev Plant Biol 57:335–360

Ruan Y, Le Ber P, Ng HH, Liu ET (2004) Interrogating the transcriptome. Trends Biotechnol 22:23–30

Ruperao P, Edwards D (2014) Identification of markers from next-generation sequence data plant genotyping. Meth Mole Biol Bioinform 1245:29–47

Saha S, Sparks AB, Rago C (2002) Using the transcriptome to annotate the genome. Nat Biotechnol 20:508–512

Saito K, Kobayashi M, Gong Z, Tanaka Y, Yamazaki M (1999) Direct evidence for anthocyanidin synthase as a 2-oxoglutarate dependent oxygenase: molecular cloning and functional expression of cDNA from some red forma of *Perilla frutescens*. Plant J 17:181–189

Sanger F, Air GM, Barrell BG, Brown NL, Coulson AR et al (1977) Nucleotide sequence of bacteriophage phi X174 DNA. Nature 265:687–695

Semagn K, Babu R, Hearne S, Olsen M (2014) Single nucleotide polymorphism genotyping using Kompetitive allele specific PCR (KASP): overview of the technology and its application in crop improvement. Mol Breed 33:1–14

Shanks JV (2005) Phytochemical engineering: combining chemical reaction engineering with plant science. AICh EJ 51:2–7

Sharma SK, Bolser D, de Boer J, Sønderkær M, Amoros W et al (2013) Construction of reference chromosome-scale pseudomolecules for potato: integrating the potato genome with genetic and physical maps. G3 (Bethesda) 3:2031–2047

Smith JSC, Chin ECL, Shu H, Smith OS, Wall SJ et al (1997) An evaluation of the utility of SSR loci as molecular markers in maize (*Zea mays* L.): comparisons with data from RFLPS and pedigree. Theor Appl Genet 95:163–170

Solekdeh GH, Komatsu S (2007) Crop proteomics: aim at sustainable agriculture of tomorrow. Proteomics 7:2976–2996

Somers DJ, Isaac P, Edwards K (2004) A high-density microsatellite consensus map for bread wheat (*Triticum aestivum* L.). Theor Appl Genet 109:1105–1114

Suarez MC, Bernal A, Gutierrez J, Tohme J, Fregene M (2000) Developing expressed sequence tags (ESTs) from polymorphic transcript-derived fragments (TDFs) in cassava (*Manihot esculenta* Crantz). Genome 43:62–67

Tanksley SD, Young ND, Paterson AH, Bonierbale MW (1989) RFLP mapping in plant breeding: new tools for an old science. Nat Biotechnol 7:257–264

Urano K, Maruyama K, Ogata Y, Morishita Y, Takeda M et al (2009) Characterization of the ABA-regulated global responses to dehydration in Arabidopsis by metabolomics. Plant J 57:1065–1078

Van K, McHale LK (2017) Meta-analyses of QTLs associated with protein and oil contents and compositions in soybean [Glycine max (L.) Merr.] seed. Int J Mol Sci 18:1180

Varshney RK, Graner A, Sorrels ME (2005) Genomics-assisted breeding for crop improvement. Trends Pl Sci 10:621–630

Varshney RK, Marcel TC, Ramsay L, Russell J, Roder MS et al (2007) A high density barley microsatellite consensus map with 775 SSR loci. Theor Appl Genet 114:1091–1103

Varshney RK, Nayak SN, May GD, Jackson SA (2009) Next-generation sequencing technologies and their implications for crop genetics and breeding. Trends Biotechnol 27:522–530

Varshney RK, Song C, Saxena RK, Azam S, Yu S et al (2013) Draft genome sequence of chickpea (Cicer arietinum) provides a resource for trait improvement. Nat Biotechnol 31:240–248

Vassilev D, Leunissen J, Atanassov A et al (2005) Applications of bioinformatics in plant breeding. Australas Biotechnol 19:139–152

Velculescu VE, Vogelstein B, Kinzler KW (2000) Analysing uncharted transcriptomes with SAGE. Trends Genet 16:423–425

Vos P, Hogers R, Bleeker M, Reijans M, Van De Lee T et al (1995) AFLP: a new technique for DNA fingerprinting. Nucleic Acids Res 23:4407–4414

Vukmirovic OG, Tilghman S (2000) Exploring the genome space. Nature 405:820–822

Wang K, Peng X, Ji Y, Yang P, Zhu Y et al (2013) Gene, protein and network of male sterility in rice. Front Plant Sci 92:1–10

Washburn MP, Wolters D, Yates JR (2001) Large-scale analysis of the yeast proteome by multidimensional protein identification technology. Nature Biotech 19:242–247

Weckwerth W (2003) Metabolomics in systems biology. Annu Rev Plant Biol 54:669–689

Whitelegge JP (2002) Plant proteomics: blasting out of a mud PIT. PNAS 99:11564–11566

Wilkins Marc R, Jean-Charles S, Gooley Andrew A, Appel Ron D, Ian H-S et al (1996) Progress with proteome projects: why all proteins expressed by a genome should be identified and how to do it. Biotechnol Genet Eng Rev 13:19–50

Yamazaki M, Gong Z, Fukuchi-Mizatani M, Fukui Y, Tanaka Y et al (1999) Molecular cloning and biochemical characterization of a novel anthocyanin 5-O-glucosyltransferase by mRNA differential display for plant forms regarding anthocyanin. J Biol Chem 274:7405–7411

Yan J, Warburton M, Crouch J (2011) Association mapping for enhancing maize (*Zea mays* L.) genetic improvement. Crop Sci 51:433–449

Yu H, Goh C (2000) Identification and characterization of three orchid MADS-box genes of the AP1/AGL9 subfamily during floral transition. Plant Physiol 123:458–464

Yung M, Schaffer R, Putterill J (1999) Identification of genes expressed during early Arabidopsis carpel development by mRNA differential display: characterisation of ATCEL2, a novel endo-1,4-beta-D-glucanase gene. Plant J 17:203–208

Zhang X, Huang C, Wu D, Qiao F, Li W et al (2017) High-throughput phenotyping and QTL mapping reveals the genetic architecture of maize plant growth. Plant Physiol 173:1554–1564

Zhao Q, Chen S, Dai S (2013) C4 photosynthetic machinery: insights from maize chloroplast proteomics. Front Plant Sci 4:85

# Relevance of Bioinformatics and Database in Omics Study

# 2

Rama Shankar, Vikas Dwivedi, and Gulab Chand Arya

## Abstract

Bioinformatics is an interdisciplinary branch of biological sciences that assists biologists to interpret and extract the biological information from the omics data. The biological information is further used to create different databases for annotation of an unknown molecules from the novel organism. In the cell, different types of molecules are present with diverse functions. Based on their type and functions, these molecules are divided into various categories. These molecules are majorly categorized into DNA, RNA, proteins, and metabolites. The bioinformatics tools and techniques are specific to study and analyze the variations and mechanism of these molecules. These molecules are divided into different omics for better understanding. In DNA, majorly two types of variations occur, which is categorized as genetic and epigenetic variations and known as genomics and epigenomics variations, respectively. Diversity in RNA is studied under the transcriptome category, where the level of mRNA, their regulatory molecules and modifications during synthesis and post-synthesis are examined. In addition, synthesis, modification, and interaction of proteins and metabolites are studied in proteome and metabolome categories. These studies are being analyzed by different bioinformatics tools and their respective databases are

R. Shankar (✉)
Department of Pediatrics and Human Development, Michigan State University, Grand Rapids, MI, USA
e-mail: ramashankar@hc.msu.edu

V. Dwivedi
Department of Ornamental Plants and Agricultural Biotechnology, Institute of Plant Sciences, Agricultural Research Organization, Rishon LeZion, Israel

G. C. Arya
Department of Vegetable and Field Crops, Institute of Plant Sciences, Agricultural Research Organization, Rishon LeZion, Israel



A. Kumar et al. (eds.), *Omics Technologies for Sustainable Agriculture and Global Food Security Volume 1*, https://doi.org/10.1007/978-981-16-0831-5_2

used to extract their biological information. Here, we have discussed in brief about the relevance of various bioinformatics tools and databases, which are being used for the analysis of biologically important molecules. This would provide a basic overview of the importance and application of these tools and databases in different omics study.

**Keywords**

Genomics · Epigenomics · Transcriptomics · Proteomics · Metabolomics · Database · Bioinformatics · Crop improvement

## 2.1 Introduction

Bioinformatics is an interdisciplinary branch of biological sciences that deals with applications of computational biology for the collection, storage, and analysis of biological data. In recent years, several omics projects in plants have been performed, which were contributed by a vast amount of sequencing data. These omics data generated through the traditional or high-throughput next-generation sequencing (NGS) approaches and belong to genome, transcriptome, proteome, or metabolome of the plants (Knasmüller et al. 2008). The term genome refers to the complete nuclear chromosomal DNA sequence of an organism, whereas the total messenger RNA (mRNA) content in a cell at a time is termed as trancriptome. Its level varied with different plant developmental stages and external environmental condition. The latter produce proteome, which is the result of the translation of the mRNA. During the cell metabolism, primary and secondary metabolites are generated and complete set of metabolites present in the cell are called as metabolome (Lister et al. 2009; Saito and Matsuda 2010). Besides, various inevitable modifications, such as expression of genes without changing original genetic material (DNA) of the organism occurs during lifetime and inherited to next-generation, are termed as epigenetics changes.

The data and related information obtained from the plant omics can be useful for generating high-density linkage maps, allele mining, QTL mapping, genome-wide association studies (GWAS), SNP genotyping, single sequence repeats (SSR), and a better understanding of metabolic pathways and its regulations. All these information may be helpful for better plant breeding and improvement programs.

Besides, bioinformatics with the support of highly advanced experimental evidences, various databases have been developed and curated (Shinozaki and Sakakibara 2009). These databases help to discover the novel and unknown information of novel plants and organisms. The National Center for Biotechnology Information (NCBI) is among the world's largest resource databases, storing a vast amount of data in various categories. Also, there are various other databases related to specific plants are available, such as rice genome annotation project (RGAP) database for rice (Kawahara et al. 2013), The Arabidopsis Information Resource (TAIR) for Arabidopsis (https://www.arabidopsis.org/), Phytozome (https://

phytozome.jgi.doe.gov/pz/portal.html), and OmicsDI (open source platform facilitating the access and dissemination of omics datasets) (https://www.omicsdi.org). The Phytozome and OmicsDI databases are one of the comprehensive omics databases that included information about several datasets including genomic, transcriptomic, proteomic, and metabolomic data (Goodstein et al. 2012). There is one important tool known as ODG (Omics database generator), which is a tool used for generating, querying, and analyzing multi-omics comparative databases to facilitate biological understanding (Guhlin et al. 2017). A list of various omics integration, software tools, and web applications is provided in Table 2.1

The present chapter describes the available tools and techniques used for curation, interpretation, and functional relevance of biological data using web-based resources. Further, this chapter also describes the online available databases, which can be used to extract the functional and structural information of unknown genes and proteins of novel plants. The relevant resources are also included for validating metabolic pathways. A basic overview is provided for the workflow of different omics analysis (Fig. 2.1).

## 2.2 Relevance of Bioinformatics in Genomics

DNA polymorphism is the variation of nucleotides in the genomic DNA. These modifications can be originated as a result of single nucleotide polymorphism (SNP), insertion and deletion (InDels), or simple sequence repeats (SSRs). SNPs are locations within the genome, where the original nucleotide is substituted with other nucleotide, whereas InDels are insertion and deletion of nucleotide in the genome, and these changes are inheritable from one generation to other. The length of insertion and deletion in the genomic DNA varies from one to many bases. However, three nucleotide insertion or deletion is very common (Chai et al. 2018; Jain et al. 2014). This could be an evolutionary adaptation as three nucleotides code for an amino acid. SSRs are another genetic variation that occurs in genome and known as simple sequence repeats of single nucleotide to ten nucleotides. However, during the analysis repeats of two nucleotides or more with specific repetition are considered as the SSRs (Agarwal et al. 2015; Daware et al. 2016; Parida et al. 2015; Dwivedi et al. 2017).

Identification of DNA polymorphisms is highly essential for gene mapping, QTL analysis, and marker-assisted breeding. Various techniques have been used to identify DNA polymorphisms including gel-based, like random amplified polymorphic DNA (RAPD), amplified fragment length polymorphism (AFLP), restriction fragment length polymorphism (RFLP), microsatellites, SSR, simple sequence length polymorphism (SSLP), and non-gel-based techniques, like SNPs and InDels. SNPs/InDels are the most popular non-gel-based DNA marker systems, which represent the position of nucleotide(s), where DNA sequence differs by a single or more bases. SNPs/InDels have gained importance due to their ubiquity in the genome coupled with various characteristics, such as stability, robustness, efficiency, and cost-effectiveness (Alkan et al. 2011; Kumar et al. 2012b; McCouch

**Table 2.1** Summary of multi-omics integration software tools and web applications

| Tools | Omics integrated | Domain | Functionality | Type of license |
|---|---|---|---|---|
| Omics | Transcriptomics, proteomics, and metabolomics | Medical (human) | Correlation network analysis, co-expression analysis, phenotype generation, KEGG/ Human Cyc, pathway enrichment, GO enrichment, Name to ID conversion | Open |
| COBRA | Transcriptomics, proteomics, metabolomics, and Fluxomics | Unspecified | Genome scale integrated modeling of cell metabolism and macro molecular expression | Open |
| Gaggle | Variety of omics platform bioinformatics solutions | Unspecified | Inoperability of the following tools: Bioinformatics resource manager, Cytoscape, Data Matrix Viewer, KEGG, Genome Browser, MeV, PIPE, Bio Tapestry, N-Browse | Open |
| KaPPA-view | Transcriptomics, and metabolomics | Plants | Integrates transcriptomics and metabolomics data to map pathways | Open |
| MADMAX | Metagenomics, transcriptomics, and metabolomics | Plants, medical and clinical | Integrates omics data—Statistical analysis and pathway mapping | Open |
| MapMan | Metagenomics, transcriptomics, and metabolomics | Plants | Compare data across these two species, KEGG classification, classification into KOG clusters, mapping expression responses | Open |
| MetaboAnalyst | Genomics, transcriptomics, proteomics, metabolomics, and clinical | Plants, microbial, microbiome, medical and clinical | Data processing and statistical analysis, pathway analysis, multi-omics integration | Open |
| mixOmics (R package) | Metagenomics, transcriptomics, proteomics, and metabolomics | Unspecified | Integration of data, Chemometric analysis (similarity/difference) | Open |
| Omickriging (R package) | Transcriptomics, proteomics, and metabolomics | Unspecified | Integration and visualization of omics data | Open |

(continued)

**Table 2.1** (continued)

| Tools | Omics integrated | Domain | Functionality | Type of license |
|---|---|---|---|---|
| PaintOmics | Transcriptomics, and metabolomics | 100 top species of different biological kingdoms | Integration and visualization of transcriptomics and metabolomics data | Open |
| Reactome | Genomics, transcriptomics, proteomics, and metabolomics | Unspecified | Multi-omics data visualization, metabolic map of known biological processes and pathways | Open |
| SIMCA | Metagenomics, transcriptomics, proteomics, and metabolomics | Unspecified | Integration of data, Chemometric analysis (similarity/difference) | Commercial |
| VitisNet | Metagenomics, transcriptomics, proteomics, and metabolomics | Grapes | Integration of data - visualization of connectivity | Open |
| GenBank (database) | Proteomics | Numerous (over 100,000 organisms) | Proteomics database, open access, annotated collection of all publically available nucleotide sequences and their protein transitions. | Open |
| Plant metabolic network (PMN) | Genomics, proteomics, and metabolomics | Plants | Plant-specific database containing pathways, enzymes, reactions, and compounds | Open |
| PRIDE | Proteomics | Unspecified | Proteomics database | |
| KEGG | Genomes, transcriptomics, proteomics, and metabolomics | Plants animals microbes | Collection of databases dealing with genomes, biological pathways, diseases, drugs, and chemical substances | Open and licensed |
| Yeast metabolome data (YMDB) | Metabolomics | Microbe (yeast) | Metabolite database | Open |
| VANTED | Metagenomics, transcriptomics, proteomics, and metabolomics | Unspecified | Comparison of multiple omics data sets, visualization of metabolic maps, correlation networks analysis | Open |

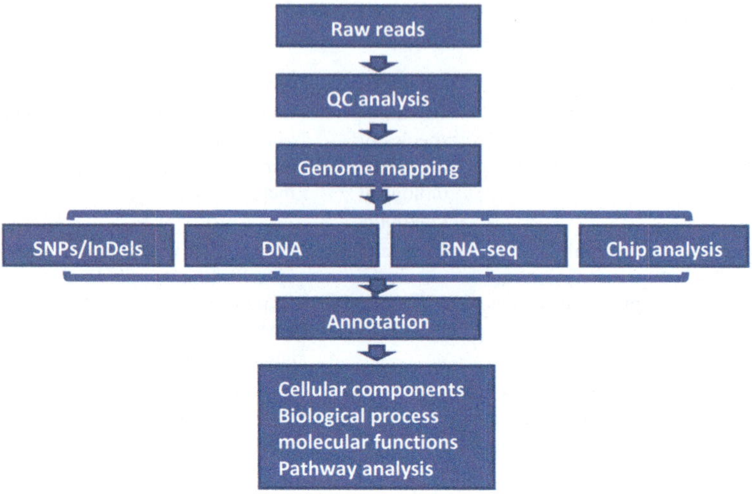


**Fig. 2.1** Basic workflow of omics analysis

et al. 2010; Rafalski 2002; Steemers and Gunderson 2007). The next-generation sequencing (NGS) is an easy and cost effective method for discovery of SNPs/InDels in a population. A large number of SNPs have been discovered from several plant species like Arabidopsis (Atwell et al. 2010), rice (Huang et al. 2010, 2011; Jain et al. 2014; McNally et al. 2009; Meyer et al. 2016; Zhao et al. 2011), maize (Kump et al. 2011; Tian et al. 2011), chickpea (Deokar et al. 2014; Thudi et al. 2014), and soybean (Hwang et al. 2014; Lam et al. 2010) via genome re-sequencing.

Since huge data of SNPs/InDels are being generated using the NGS, a large number of bioinformatics tools are available to validate the biological significance of the aforesaid changes in the genome. For the analysis and validation of SNPs/InDels various bioinformatics tools are available (Li and Wei 2015; Seal et al. 2014), in which GATK and Freebays are the two important tools to discover the SNPs/InDels from the genome mapped files (Garrison and Marth 2012; Van der Auwera et al. 2013). The genome mapping of sequence reads is performed using different tools, mainly TopHat, STAR, and Bowtie tools (Dobin et al. 2013; Trapnell et al. 2009; Wu et al. 2018). Once the DNA polymorphism is identified, it is annotated using the snpEff software (Cingolani et al. 2012). This helps to understand the effect of SNPs/InDels on various transcriptional, post-transcriptional, and post-translational modifications. These genetic modifications can be further associated with various traits using the genome-wide association (GWAS) study in plants (Marees et al. 2018). The SNPs/InDels associated with various traits can be used for the genetic engineering and crop breeding purposes to improve the crop productivity.

## 2.3 Application of Bioinformatics in Epigenomics

DNA methylation is one of the epigenetic variations that occur by addition of a methyl group to the genomic DNA. It plays a crucial role in the regulation of chromatin structure and regulates the gene expression in eukaryotes. DNA methylation mainly occurs at the cytosine and adenine nucleotides in DNA; however, methylation in cytosine is specific to higher eukaryotes. In plants, DNA methylation is occurred in three different sequence contexts, CG, CHG, and CHH (where H = A, C or T). This methylation is established and maintained by de novo methyltransferases (DRM1/2/CMT3) via RNA-directed DNA methylation (RdDM) pathway and MET1 proteins (Cao and Jacobsen 2002; Lindroth 2001). Epigenetic modifications are highly stable and heritable, and it regulates cellular and developmental modifications including agronomically important traits in the plants (Manning et al. 2006; Miura et al. 2009; Soppe et al. 2000). DNA methylation analysis has been carried out in different plants to study their role in different developmental processes and stress responses (Chinnusamy and Zhu 2009; Dowen et al. 2012; Gehring et al. 2009; Hsieh et al. 2009; Lang-Mladek et al. 2010; Mirouze et al. 2009; Saze et al. 2003; Zemach et al. 2010).

To study the genome-wide DNA methylation, various techniques have been developed (HPLC, mass spectrometry, Sssl methyltransferase tritium labeling and methyl sensitive restriction enzyme). Initially, these methods were low throughput because they could capture the DNA methylation only in few genes (Karan et al. 2012; Wang et al. 2011). Later, microarray has been proved as first high-throughput technique to study the DNA methylation (Schumacher et al. 2006). Further, next-generation sequencing (NGS) based technique has also been evolved to capture the DNA methylation at the single-base resolution and has been used to study the DNA methylation in various plants including Arabidopsis and rice (Dowen et al. 2012; Garg et al. 2015; Rajkumar et al. 2020; Wang et al. 2011). This technique provides more in-depth knowledge about the DNA methylation, its distribution, and regulation.

Bioinformatics tools such as Bismark and Methylkit are highly efficient tools to analyze the DNA methylation data. Bisulfite sequencing is widely used technique to study the DNA methylation, in which nonmethylated thymine is changed into a cytosine but methylated thymine nucleotide does not modify (Li and Tollefsbol 2011). The first step of bisulfite sequencing is NGS based sequencing. Further, the sequencing data needs to be mapped on genomic DNA. Specific sequence aligner is required to align the sequence reads on the genome. The most widely used sequence aligner is Bismark (Krueger and Andrews 2011). Further, the mapped reads are mined by another bioinformatic tool widely known as Methylkit (Akalin et al. 2012). It extracts the methylated cytosine from the data throughout the genome. This information is further used to annotate and study the biological relevance of methylation on the various biological processes and metabolic pathways using different databases.

## 2.4 Bioinformatic Tools to Identify the Transcriptomic Alterations

### 2.4.1 RNA-Seq Analysis

Transcriptome can be defined as the total mRNA in a cell at a particular time. mRNA is derived from one strand of genomic DNA. Further, it translates into a protein with the help of the ribosomes. Transcriptome of the cell can be studied by the microarray and RNA sequencing (RNA-seq) (Jain 2012; Wang et al. 2011). Microarray has low throughput and various limitations as compared to the RNA-seq.

Microarray is based on the hybridization of the DNA probe designed for every gene (Page et al. 2007). They are very specific for the genes. mRNA in one condition is labeled with the green color and mRNA in other condition is tagged with red color. These labeled mRNAs are hybridized on a chip containing DNA probes for various genes. Once the labeled mRNA hybridized with the probe, it emits a fluorescent color, which is detected by the highly sensitive camera. Further, these patterns of color overlap between two conditions and based on the intensity, the differential expression between two conditions is estimated. To analyze these data, GeneSpring GX is one of the most widely used bioinformatics tool provided by the Agilent (Agapito 2019). It is a combination of different utilities that provides powerful, accessible statistical tools for data analysis and visualization. It is designed basically for the need of biologist and enables understanding of transcriptomics, genomics, proteomics, metabolomics, and NGS data within the biological context. It allows the researchers to quick and reliable identification of the biologically significant genes and pathways.

RNA-seq is one of the most advanced techniques based on next-generation sequencing (NGS) to study the transcriptome (Børsting and Morling 2015; Jain 2012; Lister et al. 2009). It has various advantages over microarray, as it can be used to study alternative splicing, polyadenylation, and novel genes or transcript discovery (Rao et al. 2018). During the RNA-seq library preparation process, mRNA is converted into cDNA to enhance stability. The cDNA is mechanically fragmented into small fragments (100–500 nucleotides). These fragments are attached with the adopter sequences present on the sequencing chip. The attached fragments further PCR amplified using the primers based on the adopter sequences to enhance the number of fragments for each molecule. These cDNA fragments are further sequenced by the sequencing technology (Kumar et al. 2012a; Zhong et al. 2011). The sequencing platform uses the sequence by synthesis approach. Based on the sequence length, these techniques are divided into two groups, i.e. short reads and long reads (Berbers et al. 2020). Both of these groups have advantages and disadvantages. The short reads sequencing technology can provide more read depth, whereas long reads technology provides the longer reads but shallow read depth (Reinert et al. 2015).

Once the sequencing is complete, the sequencing reads are mapped on the genome sequence of the respective plant. Mapping of sequencing reads is done by various bioinformatics tools, such as Tophat, SOAP, STAR, Salmon, Bowtie (Dobin

**Table 2.2** Databases for the study of promoter sequences and regulatory elements of a gene

| Database | Description | URL |
|---|---|---|
| TRANSFAC | Transcription factor database | http://transfac.gbf.de/TRANSFAC/ |
| PlantCARE | Plant cis-acting regulatory elements database | http://sphinx.rug.ac.be:8080/PlantCARE/ |
| PLACE | Plant cis-acting regulatory elements database | http://www.dna.affrc.go.jp/htdocs/PLACE/ |
| SignalP 4.0 | Identification of signal peptides | http://www.cbs.dtu.dk/services/SignalP/ |
| TargetP | Subcellular localization of sequences | http://www.cbs.dtu.dk/services/TargetP/ |
| LOCTREE3 | Subcellular localization of sequences | https://www.rostlab.org/services/loctree3/ |
| Plant-mPLoc | Subcellular localization of sequences | www.csbio.sjtu.edu.cn/bioinf/plant-multi/ |
| PSI-Pred | Prediction of transmembrane regions of the gene | http://bioinf.cs.ucl.ac.uk/psipred/ |
| DNASTAR | Making of sequence assembly | http://www.dnastar.com/ |
| PromPredict | Promoter analysis | http://nucleix.mbu.iisc.ernet.in/prompredict/prompredict.html |
| CTDB | Transcriptome | http://www.nipgr.ac.in/ctdb.html |

et al. 2013; Kim et al. 2013; Patro et al. 2017; Trapnell et al. 2009; Xie et al. 2014). Among all, STAR is the better alignment tool and it provides the normalized count of reads mapped on each gene in every sample (Dobin et al. 2013). Normalized mapped read count is used to estimate the differential gene expression between two samples or conditions. To estimate the differential gene expression, various bioinformatics tools are being used including EdgeR, DESeq, Limma, cufflinks/Cuffdiff, RSEM, and Salmon (Ghosh and Chan 2016; Li and Dewey 2011; Love et al. 2014; Patro et al. 2017; Pollier et al. 2013; Ritchie et al. 2015; Robinson et al. 2010). Edger, DEseq, and Limma are the most used tools for identification of differentially expressed (DE) genes (Love et al. 2014; Ritchie et al. 2015; Robinson et al. 2010).

The DE genes are further used to discover the biological processes and pathways regulated by them. The biological processes were discovered by the EnrichR and BinGO tools (Kuleshov et al. 2016; Maere et al. 2005). For the annotation of DE genes, these tools used the functional annotation from the ontology databases. To discover the role of DE genes in biological pathways KEGG pathway database (https://www.genome.jp/kegg/pathway.html) is used. DE genes were also used to discover the transcription regulatory elements using different databases (Table 2.2). Among all, plant cis-acting regulatory elements database (PlantCARE) and PLACE are the most suitable and highly used database (Guo et al. 2008).

For transcriptomic studies, there are several public databases available to store the transcriptomic data, such as Genevestigator, NASCArrays, ArrayExpress, Stanford Microarray Database, Omics DI, and Gene Expression Omnibus (Bhardwaj and Somvanshi 2015). An example of the database is Chickpea Transcriptome Database (CTDB), which has information about the tools used for transcriptome sequence, transcription factor families, conserved domain(s), and molecular markers in chickpea (Verma et al. 2015) (Table 2.2).

### 2.4.2 Tools and Databases for Transcription Factor Binding Site

Chromatin immunoprecipitation (ChIP)-sequencing (ChIP-seq) is the method to analyze the protein DNA interaction. It is a combination of chromatin immunoprecipitation (ChIP) coupled with NGS to identify the binding sites of DNA associated proteins. It could be useful to discover the binding sites of any protein and has primarily been used to study the transcription factor (TF) binding sites and chromatin-associated proteins (Mundade et al. 2014).

ChIP-seq includes a few critical steps before the sequencing of the DNA-fragments attached with TF/protein. It starts with the crosslinking of protein with the DNA using formaldehyde (Hoffman et al. 2015; Klockenbusch and Kast 2010; Nadeau and Carlson 2007). However, along with the protein DNA crosslinking there are chances of contamination of RNA-protein complexes in the reaction mixture. This crosslinked sample was fragmented to get the DNA-protein crosslinked fragments and pull-down using antibody. The DNA fragments are then sequenced using the deep short-read sequencing platform. The first step in the ChIP-seq data analysis is known as the peak calling.

The most popular bioinformatics tool for peak calling is MACS (Feng et al. 2012; Zhang et al. 2008). This empirically models the shift size of ChIP-seq tags and uses it to improve the spatial resolution of predicted binding sites. Once the binding sites in the whole genome are predicted, these binding sites must be annotated to find out the respective genes, which are present at the downstream. This can be performed by HOMER and various other databases available to annotate these binding sites and related TFs (Table 2.3) (Heinz et al. 2010, 2018). It provides information about the binding sites and their regulating genes and pathways. This information can be used to identify genes and relevant pathways that can be used to implement in the crop improvement.

### 2.4.3 Tools and Databases for Analysis of Post-Transcriptional Modifications

Another important event known as alternative splicing is also studied in transcriptome analysis as the post-transcriptional event. Alternative splicing is divided into five categories such as exon skipping, mutually exclusive exon, alternative 5′ donor site, alternative 3′ acceptor site, and intron retention (Bedre et al. 2019; Eckardt 2013; Shang et al. 2017; Shankar et al. 2016). Intron retention is the most common alternative splicing events that happened during the transcription process under normal or any stress condition (Shankar et al. 2016). The recommended tools to identify the alternative splicing are TopHat, MapSplice, SpliceMap, HMMsplicer, STAR, and HISAT (Au et al. 2010; Dimon et al. 2010;

**Table 2.3** Database for transcription factor prediction

| | | |
|---|---|---|
| AGRIS, AtTFDB | Arabidopsis | http://arabidopsis.med.ohio-state.edu/AtTFDB/ |
| DRTF | Rice | http://drtf.cbi.pku.edu.cn/ |
| DPTF | Poplar | http://dptf.cbi.pku.edu.cn/ |
| TOBFAC | Tobacco | http://compsysbio.achs.virginia.edu/tobfac/ |
| PlantTFDB | Plant species | http://planttfdb.cbi.pku.edu.cn/22 |
| PlnTFDB | Plant species | http://plntfdb.bio.uni-potsdam.de/v3.0/20 |
| GRASSIUS, GrassTFDB | Maize, rice, sorghum, and sugarcane | http://grassius.org/grasstfdb.html |
| LegumeTFDB | Soybean, lotus japonicas, and Medicago truncatula | http://legumetfdb.psc.riken.jp/ |
| DBD | 700 species | http://dbd.mrc-lmb.cam.ac.uk/DBD/index.cgi?Home |
| PlantTFDB | 83 species | http://planttfdb.cbi.pku.edu.cn/ |

Dobin et al. 2013; Kim et al. 2015; Trapnell et al. 2009; Wang et al. 2010). These tools provide information about the alternative splicing in mRNA. Various bioinformatics tools are available for computing the differential expression of transcript isoforms produced as a result of alternative splicing (Kim et al. 2013; Patro et al. 2017). This will help to identify a specific isoform produced during the stress or different developmental stages (Akhter et al. 2018; Jiang et al. 2015; Shankar et al. 2016). A biologist to understand the deeper knowledge of plant development and stress responses will use this information.

RNA secondary structure is another post-transcriptional changes happened in the RNA during the post-transcriptional event (Ding et al. 2014; Wang et al. 2019b; Yang et al. 2018). It is known that genomic DNA is folded into specific shapes in the nucleus. Similar folding is reported in RNA also after post-transcriptional process to deliver its function or stability. It is well established that ribosomal RNA folded into distinct three-dimensional shape including internal loops and helices. It binds with the ribosomal protein and make ribosomal subunit required for protein synthesis. Various studies have been carried out to discover the mRNA secondary structure in plants using the NGS techniques (Ding et al. 2014; Wang et al. 2019b; Yang et al. 2018). It has been observed that mRNA with variations in RNA secondary structure lead to affect various transcriptional and post-transcriptional events (Li et al. 2012). There are several bioinformatics tools available, which can provide the secondary structure of the RNA (Gruber et al. 2008; Reuter and Mathews 2010; Wang et al. 2019a). It has been observed that RNA secondary structure predicted using the bioinformatics tools and structure detected using the NGS technique are very similar (Li et al. 2012).

## 2.5 Importance of Bioinformatics in Proteomics and Metabolomics

Proteins regulate various biochemical and physiological functions in the cells. The dysregulation of proteins may result in various diseases like cancer, neurodegenerative disease, and metabolic imbalance. Protein is synthesized from the mRNA during the translation process and folded into three-dimensional structure after protein synthesis. If the 3D structure is not folded properly, the protein will not be able to perform its activity and will not be able to interact with other proteins as well. The knowledge of protein–protein interactions and structure can be obtained from various databases (Table 2.4).

One of the most advanced techniques available for proteomic analysis is known as mass spectrometry (Di Falco 2018; Reinders et al. 2004). All the proteins from a sample are needed to be extracted and digested using specific proteases to generate a defined peptide. The peptides obtained are analyzed by the liquid chromatography coupled to mass spectrometry (GC-MS) (Lluveras-Tenorio et al. 2017). During the analysis, peptides eluted from the chromatography are selected and data is recorded as a mass spectrometer. The resulted tandem spectra provide information about the sequence of the peptide. These proteins are further used for functional annotation using the gene ontology (GO) terms and KEGG pathways database. The GO term provides the information about the cellular component, biological process, and molecular functions of the respective genes and proteins. The cellular component GO term provides information about the protein location in the cell compartment. The biological process GO terms provide information about the biological processes and molecular functions GO terms represent activities rather than the entities (molecules or complexes) performed by the genes or proteins (Hill et al. 2008). Similarly, the KEGG pathways database provides knowledge about the metabolic pathways regulated by these proteins. This information is further used by the research scientist to conclude the pathways regulated by these genes and used it to translate into genetic engineering and crop improvement.

There are different public databases available for MS proteomics research. These databases are Global Proteome Machine Database (GPMDB), Mass Spectrometry Interactive Virtual Environment (MassIVE), PRIDE, PeptideAtlas, PeptideAtlas SRM Experiment Library (PASSEL), and Proteomics DB. Moreover, for more integration and sharing of public databases, the Proteome Xchange consortium has been made recently to take its advantage for the scientific community (Perez-Riverol et al. 2015).

Metabolomics is another direction of omics included in the comprehensive assessment and quantification of metabolites present in the cell. Metabolites represent a diverse group of low molecular weight molecules including lipids, amino acids, peptides, nucleic acids, organic acids, vitamins, thiols, and carbohydrates. These metabolites have a different role in the biological systems and their role in various plant stress and development processes needed to be understood (Hussein and El-Anssary 2019; Bartwal et al. 2013; Jwa et al. 2006; Saito and Matsuda 2010; Shankar et al. 2016). Further, this information can be used by the biologist to

**Table 2.4** Important computational tools for predicting protein structure and protein–protein interactions

| S. No. | Software/ server | URL | Description |
|---|---|---|---|
| 1 | SWISS-MODEL | http://swissmodel.expasy.org/ | Automated protein homology modeling server |
| 2 | YASARA | http://www.yasara.org/ | Molecular modeling tool |
| 3 | ESyPred3D | http://www.unamur.be/sciences/biologie/urbm/bioinfo/easypred/ | Homology modeling with increased alignment performance |
| 4 | ROSETTA | http://boinc.bakerlab.org/resetta/ | 3D structure prediction |
| 5 | RaptorX | http://raptorx.uchicago.edu/ | Protein structure prediction |
| 6 | HHPred | http://toolkit.tuebingen.mpg.de/hhpred | Homology detection and structure prediction server |
| 7 | Phyre2 | http://www.sbg.bio.ic.ac.uk/~phyre2/html/page.cgi?id=index | 3D structure prediction |
| 8 | Bhageerah | http://www.scfbio-iitd.res.in/bhageerath/index.jsp | Energy-based protein structure prediction server |
| | 3DJigsaw | http://bmm.cancerresearchuk.org/~3djigsaw/ | Predict structure and function of protein |
| 9 | I-TASSER | http://zhanglab.ccmb.med.umich.edu/I-TASSER/ | Predict structure and function of protein |
| 10 | 3DJigsaw | http://bmm.cancerresearchuk.org/~3djigsaw/ | Predict structure and function of protein |
| 11 | MODELLER | http://salilab.org/modeller/ | Comparative modeling of protein 3D structures |
| 12 | PIPE2 | http://cgmlab.carleton.ca/PIPE2 | PIPE2 queries the protein interactions between two proteins based on specificity and sensitivity |
| 13 | HomoMINT | http://mint.bio.uniroma2.it/HomoMINT | HomoMINT predicts interaction in human based on ortholog information in model organisms |
| 14 | MirrorTree | http://csbg.cnb.csic.es/mtserver/ | The MirrorTree allows graphical and interactive study of the coevolution of two protein families and assesses their interactions in a taxonomic context |
| 15 | COG | http://www.ncbi.nlm.nih.gov/COG/ | COG shows phylogenetic classification of proteins encoded in genomes |
| 16 | PreSPI | http://code.google.com/p/prespi/ | PreSPI predicts protein interactions using a combination of domains |

(continued)

**Table 2.4** (continued)

| S. No. | Software/ server | URL | Description |
|---|---|---|---|
| 17 | InPrePPI | http://inpreppi.biosino.org/InPrePPI/index.jsp | InPrePPI predicts protein interactions in prokaryotes based on genomic context |
| 18 | STRING | http://string.embl.de | STRING database includes protein interactions containing both physical and functional associations |
| 19 | InterPreTS | http://gabrmn.uab.es/interpret/ | InterPreTS uses tertiary structure to predict interactions |
| 20 | iWARP | http://groups.csail.mit.edu/cb/iwrap/ | iWARP is a threading-based method to predict protein interaction from protein sequences |
| 21 | Coev2Net | http://groups.csail.mit.edu/cb/coev2net/ | Coev2Net is a general framework to predict, assess, and boost confidence in individual interactions inferred from a high-throughput experiment |

perform genetic engineering or plant breeding to improve the crop plants. Various methods have been developed to study the metabolites including GC, HPLC, UPLC, CE coupled to MS and NMR spectroscopy (Boizard et al. 2016; Boros et al. 2018; Garcia-Perez et al. 2020; Lluveras-Tenorio et al. 2017; Patel et al. 2017; Yang et al. 2013, 2020). This could help in separation, detection, characterization, and quantification of such metabolites and their related pathways. However, the diverse group of molecules makes it more challenging to study the metabolites using a single technique. Thus, more than one technique is used to identify the different metabolites in the plant system.

## 2.6 Challenges and Opportunity in Omics Study

Various advancements have been achieved in the field of omics study. Now we can detect the maximum number of RNA, DNA, and protein content present in the cell. However, different challenges are still persisted, which need to be answered. Even today, during library preparation of DNA or RNA sequencing, we are not able to capture all the DNA and RNA molecules. A large number of RNA and DNA have become degraded during the sample preparation. Genome re-sequencing with advanced technology is not able to cover 100% of the genome of any organism. We used to get a lot of redundancy during the mapping of the sequencing reads on the genome and/or transcriptome. This problem is more prominent in the plants with genome $\geq$2n (diploid). Study of proteomics and metabolomics are at very early stage and recent development in large scale proteomics data impose a substantial challenge for available bioinformatics tools to validate these results (Cho 2007; Hongzhan et al. 2007; Reinders et al. 2004; Schubert et al. 2017). During the proteomic analysis, a large number of challenges needed to be resolved besides sample preparation such as data assembly and database search for the functional annotation (Reinders et al. 2004; Schubert et al. 2017). We can annotate only those proteins, whose information is present in the database, but identifying a novel protein is very challenging.

To capture all the DNA and RNA new methods and techniques are being developed. Single molecule sequencing is evolving, as the new approach is developed to improve the genome coverage. The analysis for these molecules is also being improved. It provides a complete sequence information of all the mRNA expressed in a cell or tissue. This will also enable to get a deeper understanding of the post-transcriptional modifications occurred in RNA. Implementation of this method can solve the limitation of protein sequencing and quantification. During the sample preparation, one part of tissue is used to extract either DNA or RNA or proteins and metabolites. This adds the batch effect in the analysis. Now molecular signature is being analyzed from single cell, so developing methods to extract the entire molecular signature from the same cell or tissue has a great opportunity. Recently, few protocols have been developed to extract the DNA and RNA from same tissue but still need a lot of optimization. In bioinformatics analysis, all the tools and techniques come with few limitations. To solve all these limitations, novel techniques and methods are being developed. Hopefully, in future we will be able to develop more advanced technology to solve all these challenges and limitations.

## References


Agapito G (2019) Computer tools to analyze microarray data. Methods Mol Biol 1986:267–282

Agarwal G, Sabbavarapu MM, Singh VK, Thudi M, Sheelamary S, Gaur PM, Varshney RK (2015) Identification of a non-redundant set of 202 in silico SSR markers and applicability of a select set in chickpea (Cicer arietinum L.). Euphytica 205:381–394

Akalin A, Kormaksson M, Li S, Garrett-Bakelman FE, Figueroa ME, Melnick A, Mason CE (2012) methylKit: a comprehensive R package for the analysis of genome-wide DNA methylation profiles. Genome Biol 13:87

Akhter S, Kretzschmar WW, Nordal V, Delhomme N, Street NR, Nilsson O, Emanuelsson O, Sundström JF (2018) Integrative analysis of three RNA sequencing methods identifies mutually exclusive exons of MADS-box isoforms during early bud development in Picea abies. Front Plant Sci 9:1625

Alkan C, Coe BP, Eichler EE (2011) Genome structural variation discovery and genotyping. Nat Rev Genet 12:363–376

Atwell S, Huang YS, Vilhjálmsson BJ, Willems G, Horton M, Li Y, Meng D, Platt A, Tarone AM, Hu TT et al (2010) Genome-wide association study of 107 phenotypes in Arabidopsis thaliana inbred lines. Nature 465:627–631

Au KF, Jiang H, Lin L, Xing Y, Wong WH (2010) Detection of splice junctions from paired-end RNA-seq data by SpliceMap. Nucleic Acids Res 38:4570–4578

Bartwal A, Mall R, Lohani P, Guru SK, Arora S (2013) Role of secondary metabolites and Brassinosteroids in plant defense against environmental stresses. J Plant Growth Regul 32:216–232

Bedre R, Irigoyen S, Petrillo E, Mandadi KK (2019) New era in plant alternative splicing analysis enabled by advances in high-throughput sequencing (HTS) technologies. Front Plant Sci 10:740

Berbers B, Saltykova A, Garcia-Graells C, Philipp P, Arella F, Marchal K, Winand R, Vanneste K, Roosens NHC, De Keersmaecker SCJ (2020) Combining short and long read sequencing to characterize antimicrobial resistance genes on plasmids applied to an unauthorized genetically modified bacillus. Sci Rep 10:4310

Bhardwaj T, Somvanshi P (2015) Plant systems biology: insights and advancements. In: Barh D, Khan MS, Davies E (eds) PlantOmics: the omics of plant science. New Delhi, Springer, pp 791–819

Boizard F, Brunchault V, Moulos P, Breuil B, Klein J, Lounis N, Caubet C, Tellier S, Bascands J-L, Decramer S et al (2016) A capillary electrophoresis coupled to mass spectrometry pipeline for long term comparable assessment of the urinary metabolome. Sci Rep 6:34453

Boros E, Pinkhasov OR, Caravan P (2018) Metabolite profiling with HPLC-ICP-MS as a tool for in vivo characterization of imaging probes. EJNMMI Radiopharm Chem 3:2

Børsting C, Morling N (2015) Next generation sequencing and its applications in forensic genetics. Forensic Sci Int Genet 18:78–89

Cao X, Jacobsen SE (2002) Role of the Arabidopsis DRM methyltransferases in De novo DNA methylation and gene silencing. Curr Biol 12:1138–1144

Chai C, Shankar R, Jain M, Subudhi PK (2018) Genome-wide discovery of DNA polymorphisms by whole genome sequencing differentiates weedy and cultivated rice. Sci Rep 8:14218

Chinnusamy V, Zhu J-K (2009) Epigenetic regulation of stress responses in plants. Curr Opin Plant Biol 12:133–139

Cho WCS (2007) Proteomics technologies and challenges. Genomics Proteomics Bioinformatics 5:77–85

Cingolani P, Platts A, Wang LL, Coon M, Nguyen T, Wang L, Land SJ, Lu X, Ruden DM (2012) A program for annotating and predicting the effects of single nucleotide polymorphisms, SnpEff. Fly (Austin) 6:80–92

Daware A, Das S, Srivastava R, Badoni S, Singh AK, Agarwal P, Parida SK, Tyagi AK (2016) An efficient strategy combining SSR markers- and advanced QTL-seq-driven QTL mapping unravels candidate genes regulating grain weight in rice. Front Plant Sci 7:1535

Deokar AA, Ramsay L, Sharpe AG, Diapari M, Sindhu A, Bett K, Warkentin TD, Tar'an B (2014) Genome wide SNP identification in chickpea for use in development of a high density genetic map and improvement of chickpea reference genome assembly. BMC Genomics 15:708

Di Falco MR (2018) Mass spectrometry-based proteomics. Methods Mol Biol 1775:93–106

Dimon MT, Sorber K, DeRisi JL (2010) HMMSplicer: a tool for efficient and sensitive discovery of known and novel splice junctions in RNA-Seq data. PLoS One 5:e13875

Ding Y, Tang Y, Kwok CK, Zhang Y, Bevilacqua PC, Assmann SM (2014) In vivo genome-wide profiling of RNA secondary structure reveals novel regulatory features. Nature 505:696–700

Dobin A, Davis CA, Schlesinger F, Drenkow J, Zaleski C, Jha S, Batut P, Chaisson M, Gingeras TR (2013) STAR: ultrafast universal RNA-seq aligner. Bioinformatics 29:15–21

Dowen RH, Pelizzola M, Schmitz RJ, Lister R, Dowen JM, Nery JR, Dixon JE, Ecker JR (2012) Widespread dynamic DNA methylation in response to biotic stress. Proc Natl Acad Sci U S A 109:E2183–E2191

Dwivedi V, Parida SK, Chattopadhyay D (2017) A repeat length variation in myo-inositol monophosphatase gene contributes to seed size trait in chickpea. Sci Rep 7:4764

Eckardt NA (2013) The plant cell reviews alternative splicing. Plant Cell 25:3639

Feng J, Liu T, Qin B, Zhang Y, Liu XS (2012) Identifying ChIP-seq enrichment using MACS. Nat Protoc 7:1728–1740

Garcia-Perez I, Posma JM, Serrano-Contreras JI, Boulangé CL, Chan Q, Frost G, Stamler J, Elliott P, Lindon JC, Holmes E et al (2020) Identifying unknown metabolites using NMR-based metabolic profiling techniques. Nat Protoc 15:2538–2567

Garg R, Narayana Chevala V, Shankar R, Jain M (2015) Divergent DNA methylation patterns associated with gene expression in rice cultivars with contrasting drought and salinity stress response. Sci Rep 5:14922

Garrison E, Marth G (2012) Haplotype-based variant detection from short-read sequencing. ArXiv:1207.3907 [q-Bio]

Gehring M, Bubb KL, Henikoff S (2009) Extensive demethylation of repetitive elements during seed development underlies gene imprinting. Science 324:1447–1451
Ghosh S, Chan C-KK (2016) Analysis of RNA-seq data using TopHat and cufflinks. Methods Mol Biol 1374:339–361
Goodstein DM, Shu S, Howson R, Neupane R, Hayes RD, Fazo J, Mitros T, Dirks W, Hellsten U, Putnam N et al (2012) Phytozome: a comparative platform for green plant genomics. Nucleic Acids Res 40:D1178–D1186
Gruber AR, Lorenz R, Bernhart SH, Neuböck R, Hofacker IL (2008) The Vienna RNA websuite. Nucleic Acids Res 36:W70–W74
Guhlin J, Silverstein KAT, Zhou P, Tiffin P, Young ND (2017) ODG: omics database generator—a tool for generating, querying, and analyzing multi-omics comparative databases to facilitate biological understanding. BMC Bioinformatics 18:367
Guo A-Y, Chen X, Gao G, Zhang H, Zhu Q-H, Liu X-C, Zhong Y-F, Gu X, He K, Luo J (2008) PlantTFDB: a comprehensive plant transcription factor database. Nucleic Acids Res 36:D966–D969
Heinz S, Benner C, Spann N, Bertolino E, Lin YC, Laslo P, Cheng JX, Murre C, Singh H, Glass CK (2010) Simple combinations of lineage-determining transcription factors prime cis-regulatory elements required for macrophage and B cell identities. Mol Cell 38:576–589
Heinz S, Texari L, Hayes MGB, Urbanowski M, Chang MW, Givarkes N, Rialdi A, White KM, Albrecht RA, Pache L et al (2018) Transcription elongation can affect genome 3D structure. Cell 174:1522–1536
Hill DP, Smith B, McAndrews-Hill MS, Blake JA (2008) Gene ontology annotations: what they mean and where they come from. BMC Bioinformatics 9:5
Hoffman EA, Frey BL, Smith LM, Auble DT (2015) Formaldehyde crosslinking: a tool for the study of chromatin complexes. J Biol Chem 290:26404–26411
Hongzhan H, Shukla HD, Cathy W, Satya S (2007) Challenges and solutions in proteomics. Curr Genomics 8:21–28
Hsieh T-F, Ibarra CA, Silva P, Zemach A, Eshed-Williams L, Fischer RL, Zilberman D (2009) Genome-wide demethylation of Arabidopsis endosperm. Science 324:1451–1454
Huang X, Wei X, Sang T, Zhao Q, Feng Q, Zhao Y, Li C, Zhu C, Lu T, Zhang Z et al (2010) Genome-wide association studies of 14 agronomic traits in rice landraces. Nat Genet 42:961–967
Huang X, Zhao Y, Wei X, Li C, Wang A, Zhao Q, Li W, Guo Y, Deng L, Zhu C et al (2011) Genome-wide association study of flowering time and grain yield traits in a worldwide collection of rice germplasm. Nat Genet 44:32–39
Hussein RA, El-Anssary AA (2019) Plants secondary metabolites: the key drivers of the pharmacological actions of medicinal plants. In: Builders PF (ed) Herbal medicine. IntechOpen, London
Hwang E-Y, Song Q, Jia G, Specht JE, Hyten DL, Costa J, Cregan PB (2014) A genome-wide association study of seed protein and oil content in soybean. BMC Genomics 15:1
Jain M (2012) Next-generation sequencing technologies for gene expression profiling in plants. Brief Funct Genomics 11:63–70
Jain M, Moharana KC, Shankar R, Kumari R, Garg R (2014) Genomewide discovery of DNA polymorphisms in rice cultivars with contrasting drought and salinity stress response and their functional relevance. Plant Biotechnol J 12:253–264
Jiang J, Zhang C, Wang X (2015) A recently evolved isoform of the transcription factor BES1 promotes brassinosteroid signaling and development in Arabidopsis thaliana. Plant Cell 27:361–374
Jwa N-S, Agrawal GK, Tamogami S, Yonekura M, Han O, Iwahashi H, Rakwal R (2006) Role of defense/stress-related marker genes, proteins and secondary metabolites in defining rice self-defense mechanisms. Plant Physiol Biochem 44:261–273

Karan R, DeLeon T, Biradar H, Subudhi PK (2012) Salt stress induced variation in DNA methylation pattern and its influence on gene expression in contrasting rice genotypes. PLoS One 7: e40203

Kawahara Y, de la Bastide M, Hamilton JP, Kanamori H, McCombie WR, Ouyang S, Schwartz DC, Tanaka T, Wu J, Zhou S et al (2013) Improvement of the Oryza sativa Nipponbare reference genome using next generation sequence and optical map data. Rice (N Y) 6:4

Kim D, Pertea G, Trapnell C, Pimentel H, Kelley R, Salzberg SL (2013) TopHat2: accurate alignment of transcriptomes in the presence of insertions, deletions and gene fusions. Genome Biol 14:36

Kim D, Langmead B, Salzberg SL (2015) HISAT: a fast spliced aligner with low memory requirements. Nat Methods 12:357–360

Klockenbusch C, Kast J (2010) Optimization of formaldehyde cross-linking for protein interaction analysis of non-tagged integrin β 1. J Biomed Biotechnol 2010:1–13

Knasmüller S, Nersesyan A, Misík M, Gerner C, Mikulits W, Ehrlich V, Hoelzl C, Szakmary A, Wagner K-H (2008) Use of conventional and -omics based methods for health claims of dietary antioxidants: a critical overview. Br J Nutr 99:3–52

Krueger F, Andrews SR (2011) Bismark: a flexible aligner and methylation caller for bisulfite-Seq applications. Bioinformatics 27:1571–1572

Kuleshov MV, Jones MR, Rouillard AD, Fernandez NF, Duan Q, Wang Z, Koplev S, Jenkins SL, Jagodnik KM, Lachmann A et al (2016) Enrichr: a comprehensive gene set enrichment analysis web server 2016 update. Nucleic Acids Res 44:90–97

Kumar R, Ichihashi Y, Kimura S, Chitwood DH, Headland LR, Peng J, Maloof JN, Sinha NR (2012a) A high-throughput method for Illumina RNA-Seq library preparation. Front Plant Sci 3:202

Kumar S, Banks TW, Cloutier S (2012b) SNP discovery through next-generation sequencing and its applications. Int J Plant Genomics 2012:831460

Kump KL, Bradbury PJ, Wisser RJ, Buckler ES, Belcher AR, Oropeza-Rosas MA, Zwonitzer JC, Kresovich S, McMullen MD, Ware D et al (2011) Genome-wide association study of quantitative resistance to southern leaf blight in the maize nested association mapping population. Nat Genet 43:163–168

Lam H-M, Xu X, Liu X, Chen W, Yang G, Wong F-L, Li M-W, He W, Qin N, Wang B et al (2010) Resequencing of 31 wild and cultivated soybean genomes identifies patterns of genetic diversity and selection. Nat Genet 42:1053–1059

Lang-Mladek C, Popova O, Kiok K, Berlinger M, Rakic B, Aufsatz W, Jonak C, Hauser M-T, Luschnig C (2010) Transgenerational inheritance and resetting of stress-induced loss of epigenetic gene silencing in Arabidopsis. Mol Plant 3:594–602

Li B, Dewey CN (2011) RSEM: accurate transcript quantification from RNA-Seq data with or without a reference genome. BMC Bioinformatics 12:323

Li Y, Tollefsbol TO (2011) DNA methylation detection: bisulfite genomic sequencing analysis. Methods Mol Biol 791:11–21

Li L, Wei D (2015) Bioinformatics tools for discovery and functional analysis of single nucleotide polymorphisms. Adv Exp Med Biol 827:287–310

Li F, Zheng Q, Vandivier LE, Willmann MR, Chen Y, Gregory BD (2012) Regulatory impact of RNA secondary structure across the Arabidopsis transcriptome. Plant Cell 24:4346–4359

Lindroth AM (2001) Requirement of CHROMOMETHYLASE3 for maintenance of CpXpG methylation. Science 292:2077–2080

Lister R, Gregory BD, Ecker JR (2009) Next is now: new technologies for sequencing of genomes, transcriptomes, and beyond. Curr Opin Plant Biol 12:107–118

Lluveras-Tenorio A, Vinciguerra R, Galano E, Blaensdorf C, Emmerling E, Perla Colombini M, Birolo L, Bonaduce I (2017) GC/MS and proteomics to unravel the painting history of the lost Giant Buddhas of Bāmiyān (Afghanistan). PLoS One 12:e0172990

Love MI, Huber W, Anders S (2014) Moderated estimation of fold change and dispersion for RNA-seq data with DESeq2. Genome Biol 15:550

Maere S, Heymans K, Kuiper M (2005) BiNGO: a Cytoscape plugin to assess overrepresentation of gene ontology categories in biological networks. Bioinformatics 21:3448–3449

Manning K, Tör M, Poole M, Hong Y, Thompson AJ, King GJ, Giovannoni JJ, Seymour GB (2006) A naturally occurring epigenetic mutation in a gene encoding an SBP-box transcription factor inhibits tomato fruit ripening. Nat Genet 38:948–952

Marees AT, de Kluiver H, Stringer S, Vorspan F, Curis E, Marie-Claire C, Derks EM (2018) A tutorial on conducting genome-wide association studies: quality control and statistical analysis. Int J Methods Psychiatr Res 27:e1608

McCouch SR, Zhao K, Wright M, Tung C-W, Ebana K, Thomson M, Reynolds A, Wang D, DeClerck G, Ali ML et al (2010) Development of genome-wide SNP assays for rice. Breed Sci 60:524–535

McNally KL, Childs KL, Bohnert R, Davidson RM, Zhao K, Ulat VJ, Zeller G, Clark RM, Hoen DR, Bureau TE et al (2009) Genomewide SNP variation reveals relationships among landraces and modern varieties of rice. Proc Natl Acad Sci U S A 106:12273–12278

Meyer RS, Choi JY, Sanches M, Plessis A, Flowers JM, Amas J, Dorph K, Barretto A, Gross B, Fuller DQ et al (2016) Domestication history and geographical adaptation inferred from a SNP map of African rice. Nat Genet 48:1083–1088

Mirouze M, Reinders J, Bucher E, Nishimura T, Schneeberger K, Ossowski S, Cao J, Weigel D, Paszkowski J, Mathieu O (2009) Selective epigenetic control of retrotransposition in Arabidopsis. Nature 461:427–430

Miura K, Agetsuma M, Kitano H, Yoshimura A, Matsuoka M, Jacobsen SE, Ashikari M (2009) A metastable DWARF1 epigenetic mutant affecting plant stature in rice. Proc Natl Acad Sci U S A 106:11218–11223

Mundade R, Ozer HG, Wei H, Prabhu L, Lu T (2014) Role of ChIP-seq in the discovery of transcription factor binding sites, differential gene regulation mechanism, epigenetic marks and beyond. Cell Cycle 13:2847–2852

Nadeau OW, Carlson GM (2007) Protein interactions captured by chemical cross-linking: one-step cross-linking with formaldehyde. CSH Protoc 2007:4634

Page GP, Zakharkin SO, Kim K, Mehta T, Chen L, Zhang K (2007) Microarray analysis. Methods Mol Biol 404:409–430

Parida SK, Verma M, Yadav SK, Ambawat S, Das S, Garg R, Jain M (2015) Development of genome-wide informative simple sequence repeat markers for large-scale genotyping applications in chickpea and development of web resource. Front Plant Sci 6:645

Patel DP, Krausz KW, Xie C, Beyoğlu D, Gonzalez FJ, Idle JR (2017) Metabolic profiling by gas chromatography-mass spectrometry of energy metabolism in high-fat diet-fed obese mice. PLoS One 12:e0177953

Patro R, Duggal G, Love MI, Irizarry RA, Kingsford C (2017) Salmon: fast and bias-aware quantification of transcript expression using dual-phase inference. Nat Methods 14:417–419

Perez-Riverol Y, Alpi E, Wang R, Hermjakob H, Vizcaíno JA (2015) Making proteomics data accessible and reusable: current state of proteomics databases and repositories. Proteomics 15:930–950

Pollier J, Rombauts S, Goossens A (2013) Analysis of RNA-Seq data with TopHat and cufflinks for genome-wide expression analysis of jasmonate-treated plants and plant cultures. Methods Mol Biol 1011:305–315

Rafalski A (2002) Applications of single nucleotide polymorphisms in crop genetics. Curr Opin Plant Biol 5:94–100

Rajkumar MS, Shankar R, Garg R, Jain M (2020) Bisulphite sequencing reveals dynamic DNA methylation under desiccation and salinity stresses in rice cultivars. Genomics 112:3537–3548

Rao MS, Van Vleet TR, Ciurlionis R, Buck WR, Mittelstadt SW, Blomme EAG, Liguori MJ (2018) Comparison of RNA-Seq and microarray gene expression platforms for the toxicogenomic evaluation of liver from short-term rat toxicity studies. Front Genet 9:636

Reinders J, Lewandrowski U, Moebius J, Wagner Y, Sickmann A (2004) Challenges in mass spectrometry-based proteomics. Proteomics 4:3686–3703

Reinert K, Langmead B, Weese D, Evers DJ (2015) Alignment of next-generation sequencing reads. Annu Rev Genomics Hum Genet 16:133–151

Reuter JS, Mathews DH (2010) RNAstructure: software for RNA secondary structure prediction and analysis. BMC Bioinformatics 11:129

Ritchie ME, Phipson B, Wu D, Hu Y, Law CW, Shi W, Smyth GK (2015) Limma powers differential expression analyses for RNA-sequencing and microarray studies. Nucleic Acids Res 43:e47

Robinson MD, McCarthy DJ, Smyth GK (2010) edgeR: a Bioconductor package for differential expression analysis of digital gene expression data. Bioinformatics 26:139–140

Saito K, Matsuda F (2010) Metabolomics for functional genomics, systems biology, and biotechnology. Annu Rev Plant Biol 61:463–489

Saze H, Mittelsten Scheid O, Paszkowski J (2003) Maintenance of CpG methylation is essential for epigenetic inheritance during plant gametogenesis. Nat Genet 34:65–69

Schubert OT, Röst HL, Collins BC, Rosenberger G, Aebersold R (2017) Quantitative proteomics: challenges and opportunities in basic and applied research. Nat Protoc 12:1289–1294

Schumacher A, Kapranov P, Kaminsky Z, Flanagan J, Assadzadeh A, Yau P, Virtanen C, Winegarden N, Cheng J, Gingeras T et al (2006) Microarray-based DNA methylation profiling: technology and applications. Nucleic Acids Res 34:528–542

Seal A, Gupta A, Mahalaxmi M, Aykkal R, Singh TR, Arunachalam V (2014) Tools, resources and databases for SNPs and indels in sequences: a review. Int J Bioinform Res Appl 10:264–296

Shang X, Cao Y, Ma L (2017) Alternative splicing in plant genes: a means of regulating the environmental fitness of plants. Int J Mol Sci 18:432

Shankar R, Bhattacharjee A, Jain M (2016) Transcriptome analysis in different rice cultivars provides novel insights into desiccation and salinity stress responses. Sci Rep 6:23719

Shinozaki K, Sakakibara H (2009) Omics and bioinformatics: an essential toolbox for systems analyses of plant functions beyond 2010. Plant Cell Physiol 50:1177–1180

Soppe WJ, Jacobsen SE, Alonso-Blanco C, Jackson JP, Kakutani T, Koornneef M, Peeters AJ (2000) The late flowering phenotype of fwa mutants is caused by gain-of-function epigenetic alleles of a homeodomain gene. Mol Cell 6:791–802

Steemers FJ, Gunderson KL (2007) Whole genome genotyping technologies on the BeadArray platform. Biotechnol J 2:41–49

Thudi M, Upadhyaya HD, Rathore A, Gaur PM, Krishnamurthy L, Roorkiwal M, Nayak SN, Chaturvedi SK, Basu PS, Gangarao NVPR et al (2014) Genetic dissection of drought and heat tolerance in chickpea through genome-wide and candidate gene-based association mapping approaches. PLoS One 9:e96758

Tian F, Bradbury PJ, Brown PJ, Hung H, Sun Q, Flint-Garcia S, Rocheford TR, McMullen MD, Holland JB, Buckler ES (2011) Genome-wide association study of leaf architecture in the maize nested association mapping population. Nat Genet 43:159–162

Trapnell C, Pachter L, Salzberg SL (2009) TopHat: discovering splice junctions with RNA-Seq. Bioinformatics 25:1105–1111

Van der Auwera GA, Carneiro MO, Hartl C, Poplin R, Del Angel G, Levy-Moonshine A, Jordan T, Shakir K, Roazen D, Thibault J et al (2013) From FastQ data to high confidence variant calls: the Genome Analysis Toolkit best practices pipeline. Curr Protoc Bioinformatics 43:1–33

Verma M, Kumar V, Patel RK, Garg R, Jain M (2015) CTDB: an integrated chickpea transcriptome database for functional and applied genomics. PLoS One 10:e0136880

Wang K, Singh D, Zeng Z, Coleman SJ, Huang Y, Savich GL, He X, Mieczkowski P, Grimm SA, Perou CM et al (2010) MapSplice: accurate mapping of RNA-seq reads for splice junction discovery. Nucleic Acids Res 38:e178

Wang W-S, Pan Y-J, Zhao X-Q, Dwivedi D, Zhu L-H, Ali J, Fu B-Y, Li Z-K (2011) Drought-induced site-specific DNA methylation and its association with drought tolerance in rice (Oryza sativa L.). J Exp Bot 62:1951–1960

Wang L, Liu Y, Zhong X, Liu H, Lu C, Li C, Zhang H (2019a) DMfold: a novel method to predict RNA secondary structure with pseudoknots based on deep learning and improved base pair maximization principle. Front Genet 10:143

Wang Z, Wang M, Wang T, Zhang Y, Zhang X (2019b) Genome-wide probing RNA structure with the modified DMS-MaPseq in Arabidopsis. Methods 155:30–40

Wu DC, Yao J, Ho KS, Lambowitz AM, Wilke CO (2018) Limitations of alignment-free tools in total RNA-seq quantification. BMC Genomics 19:510

Xie Y, Wu G, Tang J, Luo R, Patterson J, Liu S, Huang W, He G, Gu S, Li S et al (2014) SOAPdenovo-trans: de novo transcriptome assembly with short RNA-Seq reads. Bioinformatics 30:1660–1666

Yang S-O, Lee SW, Kim YO, Sohn S-H, Kim YC, Hyun DY, Hong YP, Shin YS (2013) HPLC-based metabolic profiling and quality control of leaves of different Panax species. J Ginseng Res 37:248–253

Yang X, Yang M, Deng H, Ding Y (2018) New era of studying RNA secondary structure and its influence on gene regulation in plants. Front Plant Sci 9:671

Yang G, Liang K, Zhou Z, Wang X, Huang G (2020) UPLC-ESI-MS/MS-based widely targeted metabolomics analysis of wood metabolites in teak (Tectona grandis). Molecules 25:2189

Zemach A, Kim MY, Silva P, Rodrigues JA, Dotson B, Brooks MD, Zilberman D (2010) Local DNA hypomethylation activates genes in rice endosperm. Proc Natl Acad Sci U S A 107:18729–18734

Zhang Y, Liu T, Meyer CA, Eeckhoute J, Johnson DS, Bernstein BE, Nusbaum C, Myers RM, Brown M, Li W et al (2008) Model-based analysis of ChIP-Seq (MACS). Genome Biol 9:R137

Zhao K, Tung C-W, Eizenga GC, Wright MH, Ali ML, Price AH, Norton GJ, Islam MR, Reynolds A, Mezey J et al (2011) Genome-wide association mapping reveals a rich genetic architecture of complex traits in Oryza sativa. Nat Commun 2:467

Zhong S, Joung J-G, Zheng Y, Chen Y, Liu B, Shao Y, Xiang JZ, Fei Z, Giovannoni JJ (2011) High-throughput illumina strand-specific RNA sequencing library preparation. Cold Spring Harb Protoc 2011:940–949

# Omics Approaches for Understanding Plant Defense Response

3

Prajna Priyadarshini Das, Sumi Rana, Mehanathan Muthamilarasan, Monica Kannan, and Irfan Ahmad Ghazi

**Abstract**

Plants are the major components that contribute to energy, environment, and ecosystem, and they are also the primary producers of the food chain. Despite their importance, their sustenance in the environment is challenged by several biotic and abiotic factors. Among the biotic factors, diseases and infections cause devastative results, and in agriculture, such biotic stresses caused by pathogens lead to a severe decrease in yield and productivity, which ultimately challenges food security. Plants have also developed sophisticated molecular mechanisms to defend the pathogens, thus leading to resistance or tolerance to the given disease. Understanding the mechanism of tolerance or resistance is now imperative to gain insights into the molecular machinery underlying such defense responses, which could be further exploited to develop disease-resistant plant species. To study the response of plants to pathogens, different approaches have been developed that interrogates the system at varying levels of disease infection. These approaches are generically called "omics" approaches that enable the study of plant systems at the genome, transcriptome, proteome, and metabolome levels. The advent of tools and techniques has advanced these omics approaches, and the knowledge generated so far has been proven useful in developing elite cultivars resistant to

Prajna Priyadarshini Das, Sumi Rana and Mehanathan Muthamilarasan contributed equally with all other contributors.

P. P. Das · S. Rana · M. Muthamilarasan · I. A. Ghazi (✉)
Department of Plant Sciences, School of Life Sciences, University of Hyderabad, Hyderabad, Telangana, India
e-mail: irfan@uohyd.ac.in

M. Kannan
Proteomics Core Facility, School of Life Sciences, University of Hyderabad, Hyderabad, Telangana, India



A. Kumar et al. (eds.), *Omics Technologies for Sustainable Agriculture and Global Food Security Volume 1*, https://doi.org/10.1007/978-981-16-0831-5_3

pathogens. Transgene-based approaches and/or molecular breeding-based techniques are now being used to develop such improved varieties, whereas the introduction of genome editing tools like CRISPR/Cas9 is expected to expedite the crop improvement programs. Given this, the present chapter enumerates the use of different omics approaches, namely, transcriptomics, proteomics, and metabolomics, to delineate the molecular mechanism underlying disease response, and how this information could be integrated to advance the current understanding of plant defense response.

**Keywords**

Transcriptomics · Proteomics · Metabolomics · Plant defense · Plant-pathogen interaction · Molecular response · Disease resistance

## 3.1 Introduction

The defense response of plants to biotic stresses is not fully understood even when the fundamental processes underlying defense have been delineated (Muthamilarasan and Prasad 2013). The response of plants to each pathogen is unique, and therefore, gaining a complete insight into the molecular mechanisms regulating such tolerance or resistance is essential. The advent of "omics" has enabled studying the molecular response of plants to biotic stresses and the factors regulating the defense response. Among the different "omics" approaches, transcriptomics, proteomics, and metabolomics form the key tools to unlock the molecular mystery that the plants hold to effectively circumvent any pathogen. Genomics has enabled the sequencing of plant genomes, thus providing direct access to the genes encoded in the chromosomes. Annotation of the genes thus identified has provided information about the defense-related genes present in the genome; however, their expression as well as the regulation are unclear. Further, genes are present in families, where more than one gene constitutes a multigene family, and these gene family members are differentially regulated during growth, development, and stress response. A few gene families get upregulated during abiotic stresses, and several get induced after perceiving a biotic stress signal. To gain a better understanding of the expressed genes and their mode of regulation, transcriptomics approaches are being deployed. Transcriptomics investigates only the expressed RNAs, which could be mRNA or small RNA that play roles in defense response. Also, transcriptomics facilitates the study of splice variants that genomics overlooks due to the lack of resolution at the genome level. Further, transcriptomics enables the study of expression levels of each gene expressed at a given timepoint or tissue. Transcriptomics has played a significant role in understanding gene expression dynamics during pathogen infection and disease progression (Martin et al. 2013).

However, not all the transcripts are translated to form a functional protein as several studies have shown that the abundance of mRNA does not correlate with the amount of protein that is synthesized, and therefore, proteomics gains importance to study the dynamics of gene expression at proteome levels. Genomics and transcriptomics might provide information at the gene and expression level, but proteomics will suggest the expression of proteins, abundance, and their post-translational modifications. A functional protein would have undergone several modifications after translation, and thus, proteomics is important to study such proteins that play a direct role in defense response. In addition to proteins, metabolites are also involved in defense response, wherein the primary, secondary, and tertiary metabolites have unique roles to play in defending the pathogen. This underlines the importance of studying the metabolites using metabolomics approaches in plants that are challenged with pathogens.

Integrated omics involves the amalgamation of knowledge generated using each omics tool to identify the precise molecular machinery underlying defense response and characterize the genetic determinants (genes, alleles, or QTLs) to regulate the tolerance traits. While genomics provides the data on a complete set of genes present in the given organism, transcriptomics enables identifying expressed subset of genes during a pathogen challenge. Proteomics pinpoints the proteins specifically translated from the transcripts that are upregulated during pathogen stress, and metabolomics identifies the metabolites secreted to circumvent the pathogen. Such an integrated approach is the need for the time to gain a thorough insight into the defense machinery. Despite the efforts invested in understanding the same, not a complete picture has been captured yet as the plants deploy several unique strategies to resist the pathogen attack and disease development. Several such mechanisms are species-, or cultivar-specific, and therefore, comparative omics approaches are now seeing dawn in expediting the search for knowledge on molecular defense response. In view of this, the present chapter has been structured to enumerate the different omics approaches used to study the molecular defense with examples and summarize the way forward.

## 3.2 Transcriptomics

"Transcriptomics" is a generic term that indicates the study of "transcripts" or "RNAs" that are produced as a result of transcription. Among these, mRNAs encode for different proteins or enzymes obtained during translation, whereas other classes of RNAs play a multitude of functions, and altogether, the RNAome plays diverse roles in growth, development, and stress response. Thus, to identify and characterize the RNAs involved in molecular stress response, particularly to biotic stresses, transcriptomics is important. To facilitate this, several approaches have evolved and are majorly classified into non-sequencing and sequencing-based methods. Next-generation sequencing has transformed the study of RNAs, wherein the total transcriptome of a cell, tissue, or organ can be captured, sequenced, and annotated to identify the dynamics of gene expression (Ozsolak and Milos 2011). Given this, the forthcoming sections enumerate the non-sequencing and sequencing-based methods of dissecting the transcriptomes of plants during patho-stress.

## 3.2.1 Non-sequencing Based Transcriptomics

### 3.2.1.1 Microarray

Among the non-sequencing based transcriptomic approaches, microarray has a prominent place as it served as a highly preferred platform for large-scale analysis of gene expression at genome-wide levels (Taub et al. 1983; Pollack et al. 1999). In principle, microarray involves the synthesis of chips wherein the labelled single strands of DNA or cDNA were cross-linked to a solid surface. Presently, these probes are gene-printed on the slides to produce commercially viable microarray gene chips (Goldmann and Gonzalez 2000). Following this, hybridization is allowed to occur by adding the mRNA or single-stranded cDNA of the test sample onto the chip. The binding of the test strand with the DNA or cDNA probe in the chip develops a signal, which could be either colored or radioactive (Shalon et al. 1996). These signals are detected using analytical tools, and the data is processed to identify the gene expression along with the level of expression (Wei et al. 2004). California-based Affymetrix, Inc. are well known for developing commercial high-density oligo microarrays. The hybridization is captured as an image file which is further; (1) normalized using algorithms like RMA (Robust MultiArray Average), (2) used for identification of differentially expressed genes (using *t*-test), and (3) analysis of gene ontology, functional pathways, enrichment, etc.

Microarray has been widely used to understand the transcriptomic complexities associated with plant defense to environmental stresses (Lodha and Basak 2012). Baldwin et al. (1999) were the first probable researchers to test the use of microarray in maize (*Zea mays*) and identified 117 genes that were differentially expressed in response to the infection of a fungal pathogen, *Cochliobolus carbonum.* In *Arabidopsis,* Schenk et al. (2000) studied the transcriptome during *Alternaria brassicicola* infection (incompatible) to identify the genes showing altered expression. The study identified 168 upregulated and 39 downregulated genes in *A. thaliana* after infection. In the same year, Reymond et al. (2000) studied the gene expression in *A. thaliana* during mechanical wounding and feeding by *Pieris rapae* (cabbage butterfly insect). In this study, the expression of 150 genes was studied in wild type as well as a coronatine-insensitive *coi1-1* mutant to identify genes that play a role in defense response. The alteration in gene expression during systemic acquired resistance (SAR) was performed by Maleck et al. (2000) in *Arabidopsis.* The study identified a set of candidate genes that were specifically upregulated during SAR. In continuation of this, Wang et al. (2005) studied the gene expression levels in *NPR1* (Non-expression of Pathogenesis-Related 1) overexpressed as well as in mutant lines (*npr1*) during SAR. A comparison of the datasets showed that the SAR was induced by NPR1 by two mechanisms, viz., enhancing the expression of PR proteins, and induction of protein secretory pathway. Roots and shoots of tomato (*S. lycopersicum*) plant subjected to *Tomato spotted wilt virus* infection showed organ-specific responses (Catoni et al. 2009). In the shoot, defense and signal transduction-related genes were upregulated, whereas in roots, the genes having a role in biotic stresses were highly expressed. In barley (*Hordeum vulgare*), non-host resistance to three different fungal (host/non-host) pathogens, namely *Blumeria* (powdery mildew), *Puccinia* (rust), and *Magnaporthe* (blast) was studied using microarray (Zellerhoff et al. 2010). The study showed that the pathogen infections modulated the activation or deactivation of common metabolic or signalling pathways without a distinct difference between PR and non-host specific genes.

In rice (*Oryza sativa*), Zhou et al. (2010) have studied the global gene expression pattern during infection with a non-host bacterial pathogen, *Xanthomonas oryzae* pv. *oryzicola* (*Xoc*). Maize R gene, *Rxo1* (a dominant NBS-LRR type *R* gene), was overexpressed in transgenic rice lines prior to the infection (Zhao et al. 2005). Infection of *Xoc* in transgenic as well as non-transgenic rice lines followed by expression analysis using microarray revealed the involvement of *Rxo1* that activates a broad set of genes leading to a hypersensitive response against Xoc infection. Among these genes, *OsNPR1* and *OsPR1* were identified as probable candidates for further functional characterization. The application of microarray to study the gene expression expands beyond the well-studied species. In Italian ryegrass (*Lolium multiflorum*), this approach was used to identify the genes involved in defense response to *Xanthomonas translucens* pv. *graminis* (Wichmann et al. 2011). The study identified differential expression of at least 1200 genes, which were then mapped onto the genetic linkage map (Studer et al. 2006), and further analysis was performed. Altogether, the study pinpointed that a Low silicon protein 1 encoding gene, *Lsi1,* could be a potential candidate gene for marker-assisted selection for disease resistance (Wichmann et al. 2011).

In nature, plants encounter more than one stress at a given point of time, and studies have underlined that the response of plants to combined stresses is unique compared to their response to individual stresses (Ramegowda and Senthil-Kumar 2015). This demanded the study and understanding of the transcriptomic dynamics of plants exposed to multiple stress combinations in the natural environment. *A. thaliana* plants exposed to individual and combined challenge with drought as well as *Pseudomonas syringae* pv tomato DC3000 infection showed unique expression of twenty novel genes during the combined drought and pathogen stress (Gupta et al. 2016). One of these genes was *AtGBF3,* a G-Box Binding Factor 3 (transcription factor) that showed significant upregulation during individual as well as combined stresses. Dixit et al. (2019) had further characterized this gene by overexpression and knockout approaches in *A. thaliana*. The *AtGBF3* overexpression lines were observed to be tolerant to both individual and combined stresses, whereas the mutant plants were susceptible. A similar study was performed in chickpea (*Cicer arietinum*) to identify the genes playing roles in drought and pathogen interaction by Sinha et al. (2017). Here, the plants were challenged with drought stress and infection by *Ralstonia solanacearum* (causative agent of wild disease) both individually and in a combined pattern. The authors then correlated the downregulation of defense-related genes with *in planta* multiplication of

*R. solanacearum*. The upregulation of genes having roles in hormone signalling and lignin biosynthesis was identified to be potential candidates for functional characterization. The impact of pathogen infection during drought recovery was studied by Gupta and Senthil-Kumar (2017), wherein microarray analysis revealed the specific upregulation of proline dehydrogenase (*AtProDH1*) which could be studied further to delineate its role in imparting tolerance to dual stress. Recently, Fatima et al. (2019) performed a comparative study on how the host, *Brassica juncea,* and the non-host, *C. arietinum* respond to *A. brassicae* at morpho-pathological and molecular levels. Comprehensive microarray analyses identified several genes involved in multilayered pathogen defense. These genes could serve as putative candidates to study their precise involvement in non-host resistance and help in the development of blight-resistant transgenic lines *Brassica* sp.

The large-scale microarray datasets were deposited in open-access repositories, including NCBI-Gene Expression Omnibus (https://www.ncbi.nlm.nih.gov/geo/) and EMBL-ArrayExpress (https://www.ebi.ac.uk/arrayexpress/). The availability of this information has enabled several secondary studies and further analyses of microarray datasets. For example, Ramu et al. (2016) had performed a meta-analysis of sunflower microarray datasets to identify commonly regulated genes that were then characterized to delineate their roles in conferring tolerance to environmental stresses. Altogether, microarray remained a powerful tool to resolve the transcriptomic intricacies towards understanding the regulation of gene expression at a genome-wide level; however, the advent of next-generation sequencing has overtaken the pride of microarray. This has resulted in microarray becoming the least preferred approach as compared to NGS-based transcriptomics studies.

#### 3.2.1.2 Northern Blotting and Quantitative Real-Time PCR

Before microarray, gene expression analysis and quantification of transcripts were performed using the Northern Blot technique. RNA blot or Northern blot enabled the validation of gene expression by detecting the presence of corresponding mRNAs through autoradiographic signals (Alwine et al. 1977). To achieve this, the total RNA extracted from the sample will be subjected to mRNA isolation and agarose gel electrophoresis. The mRNA molecules in the gel will then be transferred to the nylon membrane through capillary action, and the mRNAs are cross-linked with the membrane using UV irradiation. The probes with complementary sequences to the test gene are synthesized and radio-labelled, and are allowed to hybridize with the nylon membrane containing the mRNA. After hybridization, the membrane is washed to remove unhybridized probes, and then the autoradiogram is generated and analyzed for signals. The signals on X-ray film could then be quantified using densitometry. Though Northern blotting appears to be a reliable experimental method used to validate and quantify RNAs (Kevil et al. 1997), the approach does not support a large-scale analysis of genes. A genome containing thousands of genes cannot be analyzed for its expression data using Northern blot as it is time taking, low-throughput, and labor-intensive. However, to date, the approach is used for validation of results obtained through other transcriptomics studies. Recently, the application of Northern blots in studying the microRNAs (miRNA) was also realized. For instance, *A. thaliana* infected with *Oil-seed Rape Mosaic Virus* were analyzed for the expression of stress-responsive miRNAs (Hajdarpašić and Ruggenthaler 2012). Northern blot showed that 28 miRNAs were differentially expressed during viral infection. Recently, Sharma and Prasad (2020) have performed *in planta* silencing of the *AC1* gene of *Tomato leaf curl virus* in tomato using artificial miRNA (AC1-amiR), and Northern blot was used to detect the AC1-amiR and viral AC1 expression. Interestingly, the transgenic tomato lines overexpressing AC1-amiR, when challenged with leaf curl virus, showed reduced disease symptoms and high percentage resistance ranging between ~40 and 80%. These examples underline the importance of Northern blot in the analysis of gene expression in several high-end investigations being conducted worldwide.

Two approaches advanced the Northern blotting method of transcript analysis, and they are semi-quantitative reverse-transcriptase PCR and quantitative Real-Time PCR (qRT-PCR). The former is a gel-based approach where the mRNA is reverse transcribed to produce cDNAs that are resolved on agarose gel (Bell 1995). The band intensity was measured using densitometry to quantify the transcript abundance (Overbergh et al. 2003). The efficiency and convenience of this approach held its popularity until the arrival of qRT-PCR (Higuchi et al. 1993). RT-qPCR or qRT-PCR allows the real-time detection of transcript abundance that can be translated to predict the expression levels in terms of "relative expression" or "fold-change." Studies involving the identification of genes and gene families having roles in growth, development, and stress response have started using qRT-PCR as a tool to validate their findings. Also, qRT-PCR serves as a prominent tool in studying the expression levels of genes across different samples, time-points, and treatments. Precisely, the method works by recording the intensity of light emitted by the fluorochrome integrated into the newly synthesized strand using PCR (Livak and Schmittgen 2001). Compared to Northern and semi-quantitative reverse-transcriptase PCR, qRT-PCR is rapid and sensitive with high specificity as well as scalability (Udvardi et al. 2008), as demonstrated by Czechowski et al. (2004) wherein the expression of thousands of Arabidopsis transcription factors were studied using qRT-PCR. The study showed that the genes detected by qRT-PCR were absent in *Arabidopsis* EST library, thus underlining the precision of this approach. Thus, the approach has gained momentum in functional genomics studies, and the search for suitable reference genes to quantify the expression levels has also expanded widely. Several computational algorithms are developed to identify suitable reference genes or internal controls for qRT-PCR, and this includes geNorm (Vandesompele et al. 2002), NormFinder (Andersen et al. 2004), BestKeeper (Pfaffl et al. 2004), and RefFinder (Silver et al. 2006). Recently, Pandey et al. (2019) had identified 131 genes encoding DEAD-box RNA helicase in tomato, and expression analysis of these genes in tomato cultivars using qRT-PCR showed that *SlDEAD23* and *SlDEAD35* were involved in multiple stress response. Previously, Mandal et al. (2018) had studied Armadillo repeat family (ARM) genes in tomato and has identified significant upregulation of *SlARM18* in the tomato cultivar tolerant to leaf curl disease as compared to the susceptible cultivar. Further functional characterization of the gene delineated the precise tolerance mechanism in tomato against

Tomato leaf curl disease. These are a few examples where qRT-PCR has been successfully used in studying the expression of genes and gene families; however, the approach could not provide genome-wide coverage.

## 3.2.2 Sequencing-Based Transcriptomics

### 3.2.2.1 Expressed Sequence Tags

A short sub-sequence of a cDNA (usually $<1000$ bp) is called expressed sequence tag (EST), which typically represents the gene that has been transcribed to derive a functional mRNA. ESTs enabled the identification of genes and determination of their sequences, and therefore, the advent of sequencing approaches has resulted in a massive accumulation of EST data in public databases. NCBI had a separate database for ESTs (dbEST), which subsequently merged into "Nucleotide" database. Large-scale single-pass sequencing of cDNA clones to generate ESTs was first reported in humans (Adams et al. 1993) that was later adapted to plants. In potato, Crookshanks et al. (2001) had reported around 6000 ESTs, and later, Ronning et al. (2003) generated 61,940 ESTs from different tissues of potato (*Solanum tuberosum*), including the tissues infected with late blight pathogen (*Phytophthora infestans*). The study identified several ESTs that are exclusively expressed during the incompatible interaction with *P. infestans*. Jantasuriyarat et al. (2005) had subjected rice (*Oryza sativa*) cultivars IR36, IR68, and Nipponbare to rice blast fungus (*Magnaporthe grisea*) infection and sequenced the cDNA libraries derived from mRNAs isolated at different time-points. The study identified ten highly induced and suppressed genes in contrasting cultivars, and their expressions were confirmed using Northern Blot. In *Brassica oleracea*, Cramer et al. (2006) had identified several ESTs encoding potential genes that confer tolerance to black spot disease (caused by *Alternaria brassicicola*). Among legumes, the sequencing of ESTs in chickpea has been reported. Ashraf et al. (2009) had identified 6272 ESTs in chickpea that were differentially expressed during fusarium wilt (caused by *Fusarium oxysporum ciceri*). Functional annotation of these ESTs provided insights into the regulators that modulate host-pathogen interaction. Using the Brazilian Coffee Genome Project Database, Alvarenga et al. (2010) identified 11,300 ESTs in coffee (*Coffea* spp.) genome and annotated them for their involvement in defense response. ESTs encoding for catalase, chitinase, protein with a BURP domain, and unknown proteins were predominantly found in the database, and their corresponding genes were speculated to be involved in resistance to diseases. Recently, Mahomed and van den Berg (2011) has studied the avocado rootstock infected with Phytophthora root rot (caused by *Phytophthora cinnamomi*), and identified the ESTs associated with disease response. Though these studies have made notable findings in determining the candidate genes, the EST sequence method suffered several disadvantages. Mainly, it does not provide a genome-wide coverage as only a limited number of mRNAs undergo reverse transcription to produce cDNAs. Further cloning of cDNAs into plasmids, followed by transformation and sequencing is laborious and time-consuming. Despite the disadvantages, EST sequencing helped develop several gene-based molecular markers (Varshney et al. 2005). EST sequences facilitated the identification of microsatellite motifs in the sequences and designing the primers flanking those motifs (Scott et al. 2000; Kumari et al. 2013b). These microsatellites were exploited as codominant markers that were useful in genetic diversity analyses, population structure prediction, and linkage mapping. Thus, ESTs served as the first step towards the high-throughput analysis of gene expression as well as expediting trait improvement programs using molecular breeding approaches.

#### 3.2.2.2 Tag-Based Sequencing Approaches

As cDNA or EST sequencing has failed to provide genome-wide coverage, expensive and not quantitative, several tag-based approaches were developed to study the transcriptomes. Serial analysis of gene expression (SAGE), cap analysis of gene expression (CAGE), and massively parallel signature sequencing (MPSS) were a few examples of tag-based approaches. Velculescu et al. (1995) were the first to report SAGE in which the cDNAs are digested with an anchoring enzyme (usually *Nla*III) followed by ligation of a linker. Then digestion of tagging enzyme and formation of ditags were performed, followed by PCR amplification, redigestion with the same anchoring enzyme, concatenation, and cloning into a vector. The clones are then sequenced to obtain the data. Matsumura et al. (1999) were the first to use SAGE in plants to identify the genes in rice seedlings. The same group used SuperSAGE, an advanced version of SAGE, to study the gene expression in rice subjected to *Magnaporthe grisea* infection (blast disease; Matsumura et al. 2003). Later, Lee and Lee (2003) identified the genes involved in regulating cold stress tolerance in pollen grains of *A. thaliana*. In cassava (*Manihot esculenta*), Fregene et al. (2004) used SAGE to analyze the gene expression pattern during *Cassava mosaic virus* infection.

In the case of CAGE (Kodzius et al. 2006), the total RNA was used as the input material. Here, random priming and oligo (dT) priming were performed separately, followed by full-length cDNA selection by Cap-trapper method. Following this, linker I ligation and second-strand cDNA synthesis were performed. The double-stranded cDNA is digested with *Mme*I to release the 5′ end-specific CAGE tag, facilitating the ligation of linker II to the sticky end. Now, the linkers I and II provide primer sites for PCR amplification of the tags. After amplification, the PCR product is digested with cloning enzymes, and the tags are ligated into concatemers (~500–800 bp size). The concatemers are cloned into the vector for sequencing, and the sequence data provides information about the mRNAs that were expressed in the given sample (Harbers and Carninci 2005; Kodzius et al. 2006). Brenner et al. (2000) defined the MPSS for the first time, where the mRNAs were subjected to cDNA synthesis followed by digestion with DpnII. The resultant products were fused to TAGs for PCR amplification and coupling to microbeads. This was followed by the determination of 16–20 bp sequence from each bead by hybridization of fluorescent labelled probes. The fluorescent signal from the beads adhered to the two-dimensional surface was analyzed to determine the DNA sequence parallelly from all the beads (Nobuta et al. 2007). Meyers et al. (2007)

had later optimized the procedure for studying plant-pathogen interactions, and Peiffer et al. (2008) used MPSS to study the gene expression during flower development in *A. thaliana*. SAGE, CAGE, and MPSS critically laborious, require expertise, and are time-consuming. Also, the cost of running the reactions was extremely high, and these demerits limited the use of SAGE, CAGE, and MPSS in studying the plant system.

#### 3.2.2.3 Suppression Subtractive Hybridization

Comparative transcriptomics requires identifying genes that were uniquely expressed in one dataset compared to another, and such a comparison requires capturing the complete transcriptome in both the datasets followed by subtraction of the commonly expressed genes. This approach is called SSH or suppression subtractive hybridization, which was first described by Diatchenko et al. (1996). Xiao et al. (2001) used this approach to identify the uniquely expressed genes in transgenic tomato overexpressing the resistance gene, Pto, compared to wild-type tomato. A set of 82 unique genes that have not been reported previously were identified through this study. In *Arabidopsis,* Mahalingam et al. (2003) studied the transcriptome response of plants treated with ozone, bacterial, and oomycete pathogen, and identified a broad repertoire of stress-responsive genes that confer tolerance to these stresses. Sahu et al. (2011) compared the transcriptomes of tolerant as well as susceptible tomato cultivars to Tomato leaf curl disease caused by *Tomato leaf curl New Delhi virus* and identified that 106 non-redundant transcripts were uniquely expressed in the tolerant cultivar upon virus infection. Among these, a 26S proteasomal subunit RPT4a (*SlRPT4*) gene was further characterized to delineate its non-proteolytic function and participation in defense pathway against virus infection in tomato (Sahu et al. 2016). Roohie and Umesha (2015) used SSH to identify the defense responsive genes in cabbage against black rot causing *Xanthomonas campestris* pv. *Campestris*. In wheat, Li et al. (2015) SSH was used to study the transcriptional response against stripe rust caused by *Puccinia striiformis* f. sp. *tritici,* and a resistance allele, *WCBP1* (wheat copper-binding protein) was identified. To investigate the genes underlying crown rust disease caused by *Puccinia coronata* in oat (*Avena sativa*), SSH was performed, and reported the presence of several new genes that were not identified in the previous studies (Loarce et al. 2016). In maize, Dhakal et al. (2017) had compared the transcriptomes of resistant and susceptible inbreds to *Aspergillus flavus* infection and identified 267 unigenes related to defense response. Similarly, Saabale et al. (2018) compared the transcriptomes of contrasting chickpea genotypes to infection with *Fusarium oxysporum* f. sp. *ciceris*. SSH analysis revealed several genes having roles in pathogen response. Recently, Jothiramshekar et al. (2020) had studied the response of a halophytic species, *Suaeda nudiflora* that was subjected to combined stress of salinity and elevated $CO_2$ treatments. Altogether, SSH still remains as one of the preferred approaches to studying the transcriptome dynamics, as it identifies uniquely expressed genes. However, the procedure involved in performing SSH is tedious and requires expertise as several cloning, PCR, and sequencing exercises had to be executed.

#### 3.2.2.4 RNA-Seq

EST sequencing, where the mRNAs were isolated, reverse transcribed, cloned, and sequenced, was later improvised into RNA-Seq to enable sequencing of total RNAome without the need for cloning libraries. Next-generation sequencing (NGS) offers direct sequencing of the total RNA after adapter ligation, PCR, and library preparation. These libraries can then be sequenced using advanced platforms, including Illumina HiSeq or PacBio or Oxford Nanopore. Illumina platform, though produces short reads, is very popular due to its low cost and easiness of sequencing. Also, Illumina sequencing does not require any tedious pre-preparations; however, PacBio requires pre-preparation of the samples. PacBio and Nanopore systems offer long-read sequencing, and it is most useful in sequencing the genomic DNA. RNA-seq data produced by these platforms are processed to identify the genes present in the respective sample. Several standard pipelines have been established to achieve this, and one such pipeline is NGS QC Toolkit (Patel and Jain 2012). These pipelines check the sequence data for base and sequence quality score distributions, average base content per read, GC distribution in the reads, PCR amplification issues, and over-represented sequences. Following this, preprocessing, alignment, and differential expression analysis are performed. Preprocessing includes the trimming of adapter sequences and low-quality bases, followed by the removal of rRNA sequences. Alignment of reads could either be de novo or reference-based, and after alignment, the read count (expressed in FPKM or RPKM) will be predicted to analyze the expression levels of each transcripts. Leebens-Mack et al. (2019) led the one thousand plant RNA-seq that successfully decoded the transcriptome of 1000 plant species. The complete data is available at the NCBI-SRA database (http://ftp.cngb.org/pub/Dataset/datapla4/; Carpenter et al. 2019). The RNA-seq not only enables the identification of genes, but also facilitates the development of gene-based molecular markers useful for genotyping applications. For example, RNA-seq data of contrasting cultivars could be aligned to identify sequence polymorphisms, which could then be exploited as SNP markers (Muthamilarasan et al. 2019). Further, subjecting the transcript sequences to computational tools such as the MISA microsatellite finder enables the identification of simple sequence repeats that could be utilized as microsatellite markers (Muthamilarasan et al. 2019).

In addition to this 1000 plant transcriptome analysis, NGS has also favored independent studies on specific plants challenged with different stress factors. Hundreds of reports were published in the year 2020 alone on the use of RNA-seq to understand the transcriptomic complexity in responding to various stresses (Fig. 3.1). In rice, contrasting cultivars to bacterial leaf streak were subjected to RNA-seq analysis that showed upregulation of resistance-related genes in the tolerant cultivar than the susceptible one (Lu et al. 2020). Chen et al. (2020) conducted a similar study in poplar (*Populus tomentosa*) infected with *Marssonina brunnea* by sequencing the RNA of 435 cultivars. The research identified ~7000 differentially expressed genes associated with basal defense and mined ~30,000 SNPs within the regions of pathogen-responsive genes, which could be useful for genotyping and breeding for disease tolerance. Contrasting cultivars of tea (*Camellia*

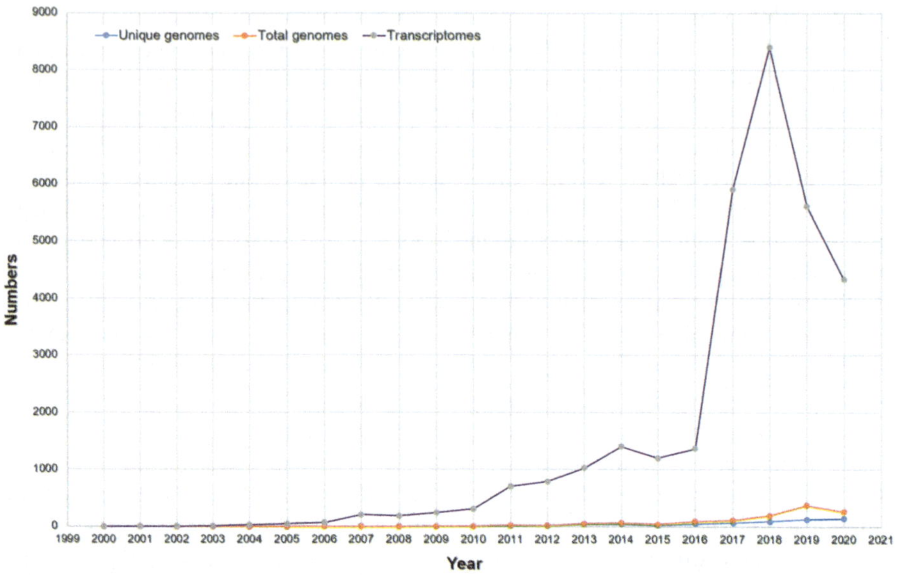


**Fig. 3.1** Plant genomes and transcriptomes sequenced until October 2020. The graph shows the data retrieved from NCBI (Genome and BioProject) to demonstrate the number of plant genomes and transcriptomes sequenced so far

*sinensis*) to the feeding of *Empoasca onukii* (green leafhopper) was studied at transcriptome level using RNA-seq by Jin et al. (2020). The study showed an elicitation of jasmonic acid-related genes in the tolerant cultivar at the early stage as compared to the susceptible cultivar, and also, the role of secondary metabolite biosynthesis was underlined in the tolerant cultivar. Transcriptomic response of tomato to potato spindle tuber viroid (PSTVd) was performed by Więsyk et al. (2020), where the plants were infected individually with highly-virulent and mildly-virulent PSTVd strains. Thousands of genes were found to be differentially expressed specifically in the plants infected with highly-virulent strain, among which the upregulation of C2C2-GATA and growth-regulating factor (GRF) transcription factor families were notable. In date palm (*Phoenix dactylifera*), Khan et al. (2020) had studied the transcriptome during *Ommatissus lybicus* attack and identified 6919 upregulated and 2695 downregulated genes during infection. The gene ontology annotation data showed that the genes are predominantly involved in the hypersensitive reaction, which could be a probable mode of defense in date palm. A Fusarium wilt-resistant cucumber was studied at transcriptome level to understand the molecular defense response against *Fusarium oxysporum* f. sp. *cucumerinum* infection (Dong et al. 2020). A total of 4116 differentially expressed genes were identified by comparing the datasets from different time-points of infection, and the annotation data showed that ethylene signalling pathways play a prominent role in conferring tolerance to fusarium wilt in cucumber. The response of peanut (*Arachis hypogaea*) to early leaf spot (caused by *Cercospora arachidicola*) was studied by

Gong et al. (2020), wherein the transcriptome sequencing identified 133 differentially expressed genes, of which a significant proportion belonged to R gene families as well as defense responsive genes.

#### 3.2.2.5 Allele Mining Approaches

Identification of novel or superior or effective alleles of the known candidate resistance genes using PCR analysis among the wild and cultivated crop populations and further confirmation of the level of novelty/superiority by expression analysis is known as allele mining. This approach is very popular in rice to identify several resistant varieties against different diseases using wild and cultivated germplasms, which can provide the molecular basis of allelic variations for any trait to establish the nucleotide changes related to novel or superior alleles. Once the superior and effective alleles are identified from available gene pools, which regulate the different plant defense responses, these can be further deployed in susceptible crops using plant breeding approach in further crop improvement programs (Ramkumar et al. 2010). This approach will also provide the degree of nucleotide diversity and conservation among the candidate genes and their transcriptional and translational control signals across the crop population. This technique has been widely used for rice blast-resistant genes *Pita* (Huang et al. 2008), *Pikh* (Ramkumar et al. 2010), *Pi54* (Kumari et al. 2013a), rice bacterial blight resistance genes *Xa27* (Bimolata et al. 2013), *Xa26, Xa21,* and *xa5* (Bimolata et al. 2015), *Xa7* (Utami et al. 2013), rice yellow mottle virus genes *RYMV1, RYMV2,* and *RYMV3* (Pidon et al. 2020), etc. from different rice cultivated varieties and wild species.

## 3.3 Proteomics

Proteins that are expressed in a cell at a given time, are collectively called a proteome, and the study of it is defined as proteomics. Proteins being workhorses of the cells are involved in several biochemical and signalling responses to plant defense mechanisms. Proteomics techniques characterize the proteins based on the following; (1) function/structure, (2) interaction with the protein, nucleic acid, lipid, and substrate, (3) post-translational modifications, (4) activity and localization, and (5) rate of synthesis and turnover rate. When exposed to biotic stresses, insects, pests, fungus, bacteria, etc., the plants elicit a cascade of events that activates a network of responses involved in immunity. Plants identify the pathogen-associated molecular patterns (PAMPs) via receptors called pattern recognition receptors (PRRs) and activates the PAMP-triggered immunity (PTI; Boller and Felix 2009; Zipfel et al. 2004). During the immune response, several genes encoding effector signalling molecules and pathogenesis-related proteins are activated. Strategically these interactions could be studied by proteomics as most of these PRRs are protein molecules and decipher the plant-host interaction and defense signalling (Ashwin et al. 2017). Comparative protein profiling of the diseased and control plants under biotic stress could be performed using either gel-based or gel-free proteomics techniques coupled with mass spectrometric analysis for identification, quantification, sequencing, and analysis.

### 3.3.1 Gel-Based Proteomics Approaches

The gel-based strategies have contributed to gain significant knowledge in understanding the protein expression profiling during biotic stresses. In the gel-based techniques, the proteins are basically separated based on their pI (isoelectric point) and molecular weight. The different types of gel-based techniques are 1-DE, 2-DE, 3-DE, and DIGE. Following these separation methods, the proteins are subjected to mass spectrometric analysis to identify, quantify, and sequence the proteins in a given sample. The crude protein extract may contain several contaminations like lipids, nucleic acid, and carbohydrate, which should be removed by detergent, pH precipitation, and enzymes, respectively. The prerequisites of the gel-based techniques are protein unfolding, solubilization, and disulfide bond breaking that can be achieved using urea, detergent, and Mercaptoethanol (or dithiothreitol and dithioerythritol), respectively.

#### 3.3.1.1 Two-Dimensional Gel Electrophoresis

In 1975, O'Farrell introduced 2DE for biochemical separation that could resolve a large number of proteins at a given time. Firstly, the proteins are separated according to their pI, called isoelectric focusing (IEF). Then, in the second dimension, they are separated based on molecular weight. Due to its significant resolving power, 10,000 spots that may correspond to a thousand protein samples with wide molecular isoforms could be separated in a single gel.

A wide range of 24 cm or 18 cm length linear immobilized pH 3–11 gradient IPG strip may be selected to study the global proteome expression levels between the control and the different stages of stress conditions. IPG strips were equilibrated before the second dimension by placing them in separate tubes containing support film near the tube wall and adding the SDS equilibration buffer solution. Incubation in DTT and followed by iodoacetamide. Mild horizontal shaking is recommended for reduction and alkylation steps. The equilibrated strips could be layered on acrylamide gels and further sealed using agarose. The second dimension of gel electrophoresis is the separation of proteins based on the molecular weight. The sodium dodecyl sulfate-polyacrylamide gel electrophoresis (SDS-PAGE) is run, and the gels were treated with fixative solution methanol:water:acetic acid overnight and further are stained with appropriate staining dyes like Coomassie brilliant blue. Destaining is done using methanol until the clearance of the background and the gel could be documented by image scanner having either MagicScan software or with Typhoon Trio+ if labelled.

2DE has extensively been used in comparative proteomic studies during pathogen attacks in crop plants, and such studies have identified several candidate defense-related proteins. In Arabidopsis, 2-DE was used to check the expression of proteins post infection with *Plutella xylostella* that revealed ROS's role in plants against the pathogen (Collins et al. 2010). *Heterodera glycines*, a nematode, make soybean a host. A study was conducted by 2-DE on the proteomics of susceptible and resistant cultivars of soybean that revealed that about 370 proteins were related to the susceptible cultivar and about 365 related to resistant cultivar (Liu et al. 2019). In Arabidopsis, a study conducted by Yin et al. (2012) studied the post-infestation by whitefly using 2-DE and checked the expression of proteins, wherein expression of 20 proteins showed drastic change. The study thus revealed the downregulation of the protease enzymes and upregulation of redox-associated enzymes. Protein profiling of apoplastic fluid (APF) proteins from coffee was performed using 2-DE. Interaction of *Coffea arabica* cultivars that were susceptible and resistant to *Hemileia vastatrix* were chosen for the study. About 210 proteins have shown differential expression in control, susceptible and resistant plants (Guerra-Guimarães et al. 2015). To understand the molecular mechanism of *Blumeria graminis* infection in wheat, expression profiling was performed using 2-DE in the susceptible wheat cultivar following infection (Li et al. 2017). In 2019, Martins' group studied the proteomics of interaction between *Meloidogyne arenaria* (Peanut root-knot nematode) and *Arachis stenosperma* (wild peanut) using 2-DE and found about 222 differentially abundant proteins (DAP) by comparing the control and nematode inoculated root (Martins et al. 2020). In wheat, *Tilletia indica* causes the Karnal burnt disease. 2-DE was used to study the change in expression profiles of the disease infected and control plants (Pandey et al. 2019). Recently, Kumar et al. (2020) had identified about 94 differentially abundant proteins using 2-DE in Arabidopsis, in wild type and *fld* mutant (SAR compromised). These proteins DAPs might have a role in developing systemic acquired resistance.

#### 3.3.1.2 Three-Dimensional Gel Electrophoresis

Difficulty in quantifying proteins that showed co-migration was one of the significant limitations of 2-DGE that led to the introduction of 3DGE (Colas et al. 2010). In this, the third time separation of co-migratory proteins is performed in different buffer systems containing ion carriers (Colignon et al. 2013). Post-translational modifications could be identified with this in addition to its increased accuracy (Rabilloud 2013). 3DGE could be exploited to understand and analyze the complex plant proteome and decipher phyto-pathoproteomics correlation (Ashwin et al. 2017). The major drawback of this technique is that it cannot identify low copy number proteins and has low reproducibility. Due to this drawback, the approach has not been used in studying the molecular defense response of crop plants.

#### 3.3.1.3 Difference Gel Electrophoresis

In DIGE, covalent labelling of each sample is done using a collection of fluorophores, viz. Cy-2, 3 and 5 that gets tagged to lysine and cysteine residues, followed by gel separation (Unlu et al. 1997). The labelling dyes have substantial dynamic range, sensitivity, and linearity that help in the comparative analysis of the proteome. Only a few mass spectrometers can match the sensitivity of that of DIGE. Hence, the separation of proteins using the Cy-labelled dyes has come into regular practice. Overstaining of gels with Coomassie blue, post-Cy-labelled fluorescent imaging is done, and then both the images are compared and excised for further mass spectrometric analysis (Thelen and Peck 2007). DIGE can overcome sensitivity problems and detect proteins that are less abundant. It is the best suitable approach for comparative proteomics to understand the differential protein expression levels. The major drawback of DIGE is that the protein sample would not be labeled in the absence of lysine residue. In this case, high-resolution laser scanners are used for visualization, and for this, the fluorophores used are specific for laser scanners, which can be pretty expensive (Marouga et al. 2005).

In *Arabidopsis thaliana*, DIGE was used for protein profiling post-infection with cell culture of *Fusarium sporotrichioides*. It was observed that the infection increased the activity rate of the enzymes, peroxidase, and phenylalanine ammonialyase, which are associated with the metabolism of phenolic compounds (Chivasa et al. 2006). In potato, comparative proteome analysis was performed using DIGE that showed that about 50 proteins were differentially expressed and might function in plant defense (Di Carli et al. 2010). *Zea mays* is one of the staple crops and is often exposed to *Aspergillus flavus* infection leading to yield loss. In a study conducted by Pechanova and group, protein profiling was performed using DIGE, which revealed the significance of housekeeping as well as inducible proteins to acquire resistance against the fungus (Pechanova et al. 2011). *Orobanche crenata* infects legumes and leads to yield loss in the Mediterranean countries. To study the molecular mechanism of the virus-host pathogenesis, 2D-DIGE was performed in two different resistant cultivars of pea. The result showed a change in expression in about 43 proteins when compared to the control and the infected ones (Castillejo et al. 2012). Proteome analysis was performed in the susceptible and resistant hosts post-infection with *Sugarcane Mosaic Virus*. The DIGE analysis of the virus and host interaction revealed 17 proteins that were responsive to the virus and about seven proteins with unknown functions (Wu et al. 2013). *Erwinia amylovora* causes fire blight in the *Rosaceae* family. The outer membrane of the pathogen, showing contrasting high and low virulence, was studied by protein profiling using 2D-DIGE (Holtappels et al. 2016). 2D-DIGE was used to detect the protein abundance in wheat infected with Fusarium head blight (FHB). Comparative protein analysis of the QTL, *Dfhb1* (confers resistance against FHB), in contrasting NIL of wheat identified about 80 proteins with differential expression (Eldakak et al. 2018).

#### 3.3.1.4 Mass Spectrometry

Though MS is not a gel-based technique, it is a follow-up procedure for identifying the proteins in gel. Mass spectrometry has gained significant importance in protein studies as it can reveal both the quality and the quantity of the proteins. In a mass spectrometer, biomolecules are analyzed based on the mass-to-charge ratio (m/z) of peptide ions that indicate the peptide sequence. Further, fragmentation analysis reveals the structure of the peptide. The three major components of a basic mass spectrometer are (1) source of ionization, (2) mass analyzer, and (3) a detector. Detection of the mass-to-charge ratio is done mainly by combining either of the 5 mass analyzers, namely Fourier transform ion cyclotron resonance (FTICR), time of flight (TOF), ion trap, quadrupole (Q), and orbitrap (Graham et al. 2007; Thelen and Miernyk 2012). Among these five, FTICR contains an ultra-high vacuum device that entraps the ions for a longer time and shows the highest degree of mass accuracy and sensitivity. FTICR helps develop a 3-D map of the protein sample by following two major approaches: bottom-up and top-down. In top-down, the proteins are intact, whereas in bottom-up fragmented proteins are used. This can be observed in tandem mass spectrometers that use either two or more mass analyzers, and for precursor fragmentation, in between the mass analyzers, collision-induced dissociation is used (MS/MS scan; de Hoffmann and Stroobant 2007). Detection by orbitrap is based on harmonic oscillations of the ions. Presently linear ion trap quadrupole Orbitrap (LTQ) is combined with increasing the levels of accuracy and sensitivity (Hu et al. 2005). Altogether, as discussed, a handful of mass analyzers could be used in various combinations to increase the resolution and accuracy of the proteome data.

In proteomics, MS soft ionization approaches are utilized in Matrix-assisted laser desorption ionization, i.e., MALDI (Karas and Kruger 2003) and electrospray ionization, i.e., ESI (Fenn 2002). In MALDI sample preparation, the digested protein is mixed in solution and co-crystallized on a target plate. For this, the mass analyzer used is the time of flight. 2D Protein spots selected for identification and sequencing by mass spectrometer were either manually or robotically excised followed by trypsin digestion (Shevchenko et al. 2006), with slight modifications. A laser beam targets the MALDI plate for ionization. The ionized peptides are separated according to the m/z ratio in TOF. MALDI-TOF is a robust technique and is less expensive. In Electrospray Ionization (ESI), an electrospray needle direct the sample to the ionizing source with a high potential difference from the needle (Fenn et al. 1989). Ionization occurs in a strong electric field with elevated temperature in the ion source (Mora et al. 2000; Fenn 2002). ESI can be combined with LC (Liquid chromatography) to fractionate large molecules or generate various charges ions. SPR-MS analysis, i.e., Surface Plasmon Resonance, can quantify protein interaction with nucleic acid, protein, and drugs using surface-immobilized ligands. The structural features of the interacting protein are understood by MS analysis. The sample solution is allowed to pass over the surface containing immobilized ligands, and SPR detects the interaction between the protein and ligand. This chip is further evaluated using MALDI or ESI added with TOF. SPR can help detect non-specific binding, PTMs, and protein interactions with other biomolecules (Bradbury et al. 2003). As per the need for the research, MS can be coupled with any of the above-mentioned techniques for a better understanding of the proteome and its interactions with other biomolecules.

As discussed earlier, there are different variants of MS and can be used in combinations according to the need of the research. To study the plant-pathogen interaction of *Brassica napus* and *Leptosphaeria maculans*, the proteome analysis of the infected leaves was performed at different time intervals post-infection using ESI-Q-TOF and ESI-IT MS/MS (Subramanian et al. 2005; Sharma et al. 2008). In *Oryza sativa*, the proteome analysis of the apoplastic extracts was performed post-infection with *Magnaporthe oryzae* using ESI-LC-MS/MS (Shenton et al. 2012). In wheat grains, the grain development is significantly interrupted by the *Blumeria graminis* infection. Comparative proteome analysis of the control and the infected samples has shown the differential expression of 43 proteins. This data could be used for further molecular analysis of the host-pathogen interaction (Li et al. 2017). Until now, researchers are facing difficulties in developing wheat cultivars that might show resistance to Karnal bunt. Comparative proteome analysis of the *Tilletia indica* low (TiP) and high (TiK) virulent strains was performed using MALDI-TOF/TOF. Eventually, it was found that the pathogenicity factor is oxalic acid as TiK isolates were found to be a malate dehydrogenase (Pandey et al. 2019). *Xanthomonas axonopodis* is a bacterial pathogen that infects several crop plants. With the use of MALDI-TOF, comparative proteome analysis was performed in 32 *Xanthomonas axonopodis* strains, and a mass spectrum was created. This data would further help in the study of plant-pathogen interaction (Sindt et al. 2018). Rice is often exposed to blast disease and salicylic acid (SA) signalling plays an important role in plant defense. Here, MALDI-TOF was implied in studying the phosphoproteomics during blast infection in contrasting rice cultivars resistant and susceptible to blast. The study revealed that SA increases rice resistance by regulating antioxidants (Sun et al. 2019).

#### 3.3.1.5 Studying the Post-Translationally Modified Proteins

Understanding the post-translational modifications can help in comprehending the mechanism underlying plant-pathogen interaction. However, such modifications occur at very minute levels in proteins and are hard to detect. Multidimensional separation of protein samples could be achieved by affinity chromatography coupled with reverse-phase (RP). It enriches the post-translationally modified protein samples and brings them to a level where an MS can detect them. Titanium dioxide ($TiO_2$) or dihydroxybenzoic acid (DHB) could be used to enrich phosphor-containing proteins in SAX chromatography, whereas charged peptide centric are used in SCX chromatography (Macek et al. 2009; Mohammed and Heck 2011) and ions like $Ga^{3+}$, $Al^{3+}$, $Fe^{3+}$, $Zr^{3+}$ or $Co^{2+}$ are utilized in IMAC (Ficarro et al. 2002). These techniques can also be used in combination with better enrichment of the PTMs. PTMs have a very dynamic nature. The dynamic nature and PTMs like glycosylation and phosphorylation have significantly less stoichiometric concentrations, making the enrichment step compulsory before performing MS (Thingholm et al. 2009).

In plants, Nod-like receptors act as immune receptors, but their regulation is very stringent as the upregulation may cause autoimmunity, whereas downregulation will cause pathogen susceptibility (Xu et al. 2015). Evolution of novel effector in *Meloidogyne graminicola*, MgGPP is abundantly expressed in initial stages when infecting the rice plants by targeting the endoplasmic reticulum. Interestingly, suppression of host response is when MgGPP undergoes N-glycosylation (Chen et al. 2017). *Xanthomonas oryzae* pv. *Oryzae* contains a collection of 10 *gigX* genes helping in glycosylation of flagellin that regulates virulent nature and the movement of the pathogen (Yu et al. 2018). *Cochliobolus carbonum* infects maize, and to inhibit histone deacetylases, and it secretes HC-toxin (HCT). During a study using iTRAQ, it was seen that HCT is vital for infection, and it does so by changing the histone deacetylases activity. This alteration in deacetylases' activity affects the acetylation process during the interaction of plant and pathogen (Walley et al. 2018). In rice, the sumoylation pathway-related genes like AOS1 and UBA2 were deleted to develop mutants. These mutants, when exposed to *Magnaporthe oryzae,* the sumoylation genes were not appropriately expressed and inhibited the invasion and growth of the fungus (Liu et al. 2018). Hence, it is evident that studying the PTMs of proteins is imperative to reveal the mechanism underlying plant-pathogen interaction.

### 3.3.2 Gel-Free Proteomics Approaches

Similar to gel-based approaches, several gel-free strategies to decipher the proteomes during pathogen infection have been developed. Some of the widely used shotgun or labelled quantification proteomics approaches to study the plant defense responses are ICAT, SILAC, iTRAQ, MudPIT, and protein microarrays (Gygi et al. 1999; Ross et al. 2004; Washburn et al. 2001; MacBeath 2002).

#### 3.3.2.1 Stable Isotope Labelling by Amino Acids in Cell Culture

SILAC is an MS-based shotgun quantification technique that utilizes labelling of the sample in vivo. Non-radioactive heavy isotopes are tagged to amino acids sample in the culture medium, followed by detection using tandem MS (Geiger et al. 2011). SILAC can only be performed on metabolically active samples like cultures in a suspension medium, tissue-cultured plants, or pathogens that are grown in vitro (Harsha and Pandey 2010). In comparison to other shotgun techniques, SILAC is very costly as well as tedious. Owing to these limitations, so far, only two studies of proteome analysis on plant-host interaction are performed (Phillips et al. 2011; Rowland et al. 2015). *Neurospora crassa* secretome, cultured in the microcrystalline cellulose, was characterized by Phillips' group using SILAC and absolute quantification. They showed that four proteins were responsible for the degradation of cellulose by fungi, including cellobiohydrolases, endoglucanase, and glucosidase (Phillips et al. 2011).

#### 3.3.2.2 Isotope Coded Affinity Tag

ICAT involves isotopes that are chemically tagged to the sample proteins and quantifies the difference in protein expression levels (Gygi et al. 1999). ICAT reagents contain a protein-reactive group, a biotin tag, and a linker. ICAT reagent can identify cysteine (thiol groups) amino acid in the sample. A chromatographic fractionation strategy is applied to separate ICAT labelled proteins. Then, Tandem MS is used to identify and quantify the protein samples (Shiio and Aebersold 2006). However, this approach has not been much popular among researchers who study plant-pathogen interactions.

#### 3.3.2.3 Isobaric Tag for Relative and Absolute Quantification

iTRAQ is a popular approach that involves labelling almost all the digested protein samples, and the approach quantifies a huge number of samples. This also widens proteome coverage as most of the trypsin digested protein samples will contain a primary amino group that can be tagged (Ross et al. 2004; Zieske 2006). The labels used are isobaric and are present evenly in each peptide, which increases the detection sensitivity than the other shotgun methods (Evans et al. 2012). Unlike other labelling approaches, direct labelling is done at the *N*-terminal and in lysine residues of the protein sample mixture. Owing to its less sophisticated strategy, iTRAQ has become an ideal quantification approach for plant-host interaction studies.

iTRAQ was used in studying the molecular mechanism of *Bradyrhizobium japonicum* infection in soybean roots (Nguyen et al. 2012). In *Solanum lycopersicum*, a comparative proteome analysis was done using iTRAQ during infection with *Pseudomonas syringae* (Parker et al. 2013). A combination of iTRAQ-RNA-seq in *Ziziphus jujuba* infected with phytoplasma showed differential expressions of about 37 genes responsible for multistep regulation post-infection (Ye et al. 2017). Proteome analysis for *Rhizoctonia solani* infected resistant and tolerant cultivars of cotton showed the differential expression of about 170 proteins, and most of them were associated with ROS activity (Zhang et al. 2017). In a study in *Zea mays*, iTRAQ was implied to demonstrate the damaging consequences of *Maize chlorotic mottle virus*. It was seen that about 970 proteins were differentially expressed, with about 310 downregulated and 660 upregulated (Dang et al. 2020). iTRAQ was used to develop a comparative expression profile with two virulent strains of *Nilaparvata lugens* that infect rice. The expression profile indicated that about 258 proteins were differentially expressed, out of which about 151 were upregulated (Zha and You 2020). Recently, Yang et al. studied the phenomenon of heterologous superinfection exclusion (HSE) against *Cucumber mosaic virus*–Fny strain in *Nicotiana benthamiana*. This HSE was developed by a prior infection of the plant with a mild mutant strain of the *Tobacco mosaic virus* -43A. The proteome analysis of superinfected samples was performed using iTRAQ, which showed that TMV-43A could protect the plant from CMV (Yang et al. 2020). In Chinese cabbage, iTRAQ was used to quantify the proteome during TuMV infection, and the differentially expressed proteins identified had a vital role in calcium signalling, random lipid transfer, HSP, and WRKY transcription factor (Lyu et al. 2020).

#### 3.3.2.4 Multidimensional Protein Identification Technology

In MudPIT, the tryptic peptides are subjected to separation using a strong cation exchanger and reversed-phase high-performance liquid chromatography. Post-separation, the samples are subjected to mass spectral analysis (Issaq et al. 2005; Washburn et al. 2001). It produces a huge number of peptides contained in the sample. MudPIT is a relatively fast and sensitive technique with an enhanced reproducibility rate. One of the major drawbacks of this approach is that it cannot quantify the sample information (Rose et al. 2004). In tomato, wilt and canker are caused by the pathogen *Clavibacter michiganensis* though the pathogenesis mechanism is not well understood. MudPIT was implied in delineating the molecular mechanism of the plant-host interaction that revealed that the pathogen secrets various hydrolytic enzymes, which is recognized by the LOX1 protein of the plant and activation of defense machinery (Savidor et al. 2012). Rice leaves were infected with *Magnaporthe oryzae* strains viz. KJ401 (incompatible) and KJ301 (compatible). The proteome of the infected leaf and the control was studied using MudPIT combined with MALDI-TOF MS. About 730 proteins were identified from the secretory proteome, which contained 60 percent of pathogen protein and 40 percent host proteins. The rice proteins were related to energy metabolism and ROS activation. The pathogen proteins include cell wall hydrolyzing enzymes (Gon et al. 2012). MudPIT was also used to analyze the differential expression of the protein in rice when infested with *Schizotetranychus oryzae*, a phytophagous mite. The data revealed that 11 proteins were upregulated in the control and one protein was upregulated during the later stages of infestation. This data was further correlated with RT-qPCR analysis that indicated these proteins might interfere with the metabolic processes in rice leaf (Blasi et al. 2017).

## 3.4 Metabolomics

There has been a significant improvement in omics technology, among which the metabolomics approach has successfully been applied for studying plants and microbes. The metabolomics approach is the qualitative and quantitative study of tiny endogenous molecules having a molecular weight below 1000 Da present within cells at a particular time and condition (Feng et al. 2020). Metabolites are the downstream processing end product of genes and proteins that directly affect the phenotypic character of a plant, i.e., curling, bending, yellowing, chlorosis, necrosis, wilting, etc. The metabolomics platform is one of the best ways to understand a particular trait that enables devising a better crop improvement strategy (Sharma et al. 2018). The metabolic profiling process identifies metabolites involved in the cellular metabolism of plants and pathogens. By this approach, the physiological and biochemical state of any given tissue can be determined in resistant and susceptible crop plants against the pathogen in a particular interval of time. Metabolomics has also proven to be an effective technique to determine the functional role of a specific gene in metabolic pathways (Zhu et al. 2018). The combined approaches of transcriptomics, proteomics, and metabolomics have categorized the genes of the biomarker metabolic compound. The metabolomics approach has been used in different crop species associated with unfavorable stress conditions (Zeiss et al. 2019; Schaker et al. 2017; Thomason et al. 2018; Kumar et al. 2017), and successfully selected the better trait which is used in crop improvement. In studying plant-pathogen interaction, the metabolomics study identifies defense responsive metabolites. This enables the development of metabolic engineering technology to be an excellent approach to developing disease-resistant plants (Fig. 3.2).

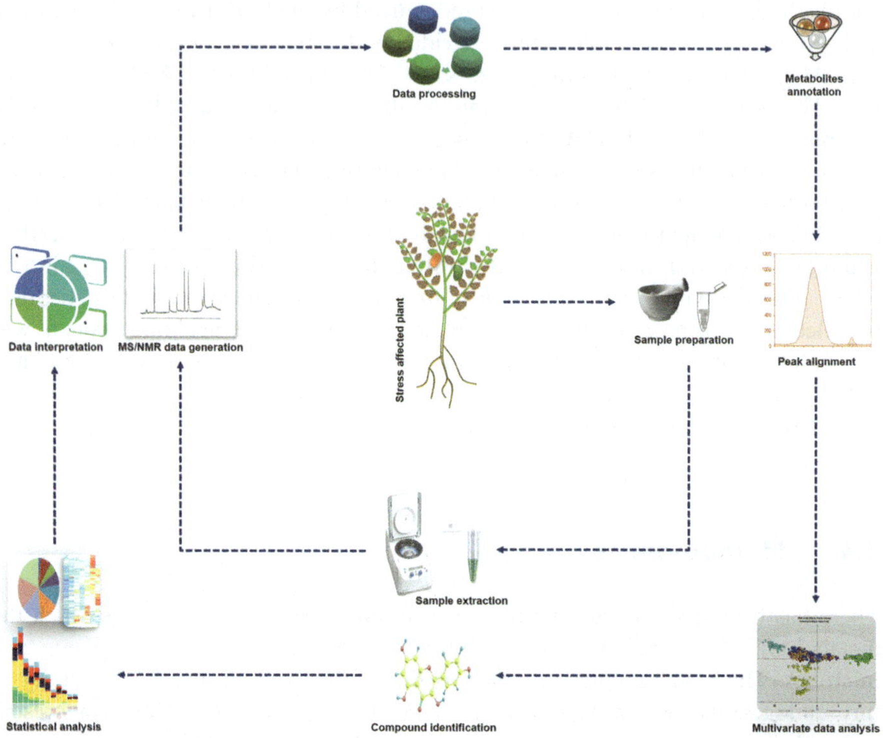


**Fig. 3.2** Metabolomics pipeline to understand the response of plants to biotic stresses. Broadly, the approach involves sample preparation, metabolomics data analyzation, compound identification, and statistical analysis

### 3.4.1 Overview of Plant Metabolites

Plant metabolites are categorized into two types, primary and secondary metabolites (Table 3.1). These two forms play different roles in plant metabolomics processes. Primary metabolites generally help in plant growth, development, and reproduction, i.e., carbohydrate, protein, amino acid, etc. In contrast, secondary metabolites play an active role in defense mechanisms to overcome biotic and abiotic stress conditions. Primary metabolites are synthesized by metabolism pathways such as the pentose phosphate pathway, nucleoside diphosphate sugar pathway, glycolysis, and tricarboxylic acid cycle. These primary metabolites help in the building of secondary metabolites. Secondary metabolites again are categorized into different classes such as fatty acids, terpenoids, phenylpropanoids, alkaloids glycosides, lignin, lignans, tannins, flavonoids, quinines, stilbenes, etc. which repel pathogens, avoid herbivores as well as protect the plants from different environmental conditions like light, temperature, water, toxic metals, and nutrient deficiency, etc. (Thomason et al. 2018).

**Table 3.1** Details of metabolomics studies performed in plants to understand their defense mechanism

| Plant | Disease | Pathogen | Metabolomics technique | Défense gene expression | Défense metabolites | Reference |
|---|---|---|---|---|---|---|
| *Phaseolus vulgaris* | Fusarium wilt | *Fusarium oxysproum* | UPLC | *CHS, CHI, IFS, IFR, F3H, DFR, ANS, LAR* | Flavonoid and isoflavonoid compound derivatives | (Chen et al. 2019b) |
| *Solanum tuberosum* | Late blight of potato | *Phytophthora infestans* | LC-ESI-LTQ-Orbitrap | *4CL, TyDC, OD, THT, PHT* | Hydroxycinnamic acid amides (HCAAs), Tyramine, Putrescine, Sinapate, Feruloylputrescine, Kaempferide 3,7-dirhamnoside, Kaempferitrin, 3-O-β-D-glucosyl-(1→2)-β-D-glucoside, phosphatidylinositol phosphate, α-Chaconine, Solanine, Calactin, Euphornin, *N*-Caffeoylputrescine, p-Coumaroylputrescine p-coumaroyltyramine, *N*-Feruloyltyramine Sinapoyltyramine, palmitate | (Pushpa et al. 2014) |
| *Asparagus officinalis* | Stem blight disease | *Phomopsis asparagi* | LC-Q-TOF-MS | *CHI3, F3'H-L, UVR8-L, RUP2-L2, SBT1, NCS1/PR10, POX29, CAT1-L, GST, SABP2, BCKDHA2, LRR-RLPK, RLPK-BAM1, PK-IRE4, IPS, INT1, ITPK3-L, PPIP5K2* | Anthocyanins, steroidal saponins | (Abdelrahman et al. 2020) |
| *Solanum lycopersicum* | Viral disease | *Tomato yellow leaf curl virus* (TYLCV) | LC-MS | *PAL1, 4CL, HCT, CCoAOMT1, CAD, ICS, LeHT1* | Rutin, Chlorogenic acid, Salicylic acid, GABA, Feruloylputrescin, Polyamine, Urea | (Sade et al. 2015) |
| *Populus deltoides* | Herbivore attack | *Hyphantria cunea* | GC-TOF-MS | *PAL2, C4H2, C3H, COMT4* | Stigmasterol, Sinapic acid, *Trans*-cinnamic acid, Caffeyl alcohol, *p*-Coumaric acid, Caffeic acid, Sinapyl alcohol | (Wang et al. 2017) |

(continued)

**Table 3.1** (continued)

| Plant | Disease | Pathogen | Metabolomics technique | Défense gene expression | Défense metabolites | Reference |
|---|---|---|---|---|---|---|
| *Camellia sinensis* | Herbivore attack | *Ectropis oblique* | UPLC-Q-TOF-MS GC-MS | *PAL, LDOX, TPS03, TPS04,* and *TPS21* | Anthocyanin, Phenylpropanoid, flavone, and flavonol, Diterpenoid, Sesquiterpenoid, and triterpenoid | (Wang et al. 2018) |
| *Glycine max* | Nematode infection | Soybean cyst nematode (SCN) | GC-MS | *P450 CYP73A100, PAL, 4CL, POD, ACO, ACS,* | Piperine, 4-vinyl phenol, Palmitic acid, Methionine | (Kang et al. 2018) |

### 3.4.2 Biochemical Effect of Metabolites on Plant-Pathogen and Plant-Environment Interaction

In nature, sessile plants are continuously exposed to several biotic stresses leading to yield loss. These biotic stresses could be pests, nematodes, bacteria, fungus, or insects. The most devastating pest, Brown plant Hopper, and the most destructive bacterial pathogen *Xanthomonas oryzae* pv. *oryzae (Xoo)* of rice causes a massive loss in productivity. However, the resistance rice cultivars have been released by breeding effort, and defense metabolic response of different rice varieties are characterized against pests and pathogen (Kang et al. 2019; Sana et al. 2010). Secondary metabolite has been involved in defense response by increasing callose deposition in the cell wall to inhibit pathogen growth in sugarcane plants against smut disease caused by *Sporisorium scitamineum* (Schaker et al. 2017). Several different metabolites are synthesized in various resistant and susceptible crop plants due to biotic and abiotic stresses. The susceptible plant associated with saline stress conditions has accumulated more ROS that interacts with DNA, proteins, lipids, and other pigments, leading to complete damage to the plant cell compared to the resistant plant (Khan et al. 2019). Different enzymes such as polyphenol peroxidase (PPO), superoxide dismutase (SOD), ascorbate peroxidase (APX), peroxidase (POD) have involved in plant-pathogen interaction; however, SOD and POD enzymatic activities were increased in a resistant variety of pepper plant infected with anthracnose disease (Padilha et al. 2019). A rapid increase in these antioxidant enzymes and salicylic acid has acted as a defense system against *Xoo* pathogen infection in rice plants. PPO plays a significant role in the pathogen defense mechanism by catalyzing phenol to quinine (Shasmita et al. 2019). Sugar is a major vital component for the growth and development of rice plants. Salicylic acid regulates sugar content and enhances the tolerance capacity to salinity stress conditions (Dong et al. 2011). Similarly, proline accumulation is one primary plant response during environmental stress conditions, especially in water stress (Anjorin et al. 2016). Sugar, protein, and plant phytohormones like cytokinin and ethylene play an essential role in plants against stress conditions, particularly in pathogen infection. The transgenic plant is protected against the harmful pathogen attack by producing cytokinin phytohormone (Bari and Jones 2009). Despite these findings, there is no clear evidence on the plants using a specific or similar pathway to defend against a pathogen. This has necessitated the development of more advanced and comprehensive tools and approaches to understand plant defense.

### 3.4.3 Prerequisites and Tools for Metabolomics Studies

Metabolomics is categorized into two groups, i.e., targeted and untargeted metabolomics. Targeted metabolomics is based on biochemically characterized and annotated metabolites, while untargeted metabolomics is based on chemically unknown metabolites—both of these methods are used to characterize the metabolites data by mass spectrometric analysis. In targeted metabolomics, a series of predefined analytes and prior knowledge determines whether the metabolites are present. The targeted method is beneficial because it has higher specificity and sensitivity than the untargeted metabolomics method. The untargeted approach helps in the separation of very complex samples without any predefined data and prior knowledge. Mass spectrometer and NMR are used for the best separation of active metabolic compounds (Wallace et al. 2020).

#### 3.4.3.1 MS-Based Methods

Separation and identification of metabolites are preferable when chromatography is coupled with mass spectroscopy (MS). The detector of MS uses samples isolated from gas chromatography, liquid chromatography, capillary electrophoresis, and Fourier Transform Ion Cyclotron. The most common physical chromatographic technique for untargeted metabolites is GC-MS, wherein samples are separated based on polarity gradient and ionized at 70 eV. Due to the involvement of the ionization process, the GC-MS approach is highly reproducible, and spectral data can be easily collected and compared with the databases (Vinaixa et al. 2016). One drawback of the GC-MS separation is that it makes the sample volatile and leaves the underivatized compounds in the unnoticed form during analysis. To overcome the situation, GC-TOF-MS has been used, which improves the separation of compounds with similar peak values as it has higher sensitivity and gives an accurate measurement of mass (Kumar et al. 2017). The LC-MS approach is used for comparative separation of both untargeted and targeted metabolites present in multiple samples without any prior knowledge. LC-MS profiling involves separating unique metabolites by mass/charge ratio and retention time, which finally gives about 100–1000 peaks (Vinaixa et al. 2016). For analyzing secondary metabolites, reverse-phase columns are used in LC-MS. It can separate a massive number of structurally similar compounds. Moreover, ultra-performance liquid chromatography (UPLC) has been developed, which is more useful for the separation of various metabolites than high-performance liquid chromatography (HPLC) due to its higher sensitivity and resolution capacity (Arrivault et al. 2009; Obata and Fernie 2012).

In recent years, CE-MS has been given significant attention to metabolomics studies. It uses global metabolic profiling of various cationic and anionic compounds. It is a useful analytical technique in which different metabolites are separated according to their electrophoretic mobility (Ramautar et al. 2017). CE-MS provides a distinct partition of both targeted and untargeted compounds. Its higher resolution capacity can separate polar, neutral, charged, and hydrophobic active metabolites based on charge and size (Ramautar and De Jong 2014). CE-MS is a less time-consuming analysis with a low-cost effect and requires minimal sample (Chen et al. 2019a).

One of the mass spectrometers based on Fourier transform technology, known as FT-ICR-MS (Fourier transform ion cyclotron-resonance mass spectrometer), analyses different fractions due to its higher sensitivity, mass accuracy, and ultra-high resolution, i.e., 1,000,000 at 400 m/z (Viant and Sommer 2013). Typically, in liquid chromatography (LC), the compound separation method is time-consuming and could be overcome using FT-ICR-MS. Despite these techniques, several other

MS-based detector chromatographic methods are used for isolation and identification of active molecules, i.e., gas chromatography coupled with time of flight MS (GC-TOF-MS), quadrupole rod tandem time of flight MS (Q-TOF-MS), triple quadrupole MS (QQQ-MS), hydrophilic interaction liquid chromatography MS (HILIC-MS), matrix-assisted laser desorption ionization MS (MALDI-MS), MALDI with the time of flight MS (MALDI-TOF-MS), hydrophilic interaction liquid chromatography MS (HILIC-MS), ion-pair LC with electrospray ionization MS (IP-LC-ESI-MS; Chen et al. 2019a, b), ESI-triple quadrupole-linear ion trap (Q TRAP)-MS (Long et al. 2019), ESI-LC-MS (Schaker et al. 2017), and ultra-high-performance liquid chromatography MS (UHPLC-MS; Tugizimana et al. 2019).

#### 3.4.3.2 NMR Profiling

Another important technique for identifying and characterizing metabolites is NMR (Nuclear Magnetic Resonance). Metabolic compounds with lower molecular weight are separated according to the nuclear atom's magnetic properties by acquiring energy transition under the magnetic field. The NMR profiling provides information about a smaller molecule with $<50$ kDa size by screening, detection, quantification, identification, and characterization. Several types of NMR spectrum, such as hydrogen (1H-NMR), carbon (13C-NMR), and phosphorus (31P-NMR), are used in the NMR technique, but the hydrogen spectrum is mostly used (Deborde et al. 2019). Several pathogens often infect tomato in nature, and the metabolite analysis in the infected leaf was studied. The study revealed that only malic acid and glucose were accumulated in the leaf tissue during viral infection, whereas upon bacterial infection, primary metabolites are accumulated (organic acids, rutin, phenyl compounds, and amino acids; Lopez-Gresa et al. 2010). Grape berries are infected by the pathogen *Botrytis cinerea*. The metabolites were studied in both control and infected berries that indicated the accumulation of amino acids, namely, alanine, arginine, proline, and glutamate. In the infected branches, accumulation of gluconate, succinate, and glycerol was observed, which might help in the growth of the pathogen. Thus, from NMR profiling post-infection, the host defense system, and the development of the pathogen were understood (Hong et al. 2012).

### 3.4.4 Bioinformatic Tools, Databases, and Data Analysis

A huge amount of data could be generated from the above techniques, and that requires extensive processing by several data processing software such as MZmine, AMDIS, MarkerLynx, AnalyzerPro, MetAlign XCMS, ChromsTof, SIEVE, MET-COFEA, MarkerView, MassProfiler, Progenesis QI, MSFACTS, etc. (Kumar et al. 2017; Chen et al. 2019a). These bioinformatics tools and software simplify a large amount of experimentally acquired data by alignment, correlation, conversion, normalization, deconvolution, noise filtration, feature detection, bucketing, etc. After processing of the metabolomics data, specific compounds are identified using several metabolites annotation database such as PMDB (Plant Metabolome Database), KEGG (Kyoto Encyclopedia of Genes and Genomes),

ChEBI (Chemical Entities of Biological Interest), CAS (Chemical Abstracts Service), DNP (Dictionary of natural products), NIST (National Institute of Standards and Technology), Chemspider, PlantCyc, Knapsack, PubChem, METLIN, GOLM, MetWare, etc. (Udayakumar et al. 2012; Lazar et al. 2015).

The metabolomics data identified from the above database can be analyzed using bioinformatics-based statistical analysis, unsupervised analysis, and supervised analysis. However, unsupervised and supervised analyses are included under multivariate data analysis (Bujak et al. 2015) such as principal component analysis (PCA), partial least square discriminant analysis (PLS-DA), hierarchical cluster analysis (HCA), orthogonal partial least squares discriminant analysis (OPLS-DA), the multiple univariate data analysis (MUDA), the linear discriminant analysis (LDA), and neural networks (NN), principal component regression (PCR), partial least squares regression (PLSR), Sammon mapping multidimensional scaling (MDS), mixture discriminant analysis (MDA), projection pursuit linear discriminant analysis (LDA), flexible discriminant analysis (FDA), quadratic discriminant analysis (QDA), variable importance in projection (VIP) plots, correlation map, K-means clustering, heat map, boxplot, chemometric modelling, and metabolomic pathways, etc. Several statistical tools, i.e., MetPA, MetaboAnalyst, Cytoscape, are used for multivariate analysis of the identified compounds. These analytical tools help the more accessible selection, characterization of metabolic markers, and significant analysis of metabolic pathways associated with defense response (Schaker et al. 2017; Tugizimana et al. 2019).

### 3.4.5 Metabolic Phenotype Study from mQTL to mGWAS

The plant produces several defense biomarker metabolites during stress conditions, which directly links to their phenotypic character. So, these key metabolites act as interlinking elements between genome sequence and phenotypic characteristics. For understanding the defense metabolic trait within the plant genome, some advanced metabolomics techniques such as metabolic Quantitative Trait Locus (mQTL) analysis and Genome-Wide Association Study (GWAS) are used (Hong et al. 2016). Metabolic based Genome-Wide Association Study (mGWAS) is used to identify the specific novel uncharacterized genes of the untargeted signature metabolites. The chromatographic metabolomics product was applied to mGWAS analysis to understand the genetic control mechanism of defense bioactive compounds, and new regulating genes have been discovered in *Arabidopsis* (Tong et al. 2018). The result of mGWAS is validated through linkage map mQTL analysis. Similarly, numerous secondary metabolites in maize plants help detect mQTL. It is more accessible because of RNA sequencing as well as SNP data developed by genomics, transcriptomics, and metabolomics technology (Wen et al. 2014). For mQTL detection, R/QTL software is used by interval mapping and composite interval mapping. The functional annotation of specific defense genes associated with QTL loci is studied by the VitisNet network database. Using the mQTL analysis, five resistance motifs have been found on chromosome 18 of grapevine species, which provided resistance to several diseases caused by pathogens (Teh et al. 2019). QTL mapping gives information about the complex and specific trait, whereas combining the mGWAS-transcriptomics network can identify novel genes and pathways of the metabolic defense process. The overall result helps breeder for selection of unique genotype of plant for biomarker metabolites assisted breeding program.

Except for these, there is a combination of metabolomics approaches with several advanced biotechnological experiments such as RNAi and gene knockout. Reverse genetic tools help in deciphering the function of the target gene in the biochemical reaction of the metabolic pathways. Metabolomics of mutant and transgenic *Arabidopsis* plant was analyzed by gene knock-out technology, and different bioactive compounds have been identified from each plant (Fukushima et al. 2014). Furthermore, genome editing technology can also study stress resistance genes or mutual interactive defense responsive genes by generating multiple mutant plants. Their metabolic profiling data can give more information about the biochemical reaction in the defense mechanism.

### 3.4.6 Metabolic Engineering

Metabolic engineering is an approach in which cellular networks are altered and modified to achieve the production of a desired metabolic compound(s). Cellular metabolomics activity is increased by enzymatic modifications involved in the regulation of plant defense function. The metabolic engineering technique provides resistance to transgenic plants against stress conditions and enhances nutritional properties. The process involves a full analytical study of the biosynthesis pathway of the desired metabolite and understanding the specific step in the path that could be altered, followed by the change in the natural biosynthesis pathway without any harmful side effects on the plant. The target gene is modified or manipulated in the biosynthetic pathway by RNA interference, antisense gene, gene knockout, CRISPR-Cas9 system, mutagenesis, and other recombinant approaches (Chownk et al. 2019). Many biotic and abiotic stress-tolerant plants have been developed by changing the gene expression of target metabolic compounds. Not only specific genes but also transcription factors could be modulated to enhance stress-tolerant plants. These transcription factors have disease resistance and stress management function, e.g., *MAPK* (Mitogen-activated protein kinase), *DREB1* (Dehydration responsive element binding protein), and *WRKY*, etc. Secondary metabolites such as terpenoids, phenylpropanoids, flavonoids, etc. are accumulated in stress-resistant plants. So, technically, if the secondary metabolites are targeted, the stress tolerance power of plants will also be improved. Genes involved in secondary metabolites have been cloned and expressed in other (transgenic) plants to achieve stress tolerance. So, by this metabolic engineering system of defense metabolites biosynthesis pathways, several transgenic plants have been developed that possess considerable tolerance to a broad spectrum of pathogens (Ganjewala et al. 2019).

### 3.4.7 Metabolic Engineering of Terpenoid Compounds

Aphids are pests that cause yield loss in several crops. So aphid resistance gene (*E*)-β-Farnesene (EβF) synthases have been engineered in different transgenic crop plants, which converted farnesyl diphosphate to EβF that repulses aphid attraction towards plant (Yu et al. 2012). The transgenic tobacco plant is engineered with terpene synthase gene *GhTPS12* derived from a cotton plant that produces a terpenoid defense compound, linalool, which reduces oviposition by moth *Helicoverpa armigera* in transgenic tobacco plant as compared to wild plant (Huang et al. 2018). Similarly, the acetoacetyl-CoA thiolase (*AACT*) gene has been overexpressed in the alfalfa plant to cope with salt stress conditions (Ganjewala et al. 2019). *FtbHLH3* gene is a drought tolerance gene induced by polyethylene glycol, and abscisic acid has been characterized from *Fagopyrum tataricum* plant and overexpressed in *Arabidopsis thaliana* to increase the drought tolerance mechanism (Yao et al. 2017).

Polyphenol oxidase (*PPO*) expressed in the tomato plant has been shown to elicit plant defense mechanisms against several pests and pathogens. *PPO* overexpressed transgenic tomato plant has increased resistance to the bacterial pathogen, *Pseudomonas syringae,* wherein reduced bacterial growth on infected leaves compared with the control plant has been observed (Li and Steffens 2002). Phytophthora root and stem rot disease of soybean plant has been reduced by introducing the harpin protein-encoding gene, i.e., *hrpZpsta* from *Pseudomonas syringae* into the soybean plant by *Agrobacterium* transformation. The transgenic plants showed upregulation of several enzymes such as *PPO*, *PAL*, peroxidase, and superoxide dismutase during pathogen infection (Du et al. 2018). *CaPAL1* gene from *Capsicum annuum* was overexpressed in *Arabidopsis thaliana,* which increased resistance to pathogens such as *Pseudomonas syringae* pv. *tomato* (*Pst*), *Hyaloperonospora arabidopsidis* by accumulating salicylic acid (Kim and Hwang 2014).

Secondary metabolites synthesized from the flavonoid biosynthesis pathway are responsible for plant stress response mechanisms. Plants and microbes have been used as model organisms for the metabolic engineering program. Overexpression of *isoflavone 7-O methyltransferase* gene produces 4-O-methylisoflavonoid phytoalexin, which has enhanced disease resistance alfalfa plant against the fungal pathogen, *Phoma medicaginis* (He and Dixon 2000). Similarly, development of resistance in pea plant against *Nectria haematococca* was achieved by downregulating the expression of 6-α-hydroxymaackiain-3-*O*-methyltransferase (Wu and VanEtten 2004). *F3H* gene of tea plant has been overexpressed in the tobacco plant, which provided resistance under biotic and abiotic stress conditions (Chownk et al. 2019). *Nicotiana tabacum* transgenic plant with *NtMYB4* gene suppression has increased flavonoid compound rutins accumulation and improved salt tolerance response and also suggested to confer resistance towards a few pathogens (Chen et al. 2019c). Integration of the omics techniques mentioned above could help in developing elite varieties of any crop. Due to the improved biotic and abiotic stress-tolerant plant by metabolic engineering technique, the productivity of crops can be increased that would address the nutritional food security to the rapidly rising world population (Tatsis and Connor 2016).

### 3.4.8 Biological Assays

Bioassays evaluate the effect of the transgene in the transgenic plant. Bioassay approaches are inexpensive and straightforward, which can be experimented in both in vitro and in vivo conditions to characterize the transgenic plants. These techniques are also applied to characterize disease resistance and susceptibility of plants against pathogens. For instance, *PPO* activity assay was performed in *PPO* overexpressed transgenic tomato plants that revealed the bacterial disease resistance which was achieved by increased activity of Polyphenol oxidase (*PPO*) (Li and Steffens 2002). Aphid resistant gene inserted in transgenic *Arabidopsis* plant was taken for aphid behavioral assay in which the agitation response of nymphs depended on defense metabolite concentration and duration of exposure to defense metabolite. It helped to determine the aphid resistant potential of the transgenic plant (Bhatia et al. 2015). In vitro and in vivo bioassays were performed to evaluate antifungal transgenes expressed in transgenic plants (Koppad and Panneerselvam 2016). The nematode mortality assessment has determined a comparative resistant and susceptible nature of nematode infected soybean plant. Different mortality rates of nematode incubated with important metabolic compounds were analyzed by SPSS software and found anti-nematode compounds (Kang et al. 2018). Transgenic *Arabidopsis* plant engineered with the *PAL1 gene,* which provides pathogen tolerance to plant was analyzed by *PAL* activity assay to find out the overexpressed enzymatic activity (Kim and Hwang 2014). Drought tolerant transgenic plants have increased antioxidant flavonoid content and decreased the ROS level (Rao et al. 2020).

## 3.5 Conclusions and Future Perspectives

Transcriptomics is a dynamic field that provides more insights into gene expression, their regulation, and functional characterization, which altogether sheds light on the precise roles of genes. Further, transcriptomics helps to overcome the demerits of studying genomics alone to deduce a gene function. For instance, genomics will not provide information on the alternate splicing, post-transcriptional regulation, and differential expression of genes; however, transcriptomics deals with these aspects to provide further insights. Being a vivacious field of research, transcriptomics has seen a tremendous growth year-by-year in terms of the introduction of new tools and strategies to improvise and accelerate the study of genes and gene expressions. The sequencing-based approaches are in the limelight at present, as they offer a high-throughput analysis of the transcriptome at a genome-wide scale. RNA-seq is highly favored to date and has developed branches for analyzing specific RNA datasets, including strand-specific RNA-seq, bulked segregant RNA-seq, double-stranded RNA-seq, differential RNA-seq, single-cell RNA-seq, etc. Thus, RNA-seq serves as a versatile platform to dissect the intricate complexities that exist in transcriptomes and their regulations. The field is expected to expand its horizon further to gain a better and precise understanding of each gene present in a given genome.

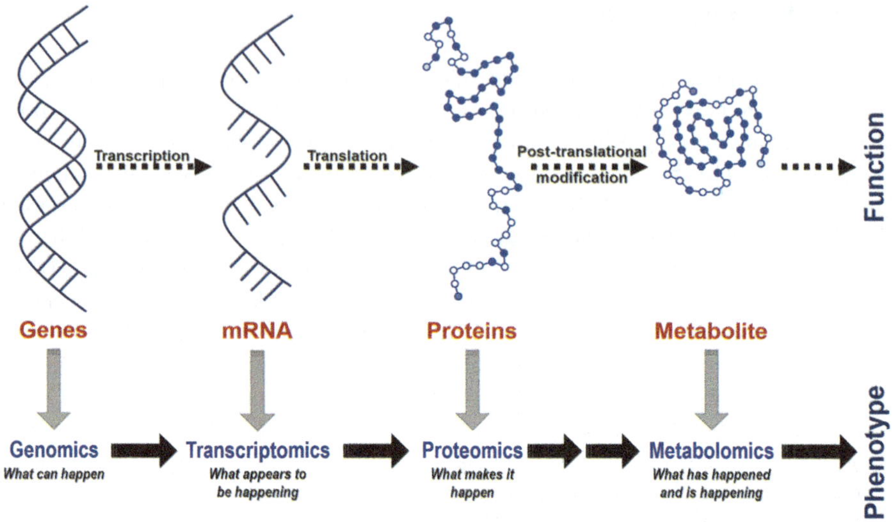


**Fig. 3.3** Schematic representation of the different omics approaches that provide insights into different levels of information encoded in gene, transcript, protein, and metabolite

Proteomic approaches help decipher the complex relationship between stress tolerance and crop productivity, enabling the development of novel plant breeding strategies with high yield and better adapted to withstand environmental changes. The advancement in the MS-based strategies gives a wealth of knowledge of the plant signalling proteomics during host-pathogen interaction. It generates enormous data sets regarding the protein; sequence, quality and quantity, modification, and structure during plants' growth and development. The study of PTMs has become crucial in plant–pathogen interaction as modification in the amino acids may help decipher pathogenesis and plant defense mechanisms. Combining multiple quantitative proteomic techniques is highly beneficial. They yield complementary datasets that improve the understanding of plant defense mechanisms and provide an in-depth characterization of proteins with respect to their abundance and in stress management and plant immunity.

Metabolomics studies the tiniest biomolecules and deciphers the vast signalling pathways related to plant-host interaction and plant defense mechanisms. In order to discriminate between host and pathogen metabolites, which is a significant issue, either one could be labelled with heavy isotopes. Due to the lack of plant-pathogen databases, it is very difficult to identify unknown compounds. Annotation of metabolites is difficult as plant–pathogen interaction is a complex process, and also, it is challenging to study the secondary metabolism pathway, hence making metabolite identification a very tedious job. These omics technologies, when used alone, they do not give enough information on plant–pathogen interaction. Thus, the integration of omics technologies is preferred to decipher the whole signaling pathway, from the genotype to the phenotype (Fig. 3.3). Linking genomics data to proteomics data is easier than linking the genotype data to the phenotype. There is no direct pathway with which the metabolites will alter the phenotype. Collaborative integration of genomics and proteomics data to the metabolomics data will fast-track the research in plant-host interaction. Omics data integration is a difficult task. Although few platforms give multi-omics data access, it is not user-friendly and needs expertise in bioinformatics. Plant-pathogen interaction could be thoroughly studied using multi-omics approaches, and pathogen-resistant/tolerant crops could be developed using various genome editing tools.

**Acknowledgments** IAG is thankful to Science and Engineering Research Board (SERB), DST for the financial support given to the lab for the project EMR/2016/006109. IAG and MM acknowledges DST-FIST Level-I and II support, UGC, DBT-CREBB, UGC-UoH-UPE Phase I and II, DST-PURSE Phase I and II and UGC-SAP-CAS for supporting different infra structural facilities to Department of Plant Sciences and School of Life Sciences, University of Hyderabad.

## References

Abdelrahman M, Nakabayashi R, Mori T et al (2020) Comparative metabolome and transcriptome analyses of susceptible Asparagus officinalis and resistant Wild A. kiusianus reveal insights into stem blight disease resistance. Plant Cell Physiol 61:1464–1476

Adams MD, Soares MB, Kerlavage AR et al (1993) Rapid cDNA sequencing (expressed sequence tags) from a directionally cloned human infant brain cDNA library. Nat Genet 4:373–380

Alvarenga SM, Caixeta ET, Hufnagel B et al (2010) In silico identification of coffee genome expressed sequences potentially associated with resistance to diseases. Genet Mol Biol 33:795–806

Alwine JC, Kemp DJ, Stark GR (1977) Method for detection of specific RNAs in agarose gels by transfer to diazobenzyloxymethyl-paper and hybridization with DNA probes. Proc Natl Acad Sci U S A 74:5350–5354

Andersen CL, Ledet-Jensen J, Ørntoft T (2004) Normalization of real-time quantitative RT-PCR data: a model based variance estimation approach to identify genes suited for normalization – applied to bladder- and colon-cancer data-sets. Cancer Res 64:5245–5250

Anjorin FB, Adejumo SA, Agboola L et al (2016) Proline, soluble sugar, leaf starch and relative water contents of four maize varieties in response to different watering regimes. Cercet Agron Mold 49(3):51–62

Arrivault S, Guenther M, Ivakov A et al (2009) Use of reverse-phase liquid chromatography, linked to tandem mass spectrometry, to profile the Calvin cycle and other metabolic intermediates in Arabidopsis rosettes at different carbon dioxide concentrations. Plant J 59(5):826–839

Ashraf N, Ghai D, Barman P et al (2009) Comparative analyses of genotype dependent expressed sequence tags and stress-responsive transcriptome of chickpea wilt illustrate predicted and unexpected genes and novel regulators of plant immunity. BMC Genomics 10:415

Ashwin NMR, Barnabas L, Ramesh S et al (2017) Advances in proteomic technologies and their scope of application in understanding plant–pathogen interactions. J Plant Biochem Biotechnol 26:371–386

Baldwin D, Crane V, Rice D (1999) A comparison of gel-based, nylon filter and microarray techniques to detect differential RNA expression in plants. Curr Opin Plant Biol 2:96–103

Bari R, Jones JDG (2009) Role of plant hormones in plant defence responses. Plant Mol Biol 69 (4):473–488. https://doi.org/10.1007/s11103-008-9435-0

Bell M (1995) Analysis of plant gene expression by reverse transcription-PCR. In: Jones H (ed) Plant gene transfer and expression protocols. Methods in molecular biology, vol 49. Springer, Totowa, NJ. https://doi.org/10.1385/0-89603-321-X:291

Bhatia V, Maisnam J, Jain A et al (2015) Aphid-repellent pheromone E-β-farnesene is generated in transgenic Arabidopsis thaliana over-expressing farnesyl diphosphate synthase2. Ann Bot 115 (4):581–591. https://doi.org/10.1093/aob/mcu250

Bimolata W, Kumar A, Sundaram RM et al (2013) Analysis of nucleotide diversity among alleles of the major bacterial blight resistance gene *Xa27* in cultivars of rice (*Oryza sativa*) and its wild relatives. Planta 238:239–305

Bimolata W, Kumar A, Reddy MSK et al (2015) Nucleotide diversity analysis of three major bacterial blight resistance genes in rice. PLoS One 10:e0120186

Blasi ÉAR, Buffon G, Rativa AGS et al (2017) High infestation levels of Schizotetranychus oryzae severely affects rice metabolism. J Plant Physiol 219:100–111

Boller T, Felix G (2009) Arenaissance of elicitors: perception of microbe-associated molecular patterns and danger signals by pattern-recognition receptors. Annu Rev Plant Biol 60:379–406

Bradbury A, Velappan N, Verzillo V et al (2003) Antibodies in proteomics I: generating antibodies. Trends Biotechnol 21:275–281. https://doi.org/10.1016/S0167-7799(03)00112-4

Brenner S, Johnson M, Bridgham J et al (2000) Gene expression analysis by massively parallel signature sequencing (MPSS) on microbead arrays. Nat Biotechnol 18:630–634. Erratum in: Nat Biotechnol 18:1021

Bujak R, Struck-Lewicka W, Markuszewski MJ et al (2015) Metabolomics for laboratory diagnostics. J Pharm Biomed Anal 113(778051):108–120. https://doi.org/10.1016/j.jpba.2014.12.017

Carpenter EJ, Matasci N, Ayyampalayam S et al (2019) Access to RNA-sequencing data from 1173 plant species: the 1000 plant transcriptomes initiative (1KP). GigaScience 8:giz126

Castillejo MÁ, Fernández-Aparicio M, Rubiales D (2012) Proteomic analysis by two-dimensional differential in gel electrophoresis (2D DIGE) of the early response of Pisum sativum to Orobanche crenata. J Exp Bot 63:107–119

Catoni M, Miozzi L, Fiorilli V et al (2009) Comparative analysis of expression profiles in shoots and roots of tomato systemically infected by tomato spotted wilt virus reveals organ-specific transcriptional responses. Mol Plant-Microbe Interact 22:1504–1513

Chen JS, Lin BR, Huang QL, Hu LL, Zhuo K, Liao JL (2017) A novel Meloidogyne graminicola efector, MgGPP, is secreted into host cells and undergoes glycosylation in concert with proteolysis to suppress plant defenses and promote parasitism. PLoS Pathog 13:4

Chen F, Ma R, Chen XL (2019a) Advances of metabolomics in fungal pathogen–plant interactions. MDPI AG 9(8):169. https://doi.org/10.3390/metabo9080169

Chen S, Wu F, Li Y et al (2019b) NtMYB4 and NtCHS1 are critical factors in the regulation of flavonoid biosynthesis and are involved in salinity responsiveness. Front Plant Sci 10:178. https://doi.org/10.3389/fpls.2019.00178

Chen L, Wu Q, He W et al (2019c) Combined de novo transcriptome and metabolome analysis of common bean response to Fusarium oxysporum f. sp. phaseoli infection. Int J Mol Sci 20:6278

Chen S, Zhang Y, Zhao Y et al (2020) Key genes and genetic interactions of plant-pathogen functional modules in poplar infected by Marssonina brunnea. Mol Plant-Microbe Interact 33:1080–1090

Chivasa S, Hamilton JM, Pringle RS et al (2006) Proteomic analysis of differentially expressed proteins in fungal elicitor-treated Arabidopsis cell cultures. J Exp Bot 57:1553–1562. https://doi.org/10.1093/jxb/erj149

Chownk M, Thakur K, Yadav SK (2019) Retrospect and prospects of plant metabolic engineering. J Plant Biochem Biotechnol 28:1–13. https://doi.org/10.1007/s13562-018-0473-7

Colas I, Koroleva O, Shaw PJ (2010) Mass spectrometry in plant proteomic analysis. Plant Biosyst 144(3):703–714. https://doi.org/10.1080/11263501003764392

Colignon B, Raes M, Dieu M et al (2013) Evaluation of three-dimensional gel electrophoresis to improve quantitative profiling of complex proteomes. Proteomics 13:2077–2082

Collins RM, Afzal M, Ward DA et al (2010) Differential proteomic analysis of Arabidopsis thaliana genotypes exhibiting resistance or susceptibility to the insect herbivore, Plutella xylostella. PLoS One 8:e10103

Cramer RA, LA Rota CM, Cho Y et al (2006) Bioinformatic analysis of expressed sequence tags derived from a compatible Alternaria brassicicola-Brassica oleracea interaction. Mol Plant Pathol 7:113–124

Crookshanks M, Emmersen J, Welinder KG et al (2001) The potato tuber transcriptome:analysis of 6077 expressed sequence tags. FEBS Lett 506:123–126

Czechowski T, Bari RP, Stitt M et al (2004) Real-time RT-PCR profiling of over 1400 Arabidopsis transcription factors: unprecedented sensitivity reveals novel root- and shoot-specific genes. Plant J 38:366–379

Dang M, Cheng Q, Hu Y et al (2020) Proteomic changes during MCMV infection revealed by iTRAQ quantitative proteomic analysis in maize. Int J Mol Sci 21

de Hoffmann E, Stroobant V (2007) Mass spectrometry: principles and applications. Wiley. https://books.google.co.in/books?id=pqhr5XQZDtUC

Deborde C, Fontaine JX, Jacob D et al (2019) Optimizing 1D 1H-NMR profiling of plant samples for high throughput analysis: extract preparation, standardization, automation and spectra processing. Metabolomics 15(3):1–12. https://doi.org/10.1007/s11306-019-1488-3

Dhakal R, Chai C, Karan R et al (2017) Expression profiling coupled with in-silico mapping identifies candidate genes for reducing aflatoxin accumulation in maize. Front Plant Sci 8:503

Di Carli M, Villani ME, Bianco L, Lombardi R et al (2010) Proteomic analysis of the plant−virus interaction in cucumber mosaic virus (CMV) resistant transgenic tomato. J Proteome Res 9 (11):5684–5697

Diatchenko L, Lau YF, Campbell AP et al (1996) Suppression subtractive hybridization: a method for generating differentially regulated or tissue-specific cDNA probes and libraries. Proc Natl Acad Sci U S A 93:6025–6030

Dixit SK, Gupta A, Fatima U et al (2019) AtGBF3 confers tolerance to Arabidopsis thaliana against combined drought and Pseudomonas syringae stress. Environ Exp Bot 168:103881

Dong CJ, Wang XL, Shang QM (2011) Salicylic acid regulates sugar metabolism that confers tolerance to salinity stress in cucumber seedlings. Sci Hortic 129(4):629–636. https://doi.org/10.1016/j.scienta.2011.05.005

Dong J, Wang Y, Xian Q et al (2020) Transcriptome analysis reveals ethylene-mediated defense responses to Fusarium oxysporum f. sp. cucumerinum infection in Cucumis sativus L. BMC Plant Biol 20:334

Du Q, Yang X, Zhang J et al (2018) Over-expression of the Pseudomonas syringae harpin-encoding gene hrpZm confers enhanced tolerance to Phytophthora root and stem rot in transgenic soybean. Transgenic Res 27(3):277–288. https://doi.org/10.1007/s11248-018-0071-4

Eldakak M, Das A, Zhuang Y et al (2018) A quantitative proteomics view on the function of Qfhb1, a major QTL for fusarium head blight resistance in wheat. Pathogens 7:1–27. https://doi.org/10.3390/pathogens7030058

Evans C, Noirel J, Ow SY et al (2012) An insight into iTRAQ: where do we stand now? Anal Bioanal Chem 404:1011–1027

Fatima U, Bhorali P, Senthil-Kumar M et al (2019) Morpho-pathological and global transcriptomic analysis reveals the robust nonhost resistance responses in chickpea interaction with Alternaria brassicae. Mol Plant-Microbe Interact 32:1598–1613

Fenn JB (2002) Electrospray ionization mass spectrometry: how it all began. J Biomol Tech 13:101–118

Fenn JB, Mann M, Meng CK et al (1989) Electrospray ionization for mass spectrometry of large biomolecules. Science 246:64–71. https://doi.org/10.1126/science.2675315

Feng Z, Ding C, Li W, Wang D et al (2020) Applications of metabolomics in the research of soybean plant under abiotic stress. Food Chem 310:125914

Ficarro SB, McCleland ML, Stukenberg PT et al (2002) Phosphoproteome analysis by mass spectrometry and its application to Saccharomyces cerevisiae. Nat Biotechnol 20:301–305

Fregene M, Matsumura H, Akano A et al (2004) Serial analysis of gene expression (SAGE) of host–plant resistance to the cassava mosaic disease (CMD). Plant Mol Biol 56:563–571

Fukushima A, Kusano M, Mejia RF et al (2014) Metabolomic characterization of knockout mutants in arabidopsis: development of a metabolite profiling database for knockout mutants in arabidopsis. Plant Physiol 165(3):948–961. https://doi.org/10.1104/pp.114.240986

Ganjewala D, Kaur G, Srivastava N (2019) Metabolic engineering of stress protectant secondary metabolites to confer abiotic stress tolerance in plants. Springer, Singapore. https://doi.org/10.1007/978-981-15-0690-1_11

Geiger D, Seedorf S, Schulze T et al (2011) Managing the Crowd: towards a taxonomy of crowdsourcing processes. 17th Americas Conference on Information Systems, AMCIS 2011. 5

Goldmann T, Gonzalez JS (2000) DNA-printing: utilization of a standard inkjet printer for the transfer of nucleic acids to solid supports. J Biochem Biophys Methods 42:105–110

Gong L, Han S, Yuan M et al (2020) Transcriptomic analyses reveal the expression and regulation of genes associated with resistance to early leaf spot in peanut. BMC Res Notes 13:381

Gon S, Wang Y, Hee K et al (2012) In-depth insight into in vivo apoplastic secretome of rice-Magnaporthe oryzae interaction. J Proteome 78:58–71

Graham RLJ, Graham C, McMullan G (2007) Microbial proteomics: a mass spectrometry primer for biologists. Microb Cell Factories 6:26

Guerra-Guimarães L, Tenente R, Pinheiro C et al (2015) Proteomic analysis of apoplastic fluid of *Coffea arabica* leaves highlights novel biomarkers for resistance against *Hemileia vastatrix*. Front Plant Sci 6:478

Gupta A, Senthil-Kumar M (2017) Transcriptome changes in Arabidopsis thaliana infected with Pseudomonas syringae during drought recovery. Sci Rep 7:9124

Gupta A, Sarkar AK, Senthil-Kumar M (2016) Global transcriptional analysis reveals unique and shared responses in Arabidopsis thaliana exposed to combined drought and pathogen stress. Front Plant Sci 7:686

Gygi SP, Rist B, Gerber SA et al (1999) Quantitative analysis of complex protein mixtures using isotope-coded affinity tags. Nat Biotechnol 17:994–999

Hajdarpašić A, Ruggenthaler P (2012) Analysis of miRNA expression under stress in Arabidopsis thaliana. Bosn J Basic Med Sci 12:169–176

Harbers M, Carninci P (2005) Tag-based approaches for transcriptome research and genome annotation. Nat Methods 2:495–502

Harsha HC, Pandey A (2010) Phosphoproteomics in cancer. Mol Oncol 4:482–495. https://doi.org/10.1016/j.molonc.2010.09.004

He XZ, Dixon RA (2000) Genetic manipulation of isoflavone 7-O-methyltransferase enhances biosynthesis of 4'-O-methylated isoflavonoid phytoalexins and disease resistance in alfalfa. Plant Cell 12(9):1689–1702. https://doi.org/10.1105/tpc.12.9.1689

Higuchi R, Fockler C, Dollinger G et al (1993) Kinetic PCR analysis: real-time monitoring of DNA amplification reactions. Nat Biotechnol 11:1026–1030

Hong M, Zhang Y, Hu F (2012) Membrane protein structure and dynamics from NMR spectroscopy. Annu Rev Phys Chem 63:1–24

Hong J, Yang L, Zhang D et al (2016) An indispensable system biology tool for plant science. Plant Metabol 17(6):767. https://doi.org/10.3390/ijms17060767

Holtappels M, Noben JP, Valcke R (2016) Virulence of Erwinia amylovora, a prevalent apple pathogen: outer membrane proteins and type III secreted effectors increase fitness and compromise plant defenses. Proteomics 16:2377–2390

Hu Q, Noll R, Hongyan L et al (2005) The Orbitrap: a new mass spectrometer. J Mass Spectrom 40:430–443

Huang CL, Hwang SY, Chiang YC et al (2008) Molecular evolution of the Pi-ta gene resistant to rice blast in wild rice (Oryza rufipogon). Genetics 179:1527–1538

Huang XZ, Xiao YT, Köllner TG et al (2018) The terpene synthase gene family in Gossypium hirsutum harbors a linalool synthase GhTPS12 implicated in direct defence responses against herbivores. Plant Cell Environ 41(1):261–274. https://doi.org/10.1111/pce.13088

Issaq HJ, Chan KC, Janini GM et al (2005) Multidimensional separation of peptides for effective proteomic analysis. J Chromatogr B 817:35–47

Jantasuriyarat C, Gowda M, Haller K et al (2005) Large-scale identification of expressed sequence tags involved in rice and rice blast fungus interaction. Plant Physiol 138:105–115

Jin S, Ren Q, Lian L et al (2020) Comparative transcriptomic analysis of resistant and susceptible tea cultivars in response to Empoasca onukii (Matsuda) damage. Planta 252:10

Jothiramshekar S, Benjamin JJ, Krishnasamy R et al (2020) Identification of salt-responsive genes from C4 halophyte Suaeda nudiflora through suppression subtractive hybridization and expression analysis under individual and combined treatment of salt and elevated carbon dioxide conditions. Physiol Mol Biol Plants 26:163–172

Kang W, Zhu X, Wang Y et al (2018) Transcriptomic and metabolomic analyses reveal that bacteria promote plant defense during infection of soybean cyst nematode in soybean. BMC Plant Biol 18(1):86. https://doi.org/10.1186/s12870-018-1302-9

Kang K, Yue L, Xia X et al (2019) Comparative metabolomics analysis of different resistant rice varieties in response to the brown planthopper Nilaparvata lugens Hemiptera: Delphacidae. Metabolomics 15(4):62. https://doi.org/10.1007/s11306-019-1523-4

Karas M, Krüger R (2003) Ion formation in MALDI: the cluster ionization mechanism. Chem Rev 103:427–440

Kevil CG, Walsh L, Laroux FS et al (1997) An improved, rapid northern protocol. Biochem Biophys Res Commun 238:277–279

Khan MS, Akther T, Ali M et al (2019) An investigation on the role of salicylic acid alleviate the saline stress in rice crop (Oryza sativa (L)). Biocatal Agric Biotechnol 18:101027. https://doi.org/10.1016/j.bcab.2019.101027

Khan AL, Asaf S, Khan A et al (2020) Transcriptomic analysis of Dubas bug (Ommatissus lybicus Bergevin) infestation to date palm. Sci Rep 10:11505

Kim DS, Hwang BK (2014) An important role of the pepper phenylalanine ammonia-lyase gene (PAL1) in salicylic acid-dependent signalling of the defence response to microbial pathogens. J Exp Bot 65(9):2295–2306

Kodzius R, Kojima M, Nishiyori H et al (2006) CAGE: cap analysis of gene expression. Nat Methods 3:211–222

Koppad AV, Panneerselvam R (2016) Simple high throughput bioassay techniques to evaluate transgenic plants expressing antifungal genes. Res J Life Sci Bioinform Pharm Chem Sci 2 (39):39–48

Kumar R, Bohra A, Pandey AK et al (2017) Metabolomics for plant improvement: status and prospects. Front Plant Sci. https://doi.org/10.3389/fpls.2017.01302

Kumar R, Barua P, Chakraborty N et al (2020) Systemic acquired resistance specific proteome of Arabidopsis thaliana. Plant Cell Rep 39:1549–1563

Kumari A, Das A, Devanna B et al (2013a) Mining of rice blast resistance gene Pi54 shows effect of single nucleotide polymorphism on phenotypic expression of alleles. Eur J Plant Pathol 137:55–65

Kumari K, Muthamilarasan M, Misra G et al (2013b) Development of eSSR-markers in Setaria italica and their applicability in studying genetic diversity, cross-transferability and comparative mapping in millet and non-millet species. PLoS One 8:e67742

Lazar AG, Romanciuc F, Socaciu MA et al (2015) Bioinformatics tools for metabolomic data processing and analysis using untargeted liquid chromatography coupled with mass spectrometry. Bulletin of University of Agricultural Sciences and Veterinary Medicine Cluj-Napoca. Anim Sci Biotechnol 72(2):11536. https://doi.org/10.15835/buasvmcn-asb:11536

Lee JY, Lee DH (2003) Use of serial analysis of gene expression technology to reveal changes in gene expression in Arabidopsis pollen undergoing cold stress. Plant Physiol 132:517–529

Leebens-Mack JH, Barker MS, Carpenter EJ et al (2019) One thousand plant transcriptomes and the phylogenomics of green plants. Nature 574:679–685

Li L, Steffens JC (2002) Overexpression of polyphenol oxidase in transgenic tomato plants results in enhanced bacterial disease resistance. Planta 215(2):239–247. https://doi.org/10.1007/s00425-002-0750-4

Li X, Liu T, Chen W et al (2015) Wheat WCBP1 encodes a putative copper-binding protein involved in stripe rust resistance and inhibition of leaf senescence. BMC Plant Biol 15:239

Li N, Jia H, Kong Z et al (2017) Identification and marker-assisted transfer of a new powdery mildew resistance gene at the Pm4 locus in common wheat. Mol Breed 37:79

Liu C, Li Z, Xing J, Yang J, Wang Z, Zhang H, Chen D, Peng YL, Chen XL (2018) Global analysis of sumoylation function reveals novel insights into development and appressorium-mediated infection of the rice blast fungus. New Phytol 219(3):1031–1047

Liu Y, Lu S, Liu K et al (2019) Proteomics: a powerful tool to study plant responses to biotic stress. Plant Methods 15:135. https://doi.org/10.1186/s13007-019-0515-8

Livak KJ, Schmittgen TD (2001) Analysis of relative gene expression data using real-time quantitative PCR and the 2(-Delta Delta C(T)) method. Methods 25:402–408

Loarce Y, Navas E, Paniagua C et al (2016) Identification of genes in a partially resistant genotype of Avena sativa expressed in response to Puccinia coronata infection. Front Plant Sci 7:731

Lodha TD, Basak J (2012) Plant–pathogen interactions: what microarray tells about it? Mol Biotechnol 50:87–97

Long L, Liu J, Gao Y et al (2019) Flavonoid accumulation in spontaneous cotton mutant results in red coloration and enhanced disease resistance. Plant Physiol Biochem 143:40–49. https://doi.org/10.1016/j.plaphy.2019.08.021

López-Gresa MP, Maltese F, Bellés JM et al (2010) Metabolic response of tomato leaves upon different plant-pathogen interactions. Phytochem Anal 21:89–94

Lu L, Yang D, Tang D et al (2020) Transcriptome analysis of different rice cultivars provides novel insights into the rice response to bacterial leaf streak infection. Funct Integr Genomics 20:681–693

Lyu S, Gao L, Zhang R et al (2020) Correlation analysis of expression profile and quantitative iTRAQ-LC-MS/MS proteomics reveals resistance mechanism against TuMV in Chinese cabbage (Brassica rapa ssp. pekinensis). Front Genet 11:1–14

Ma Z, Wang L, Zhao M et al (2020) ITRAQ proteomics reveals the regulatory response to Magnaporthe oryzae in durable resistant vs. susceptible rice genotypes. PLoS One 15:1–20. https://doi.org/10.1371/JOURNAL.PONE.0227470

MacBeath G (2002) Protein microarrays and proteomics. Nat Genet 32:526–532. https://doi.org/10.1038/ng1037

Macek B, Mann M, Olsen JV (2009) Global and site-specific quantitative phosphoproteomics: principles and applications. Annu Rev Pharmacol Toxicol 49:199–221

Mahalingam R, Gomez-Buitrago A, Eckardt N et al (2003) Characterizing the stress/defense transcriptome of Arabidopsis. Genome Biol 4:R20

Mahomed W, van den Berg N (2011) EST sequencing and gene expression profiling of defence-related genes from Persea americana infected with Phytophthora cinnamomi. BMC Plant Biol 11:167

Maleck K, Levine A, Eulgem T et al (2000) The transcriptome of Arabidopsis during systemic acquired resistance. Nat Genet 26:403–410

Mandal A, Mishra AK, Dulani P et al (2018) Identification, characterization, expression profiling, and virus-induced gene silencing of armadillo repeat-containing proteins in tomato suggest their involvement in tomato leaf curl New Delhi virus resistance. Funct Integr Genomics 18:101–111

Marouga R, David S, Hawkins E (2005) The development of the DIGE system: 2D fluorescence difference gel analysis technology. Anal Bioanal Chem 382:669–678

Martin LBB, Fei Z, Giovannoni JJ et al (2013) Catalyzing plant science research with RNA-seq. Front Plant Sci 4:66

Martins ACQ, Mehta A, Murad AM et al (2020) Proteomics unravels new candidate genes for Meloidogyne resistance in wild Arachis. J Proteome 217:103690

Matsumura H, Nirasawa S, Terauchi R (1999) Transcript profiling in rice (Oryza sativa L.) seedlings using serial analysis of gene expression (SAGE). Plant J 20:719–726

Matsumura H, Reich S, Ito A et al (2003) Gene expression analysis of plant host-pathogen interactions by SuperSAGE. Proc Natl Acad Sci U S A 100:15718–15723

Meyers BC, Haudenschild CD, Vemaraju K (2007) Use of massively parallel signature sequencing to study genes expressed during the plant defense response. In: Ronald PC (ed) Plant-pathogen interactions. Methods in molecular biology, vol 354. Humana Press, Totowa, NJ. https://doi.org/10.1385/1-59259-966-4:105

Mohammed S, Heck AJR (2011) Strong cation exchange (SCX) based analytical methods for the targeted analysis of protein post-translational modifications. Curr Opin Biotechnol 22:9–16

Mora JF, Van Berkel GJ, Enke CG et al (2000) Electrochemical processes in electrospray ionization mass spectrometry. J Mass Spectrom 35:939–952

Muthamilarasan M, Prasad M (2013) Plant innate immunity: an updated insight into defense mechanism. J Biosci 38:433–449

Muthamilarasan M, Singh NK, Prasad M (2019) Multi-omics approaches for strategic improvement of stress tolerance in underutilized crop species: a climate change perspective. Adv Genet 103:1–38

Nobuta K, Vemaraju K, Meyers BC (2007) Methods for analysis of gene expression in plants using MPSS. In: Edwards D (ed) Plant bioinformatics. Methods in molecular biology, vol vol 406. Humana Press, Totowa, NJ. https://doi.org/10.1007/978-1-59745-535-0_19

Nguyen TH, Brechenmacher L, Aldrich JT et al (2012) Quantitative phosphoproteomic analysis of soybean root hairs inoculated with Bradyrhizobium japonicum. Mol Cell Proteomics 11:1140–1155

Obata T, Fernie AR (2012) The use of metabolomics to dissect plant responses to abiotic stresses. Cell Mol Life Sci 69(19):3225–3243. https://doi.org/10.1007/s00018-012-1091-5

Overbergh L, Giulietti A, Valckx D et al (2003) The use of real-time reverse transcriptase PCR for the quantification of cytokine gene expression. J Biomol Tech 14:33–43

Ozsolak F, Milos PM (2011) RNA sequencing: advances, challenges and opportunities. Nat Rev Genet 12:87–98

Padilha HKM, Madruga NA, Aranha BC et al (2019) Defense responses of Capsicum spp. genotypes to post-harvest Colletotrichum sp. inoculation. Phytoparasitica 47(4):557–573. https://doi.org/10.1007/s12600-019-00756-9

Pandey S, Muthamilarasan M, Sharma N et al (2019) Characterization of DEAD-box family of RNA helicases in tomato provides insights into their roles in biotic and abiotic stresses. Environ Exp Bot 158:107–116

Parker J, Koh J, Yoo MJ et al (2013) Quantitative proteomics of tomato defense against Pseudomonas syringae infection. Proteomics 13:1934–1946

Patel RK, Jain M (2012) NGS QC toolkit: a toolkit for quality control of next generation sequencing data. PLoS One 7:e30619

Pechanova O, Pechan T, Williams WP, Dawn S (2011) Proteomic analysis of the maize rachis: potential roles of constitutive and induced proteins in resistance to Aspergillus flavus infection and aflatoxin accumulation. Proteomics 11:114–127

Peiffer JA, Kaushik S, Sakai H et al (2008) A spatial dissection of the Arabidopsis floral transcriptome by MPSS. BMC Plant Biol 8:43

Pfaffl MW, Tichopad A, Prgomet C et al (2004) Determination of stable housekeeping genes, differentially regulated target genes and sample integrity: BestKeeper--excel-based tool using pair-wise correlations. Biotechnol Lett 26:509–515

Phillips CM, Iavarone AT, Marletta MA (2011) Quantitative proteomic approach for cellulose degradation by Neurospora crassa. J Proteome Res 10:4177–4185

Pidon H, Cheron S, Ghesquiere A, Albar L (2020) Allele mining unlocks the identification of RYMV resistance genes and alleles in African cultivated rice. BMC Plant Biol 20:222. https://doi.org/10.1186/s12870-020-02433-0

Pollack JR, Perou CM, Alizadeh AA et al (1999) Genome-wide analysis of DNA copy-number changes using cDNA microarrays. Nat Genet 23:41–46

Pushpa D, Yogendra KN, Murphy A (2014) Identification of late blight resistance-related metabolites and genes in potato through nontargeted metabolomics. 584–595

Rabilloud T (2013) When 2D is not enough, go for an extra dimension. Proteomics 13:2065–2068

Ramautar R, De Jong GJ (2014) Recent developments in liquid-phase separation techniques for metabolomics. Bioanal Future Sci 6(7):1011–1026. https://doi.org/10.4155/bio.14.51

Ramautar R, Somsen GW, De Jong GJ (2017) CE–MS for metabolomics: developments and applications in the period 2014–2016. Electrophoresis 38(1):190–202. https://doi.org/10.1002/elps.201600370

Ramegowda V, Senthil-Kumar M (2015) The interactive effects of simultaneous biotic and abiotic stresses on plants: mechanistic understanding from drought and pathogen combination. J Plant Physiol 176:47–54

Ramkumar G, Sakthivel K, Sundaram RM et al (2010) Allele mining in crops: prospects and potentials. Biotechnol Adv 28:451–461

Ramu VS, Paramanantham A, Ramegowda V et al (2016) Transcriptome analysis of sunflower genotypes with contrasting oxidative stress tolerance reveals individual- and combined- biotic and abiotic stress tolerance mechanisms. PLoS One 11:e0157522

Rao MJ, Xu Y, Tang X, Huang Y (2020) CsCYT75B1, a citrus CYTOCHROME P450 gene, is involved in accumulation of antioxidant flavonoids and induces drought tolerance in transgenic Arabidopsis. Antioxidants 9(2):161. https://doi.org/10.3390/antiox9020161

Reymond P, Weber H, Damond M et al (2000) Differential gene expression in response to mechanical wounding and insect feeding in Arabidopsis. Plant Cell 12:707–720

Ronning CM, Stegalkina SS, Ascenzi RA et al (2003) Comparative analyses of potato expressed sequence tag libraries. Plant Physiol 131:419–429

Roohie RK, Umesha S (2015) Identification of genes associated with black rot resistance in cabbage through suppression subtractive hybridization. 3 Biotech 5:1089–1100

Rose JK, Bashir S, Giovannoni JJ et al (2004) Tackling the plant proteome: practical approaches, hurdles and experimental tools. Plant J 39:715–733

Ross PL, Huang YN, Marchese JN et al (2004) Multiplexed protein quantitation in Saccharomyces cerevisiae using amine-reactive isobaric tagging reagents. Mol Cell Proteomics 3:1154–1169

Rowland E, Kim J, Bhuiyan NH et al (2015a) The Arabidopsis chloroplast stromal N-terminome; complexities of N-terminal protein maturation and stability. Plant Physiol. https://doi.org/10.1104/pp.15.01214

Rowland E, Kim J, Bhuiyan NH et al (2015b) The Arabidopsis chloroplast stromal N-terminome: complexities of amino-terminal protein maturation and stability. Plant Physiol 169:1881–1896

Saabale PR, Dubey SC, Priyanka K et al (2018) Analysis of differential transcript expression in chickpea during compatible and incompatible interactions with Fusarium oxysporum f. sp. ciceris race 4. 3 Biotech 8:111

Sade D, Shriki O, Cuadros-Inostroza A et al (2015) Comparative metabolomics and transcriptomics of plant response to Tomato yellow leaf curl virus infection in resistant and susceptible tomato cultivars. Metabolomics 11:81–97

Sahu PP, Rai NK, Chakraborty S et al (2011) Tomato cultivar tolerant to tomato leaf curl New Delhi virus infection induces virus-specific siRNA accumulation and defense associated host gene expression. Mol Plant Pathol 11:531–544

Sahu PP, Sharma N, Puranik S et al (2016) Tomato 26S proteasome subunit RPT4a regulates ToLCNDV transcription and activates hypersensitive response in tomato. Sci Rep 6:27078

Sana TR, Fischer S, Wohlgemuth G (2010) Metabolomic and transcriptomic analysis of the rice response to the bacterial blight pathogen Xanthomonas oryzae pv. oryzae. Metabolomics 6 (3):451–465. https://doi.org/10.1007/s11306-010-0218-7

Savidor A, Teper D, Gartemann KH et al (2012) The Clavibacter michiganensis subsp. michiganensis-tomato interactome reveals the perception of pathogen by the host and suggests mechanisms of infection. J Proteome Res 11(2):736–750

Schaker PDC, Peters LP, Cataldi TR et al (2017) Metabolome dynamics of smutted sugarcane reveals mechanisms involved in disease progression and whip emission. Front Plant Sci 8:882. https://doi.org/10.3389/fpls.2017.00882

Schenk PM, Kazan K, Wilson I et al (2000) Coordinated plant defense responses in Arabidopsis revealed by microarray analysis. Proc Natl Acad Sci U S A 97:11655–11660

Scott KD, Eggler P, Seaton G et al (2000) Analysis of SSRs derived from grape ESTs. Theor Appl Genet 100:723–726

Shalon D, Smith SJ, Brown PO (1996) A DNA microarray system for analyzing complex DNA samples using two-color fluorescent probe hybridization. Genome Res 6:639–645

Sharma N, Hotte N, Rahman MH et al (2008) Towards identifying Brassica proteins involved in mediating resistance to Leptosphaeria maculans: a proteomicsbased approach. Proteomics 8:3516–3535

Sharma N, Prasad M (2020) Silencing AC1 of tomato leaf curl virus using artificial microRNA confers resistance to leaf curl disease in transgenic tomato. Plant Cell Rep. https://doi.org/10.1007/s00299-020-02584-2

Sharma K, Sarma S, Bohra A, Mitra A, Sharma NK, Kumar A (2018) Plant metabolomics: an emerging technology for crop improvement. In: Çelik O (ed) New visions in plant science, 1st edn. IntechOpen, London, pp 65–79

Shasmita MD, Mohapatra PK et al (2019) Priming with salicylic acid induces defense against bacterial blight disease by modulating rice plant photosystem II and antioxidant enzymes activity. Physiol Mol Plant Pathol 108:101427. https://doi.org/10.1016/j.pmpp.2019.101427

Shenton MR, Berberich T, Kamo M et al (2012) Use of intercellular washing fluid to investigate the secreted proteome of the rice-Magnaporthe interaction. J Plant Res 125:311–316

Shevchenko A, Tomas H, Havlis J et al (2006) In-gel digestion for mass spectrometric characterization of proteins and proteomes. Nat Protoc 1:2856–2860

Shiio Y, Aebersold R (2006) Quantitative proteome analysis using isotope-coded affinity tags and mass spectrometry. Nat Protoc 1:139–145

Sindt NM, Robison F, Brick MA et al (2018) MALDI-TOF-MS with PLS modeling enables strain typing of the bacterial plant pathogen Xanthomonas axonopodis. J Am Soc Mass Spectrom 29:413–421

Silver N, Best S, Jiang J et al (2006) Selection of housekeeping genes for gene expression studies in human reticulocytes using real-time PCR. BMC Mol Biol 7:33

Sinha R, Gupta A, Senthil-Kumar M (2017) Concurrent drought stress and vascular pathogen infection induce common and distinct transcriptomic responses in chickpea. Front Plant Sci 8:333

Studer B, Boller B, Herrmann D et al (2006) Genetic mapping reveals a single major QTL for bacterial wilt resistance in Italian ryegrass (Lolium multiflorum Lam.). Theor Appl Genet 113:661–671

Subramanian B, Bansal VK, Kav NN (2005) Proteome-level investigation of Brassica carinata derived resistance to Leptosphaeria maculans. J Agric Food Chem 53:313–324

Sun R, Qin S, Zhang T et al (2019) Comparative phosphoproteomic analysis of blast resistant and susceptible rice cultivars in response to salicylic acid. BMC Plant Biol 19:1–15

Tatsis EC, Connor SEO (2016) ScienceDirect new developments in engineering plant metabolic pathways. Curr Opin Biotechnol 42:126–132. https://doi.org/10.1016/j.copbio.2016.04.012

Taub FE, DeLeo JM, Thompson EB (1983) Sequential comparative hybridizations analyzed by computerized image processing can identify and quantitate regulated RNAs. DNA 2:309–327

Teh SL, Rostandy B, Awale M et al (2019) Genetic analysis of stilbenoid profiles in grapevine stems reveals a major mQTL hotspot on chromosome 18 associated with disease-resistance motifs. Hortic Res 6:121. https://doi.org/10.1038/s41438-019-0203-x

Thelen JJ, Peck SC (2007) Quantitative proteomics in plants: choices in abundance. Plant Cell 19:3339–3346

Thelen JJ, Miernyk JA (2012) The proteomic future: where mass spectrometry should be taking us. Biochem J 444:169–181

Thingholm TE, Jensen ON, Larsen MR (2009) Analytical strategies for phosphoproteomics. Proteomics 9:1451–1468

Thomason K, Babar MA, Erickson JE et al (2018) Comparative physiological and metabolomics analysis of wheat (Triticum aestivum L.) following post-anthesis heat stress. PLoS One. https://doi.org/10.1371/JOURNAL.PONE.0197919

Tong H, Tenenboim H, Wu S et al (2018) Mapping the Arabidopsis metabolic landscape by untargeted metabolomics at different environmental conditions. Natl Libr Med. https://doi.org/10.1016/j.molp.2017.08.012

Tugizimana F, Djami-Tchatchou AT, Steenkamp PA et al (2019) Metabolomic analysis of defense-related reprogramming in sorghum bicolor in response to Colletotrichum sublineolum infection reveals a functional metabolic web of phenylpropanoid and flavonoid pathways. Front Plant Sci 9:1–20. https://doi.org/10.3389/fpls.2018.01840

Udayakumar M, Chandar DP, Arun N et al (2012) PMDB: plant metabolome database-a metabolomic approach. Med Chem Res 21(1):47–52. https://doi.org/10.1007/s00044-010-9506-z

Udvardi MK, Czechowski T, Scheible WR (2008) Eleven golden rules of quantitative RT-PCR. Plant Cell 20:1736–1737

Unlu M, Morgan ME, Minden JS (1997) Difference gel electrophoresis: a single gel method for detecting changes in protein extracts. Electrophoresis 18:2071–2077

Utami DW, Lestari P, Koerniati S (2013) A relative expression of *Xa7* gene controlling bacterial leaf blight resistance in Indonesian local rice population (*Oryza sativa* L.). J Crop Sci Biotechnol 16:1–7

Vandesompele J, De Preter K, Pattyn F et al (2002) Accurate normalization of real-time quantitative RT-PCR data by geometric averaging of multiple internal control genes. Genome Biol 3: RESEARCH0034

Varshney RK, Graner A, Sorrells ME (2005) Genic microsatellite markers in plants, features and applications. Trends Biotechnol 23:48–55

Velculescu VE, Zhang L, Vogelstein B et al (1995) Serial analysis of gene expression. Science 270:484–487

Viant MR, Sommer U (2013) Mass spectrometry based environmental metabolomics: a primer and review. Metabolomics 9(1):144–158. https://doi.org/10.1007/s11306-012-0412-x

Vinaixa M, Schymanski EL, Neumann S et al (2016) Mass spectral databases for LC/MS- and GC/MS-based metabolomics: state of the field and future prospects. TrAC 78:23–35. https://doi.org/10.1016/j.trac.2015.09.005

Wallace ED, Todd DA, Harnly JM et al (2020) Identification of adulteration in botanical samples with untargeted metabolomics. Anal Bioanal Chem. https://doi.org/10.1007/s00216-020-02678-6

Walley JW, Shen ZX, McReynolds MR, Schmelz EA, Briggs SP (2018) Fungalinduced protein hyperacetylation in maize identifed by acetylome profling. Proc Natl Acad Sci U S A 115 (1):210–215

Wang D, Weaver ND, Kesarwani M et al (2005) Induction of protein secretory pathway is required for systemic acquired resistance. Science 308:1036–1040

Wang L, Qu L, Hu J et al (2017) Metabolomics reveals constitutive metabolites that contribute resistance to fall webworm (Hyphantria cunea) in Populus deltoides. Environ Exp Bot 136:31–40

Wang W, Li Y, Dang P et al (2018) Rice secondary metabolites: structures, roles, biosynthesis, and metabolic regulation. Molecules 23(12):3098. https://doi.org/10.3390/molecules23123098

Washburn M, Wolters D, Yates J (2001) Large-scale analysis of the yeast proteome by multidimensional protein identification technology. Nat Biotechnol 19:242–247

Wei C, Li J, Bumgarner RE (2004) Sample size for detecting differentially expressed genes in microarray experiments. BMC Genomics 5:87

Wen W, Li D, Li X et al (2014) Metabolome-based genome-wide association study of maize kernel leads to novel biochemical insights. Nat Commun 5(1):1–10. https://doi.org/10.1038/ncomms4438

Wichmann F, Asp T, Widmer F et al (2011) Transcriptional responses of Italian ryegrass during interaction with Xanthomonas translucens pv. graminis reveal novel candidate genes for bacterial wilt resistance. Theor Appl Genet 122:567–579

Więsyk A, Lirski M, Fogtman A et al (2020) Differences in gene expression profiles at the early stage of Solanum lycopersicum infection with mild and severe variants of potato spindle tuber viroid. Virus Res 286:198090

Wu Q, VanEtten HD (2004) Introduction of plant and fungal genes into pea (Pisum sativum L.) hairy roots reduces their ability to produce pisatin and affects their response to a fungal pathogen. Mol Plant-Microbe Interact 17(7):798–804. https://doi.org/10.1094/MPMI.2004.17.7.798

Wu L, Han Z, Wang S, Wang X et al (2013) Comparative proteomic analysis of the plant-virus interaction in resistant and susceptible ecotypes of maize infected with sugarcane mosaic virus. J Proteome 89:124–140

Xiao F, Tang X, Zhou JM (2001) Expression of 35S::Pto globally activates defense-related genes in tomato plants. Plant Physiol 126:1637–1645

Xu F, Huang Y, Li L, Gannon P, Linster E, Huber M, Kapos P, Bienvenut W, Polevoda B, Meinnel T, Hell R, Giglione C, Zhang YL, Wirtz M, Chen S, Li X (2015) Two N-terminal acetyltransferases antagonistically regulate the stability of a nod-like receptor in arabidopsis. Plant Cell 27(5):1547–1562

Yang X, Das PP, Oppenheimer P et al (2020) iTRAQ-based protein analysis provides insight into heterologous superinfection exclusion with TMV-43A against CMV in tobacco (Nicotiana benthamiana) plants. J Proteome 229:103948

Yao PF, Li CL, Zhao XR et al (2017) Overexpression of a Tartary buckwheat gene, FtbHLH3, enhances drought/oxidative stress tolerance in transgenic Arabidopsis. Front Plant Sci 8:625. https://doi.org/10.3389/fpls.2017.00625

Ye X, Wang H, Chen P et al (2017) Combination of iTRAQ proteomics and RNA-seq transcriptomics reveals multiple levels of regulation in phytoplasma-infected Ziziphus jujuba Mill. Hortic Res 4:1–13

Yin H, Yan F, Ji J et al (2012) Proteomic analysis of Arabidopsis thaliana leaves infested by tobacco whitefly Bemisia tabaci (Gennadius) B biotype. Plant Mol Biol Report 30:379–390. https://doi.org/10.1007/s11105-011-0351-0

Yu XD, Pickett J, Ma YZ, Bruce T et al (2012) Metabolic engineering of plant-derived (E)-β-farnesene synthase genes for a novel type of aphid-resistant genetically modified crop plants. J Integr Plant Biol 54(5):282–299. https://doi.org/10.1111/j.1744-7909.2012.01107.x

Yu C, Chen H, Tian F, Yang F, Yuan X, Yang CH, He C (2018) A ten gene-containing genomic island determines fagellin glycosylation: implication for its regulatory role in motility and virulence of Xanthomonas oryzae pv. Oryzae. Mol Plant Pathol 19(3):579–592

Zeiss DR, Mhlongo MI, Tugizimana F et al (2019) Metabolomic profiling of the host response of tomato (Solanum lycopersicum) following infection by Ralstonia solanacearum. Int J Mol Sci 20(16):3945. https://doi.org/10.3390/ijms20163945

Zellerhoff N, Himmelbach A, Dong W et al (2010) Nonhost resistance of barley to different fungal pathogens is associated with largely distinct, quantitative transcriptional responses. Plant Physiol 152:2053–2066

Zha W, You A (2020) Comparative iTRAQ proteomic profiling of proteins associated with the adaptation of brown planthopper to moderately resistant vs. susceptible rice varieties. PLoS One 15:1–13

Zhang M, Cheng ST, Wang HY et al (2017) iTRAQ-based proteomic analysis of defence responses triggered by the necrotrophic pathogen Rhizoctonia solani in cotton. J Proteome 152:226–235

Zhao B, Lin X, Poland J et al (2005) A maize resistance gene functions against bacterial streak disease in rice. Proc Natl Acad Sci U S A 02:15383–15388

Zhou YL, Xu MR, Zhao MF et al (2010) Genome-wide gene responses in a transgenic rice line carrying the maize resistance gene Rxo1 to the rice bacterial streak pathogen, Xanthomonas oryzae pv. oryzicola. BMC Genomics 11:78–88

Zhu L, Zhou Y, Li X et al (2018) Metabolomics analysis of soybean hypocotyls in response to phytophthora sojae infection. Front Plant Sci 871:1–13. https://doi.org/10.3389/fpls.2018.01530

Zieske LR (2006) A perspective on the use of iTRAQ™ reagent technology for protein complex and profiling studies. J Exp Bot 57:1501–1508

Zipfel C, Robatzek S, Navarro L et al (2004) Bacterial disease resistance in Arabidopsis through flagellin perception. Nature 428:764–767

# The Early Blight of Tomato: Omics Interventions Toward Controlling Disease Spread and Development

# 4

Abhay K. Pandey, Ajit Kumar Savani, and Pooja Singh

**Abstract**

*Alternaria* spp. is a polyphagous necrotrophic pathogen and infects many crops. In tomato, two species of *Alternaria*, namely *A. solani* and *A. alternata* cause infection. Particularly, *A. solani* infects leaves/stem and causes early blight (EB), which is a major yield-limiting disease of tomato worldwide, while *A. alternata* only infects fruit and stem leading to canker disease. This virulent pathogen causes severe damages to both fruits and plants of tomato. In the past decades, this disease was managed through an integrated approach using chemicals and bio-fungicides as well as through host-plant resistance. In the era of molecular biology, the ongoing efforts to reduce the pathogenic nature of *Alternaria* species, integration of omics technologies such as genomics, transcriptomics, proteomics, and metabolomics have recently been an advanced approach for understanding the pathogenesis and defense mechanisms involved in *Alternaria* and tomato plant interaction. The studies of omics will offer a basis for improving breeding programs through genetic manipulation that will ultimately lead to the possible protection of tomatoes from EB infection. In this chapter, we have described the disease symptoms, epidemiology, and current integrated management practices for EB along with knowledge gaps. In addition, an attempt is made to highlight

A. K. Pandey (✉)
Department of Mycology and Microbiology, TRA-North Bengal Regional R & D Center, Nagrakata, West Bengal, India

Department of Botany, DDU Gorakhpur University, Gorakhpur, Uttar Pradesh, India

A. K. Savani
Department of Plant Pathology, College of Agriculture, Assam Agricultural University, Jorhat, Assam, India

P. Singh
Department of Botany, DDU Gorakhpur University, Gorakhpur, Uttar Pradesh, India



A. Kumar et al. (eds.), *Omics Technologies for Sustainable Agriculture and Global Food Security Volume 1*, https://doi.org/10.1007/978-981-16-0831-5_4

the current research progress in tomato plant responses against EB stress using omics tools. We also deliberate the break that recent technologies of omics can provide to investigate tomato–EB pathogen interaction to project potential management strategies through crop improvement.

**Keywords**

*Alternaria* spp. · *Solanum lycopersicum* · Chemical and biological control · Genomics · Transcriptomics · Proteomics · Metabolomics · Disease control

## 4.1 Introduction

Worldwide, tomato, *Solanum lycopersicum* L. is one of the most important vegetables cultivated for its edible fruits, grown for various purposes such as for use as fresh as well as several industrial purposes (Islam et al. 2013). According to a report of FAO (2018), the total world production of tomato was 182 million tones (MT), with China as the largest producer of tomato producing 61.5% MT of tomato annually, followed by India (19.4 MT), USA (12.6 MT), Turkey (12.2 MT), and Egypt (6.6 MT). Tomato is a rich source of 17% of vitamin C of the daily value, contains 4% carbohydrates, and <1% each of protein and fat (FAO 2018). However, the worldwide production of tomatoes is constrained by several biotic and abiotic stresses, which adversely affect the quantity, quality, and profitability (Engindeniz and Ozturk 2013).

In biotic stresses, the diseases caused by fungal pathogens are particularly crucial in terms of production and quality (Sain and Pandey 2016). During the cropping periods, tomato plants are attacked by several roots and foliar fungal diseases. The wilt caused by *Rhizoctonia solani* and *Fusarium oxysporum* f. sp. *lycopersici* and damping-off by *Pythium aphanidermatum* are the major root rot diseases, while early blight incited by *Alternaria solani* or *A. alternata, Septoria* leaf spot by *Septoria lycopersici,* and late blight by *Phytophthora infestans* are the major foliar fungal diseases (Agrios 2005).

Among these diseases, early blight (EB) is one of the most severe diseases of tomato, causing 50–90% loss of the total production worldwide under favorable condition (Iqbal et al. 2019). For the management of this disease, growers rely on the use of chemicals (Mizubuti et al. 2007) and biological fungicides. But, the bio-fungicides are slow in their activity, and due to the retention of chemical fungicide residues in the vegetables, their use should be minimized (Stangarlin et al. 2011) and necessitates an alternative for disease management. In addition, small farmers growing tomatoes do not practice protective gears during the application of chemical fungicides and are not aware of the dilution instructions, thus compromising their own safety (Damalas and Koutroubas 2015). Therefore, these requirements have become more severe, especially in the amounts of chemical residues remaining in the fresh vegetables (European Commission 2012).

During recent years, using omics technology for the management of diseases of tomatoes has been found helpful to reduce the fungicidal risks problems (AbuQamar et al. 2016). Understanding the host responses and mechanisms toward a particular disease by deploying omics technologies is essential to improve the defense mechanism of tomato plants through breeding programs or by emerging ad hoc biotechnology strategies. Particularly, there is a great interest to improve tomato crops that could be free from EB, due to its global relevance as fresh and processed produce. Available literature revealed that little work has been done on the role of omics technology such as genomics or transcriptomics, proteomics, and metabolomics in understanding the *Alternaria* × tomato interaction and the management of EB. This chapter captures the latest significant studies in epidemiology, host range, and current integrated disease management strategies. In addition, we focus on the modern approaches regarding recent omics interventions for the potential management of EB disease along with knowledge gaps to deliver a role for the exploitation of candidate genes of interest and their additional analyses, offering trait-specific markers suitable for the improvement of tomato.

## 4.2 Disease Symptoms and the Biology of Causal Organism

Different pathogenic species of *Alternaria* can be distinguished by the symptoms produced on different plant parts. Initially, symptoms appear on the lower leaves as concentric rings in dark brown spots, which is the primary characteristic symptom of this disease (Fig. 4.1a, b). During humid weather, the disease progresses upwards, the areas affected by pathogen merge and form dark brown patches on the whole leaves. Under severe conditions, infected leaves may shrink and fall prematurely, resulting in early defoliation. On fruits, the infection takes place at the stalk end in the form of dark brown spots near the place of attachment with the fruit (Fig. 4.1c).

Worldwide, five different species of *Alternaria*, namely *A. alternata, A. linariae* (syn. *A. tomatophila*), *A. solani*, *A. tenuissima*, and *A. grandis* have been identified as the causal agents of EB of tomato (Bessadat et al. 2017). However, *A. solani* (Ell. And Mart) and *A. alternata* (Fr.) are the prevalent species. The mycelium of *A. solani* consists of branched, septate, light brown hyphae, which with age become darker. Conidiophores are relatively shorter, i.e. 50–90μm with dark color. *Alternaria* conidia are typically beaked, muriform, dark, and borne single or in chains, with 5–10 transverse septa and some time in each conidium a few longitudinal septa are present (Fig. 4.1d). *Alternaria alternata* possesses much fluffy margin with off white color colonies, which turn into dusky neutral gray within 96 h. Later these colonies become nearly grayish black.

**Fig. 4.1** Early blight symptoms of tomato on leaves (**a**), stem (**b**), fruits (**c**), and conidia of *Alternaria solani* (**d**)

## 4.3 Host Range and Pathogen Variability

The *Alternaria* species has a wide host range. It infects both arable crops such as crucifers, solanaceous crops, leafy vegetables (Loganathan et al. 2016) and plantation crops like tea, coconut, etc. (Rao and Subrahamanyam 1976). Based on pathogenicity tests on tomatoes, both *A. alternata* and *A. solani* isolates have been classified under the virulent category (Loganathan et al. 2016). Few species of *A. solani* (non-pathogenic) have been found to promote growth in chili plants instead of its pathogenic nature (Mauricio-Castillo et al. 2020). It is also reported the *A. solani* isolated from the different hosts exhibited pronounced variability in their pathogenicity. Also, the growth of isolates was influenced by the type of nutrients provided in the media, and among the different sources of nutrition provided, V8 juice agar supported the sporulation of the fungus (Pasche et al. 2004; Kumar et al. 2008). Several researchers reported the effect of lights such as blue or UV light on the sporulation of *A. solani* and other species, *A. tegetica*, *A. alternata,* and *A. kikuchiana* (Prasad and Dutt 1974; Cotty 1987; Fourtouni et al. 1998).

## 4.4 Epidemiology and Disease Development

The *Alternaria* species infect tomatoes are overwintered in diseased plant debris. It can survive in, or on the soil, atleast one of perhaps several years. The pathogen is seed-borne (Khulbe and Sati 1987; Shahida and Abdul 1995) and can be introduced through the infected seeds. Primary infection takes place first on lower leaves, and conidia are formed in crop debris left in the soil. The conidia developed on the primary spot helps in the secondary spread of the disease. These conidia are blown by wind or water or insects through the neighboring leaves/plants. The infection generally occurs through stomata, but *Alternaria* spp. are also capable of direct penetration.

The disease severity was reported maximum in crops sown during June–July compared to September–October and January–April planted crops (Data and Mayee 1981). Prevalence of high humidity and soil moisture favors the disease development during July, August, and September months. The optimum temperature required for the growth of *Alternaria* spp. is 28–30 °C for *A. solani* and 20–25 °C for *A. alternata* (Sahi 1990; Singh 1995). Once the infection has occurred, conidial dispersion continues throughout the growing season. Datar and Mayee (1982) reported the maximum dispersal of conidia occurs during the advanced stage of the symptom development and particularly between 9 am and 12 pm.

Among the fungal diseases, EB incited by *A. solani* or *A. alternata* is one of the major severe concerns due to substantial yield losses in tomatoes. This ascomycete pathogen usually infects tomato, potato, and eggplant. The disease is promoted by warm temperature with long periods of leaf wetness, dew, rainfall, and dense cropping. During the fruiting period, tomato plants become more susceptible to this pathogen (Cerkauskas 2005; Momel and Pemezny 2006). Although the disease is termed as EB, it may occur at all stages of development. Early blight occurs in three phases, leaf spots, fruit rot, and stem canker. Still, the foliar phase is more destructive and accountable for significant economic losses sustained by tomato producers (Chaerani and Voorrips 2006). The EB fungus can survive for several days on the infected seeds, but it is still speculative that in the next season, whether the seed-borne inoculum serves as a source of primary infection (Datar and Mayee 1982).

## 4.5 Existing Disease Mitigation Strategies

For the long term management of this disease, integrated disease management (IDM) strategies such as crop rotation, breeding of resistant cultivars of tomato, use of chemical and bio-fungicides have been practiced. Since *Alternaria* is both seed and soil-borne pathogen, both seed treatment and foliar application are recommended for disease management. Chemical and biological controls are the frequently adopted control measure for EB.

### 4.5.1 Chemical Fungicides

As far as chemical fungicides are concerned, mancozeb, hexaconazole, and zineb are effective at different concentrations against EB in both in vitro and in vivo conditions (Raza et al. 2016). At present, mancozeb is the most frequently used fungicide against EB (Singh et al. 2020). Majumder et al. (2016) reported that ED50 (effective dose) of nanoformulation of mancozeb against *A. solani* was in the range of 1.31–2.79 mg/L. In addition, mancozeb has also reduced the disease incidence of EB in the Pusa Ruby variety of tomato (Kumar and Srivastava 2013; Gondal et al. 1993). Besides, mancozeb, hexaconazole (0.05%), and azoxystrobin have also significantly managed the EB (Kumar et al. 2007). However, in the study of Arunkumar (2006) only azoxystrobin at 0.05, 0.1, and 0.15% was found to be more effective against EB than chlorothalonil, pyraclostrobin, and mancozeb. On the contrary, Singh and Singh (2006) reported that hexaconazole was more effective than chlorothalonil, azoxystrobin, mancozeb, propineb, and copper oxychloride. Recently, Farooq et al. (2019) observed that pyraclostrobin was more efficient against EB pathogen at 500 ppm, than that of hexaconazole and carbendazim. The variable range of efficacy reported for the fungicides may be due to the different isolates of the pathogens or active ingredients present in the chemical fungicides.

In addition, the fungicide resistance has also been reported for the EB pathogen, *A. solani* due to the higher pathogenic and genetic variability among different isolates, isolated from various agro-climatic regions (Pasche et al. 2004) and it could also break down the genetic resistance of the host (van derWaals et al. 2004). Therefore, to reduce the risk of chemical fungicide resistance, fungicides rotation strategies, use of different modes of action of fungicides through mixing should be executed at the regional and national level where fungicide resistance is a severe problem in EB prone areas. The increased use of fungicides to mitigate EB of tomato requires the implementation of alternative disease control practices.

### 4.5.2 Biological and Botanical Control

In recent years, to minimize the use of chemical fungicides, investigations were carried out to use the microbial biocontrol agents (MBCAs) and botanicals to combat EB where it was severe. There are several formulations of *Trichoderma* spp. and *Pseudomonas* spp. available in the markets that can be used against EB, and their efficacy has been confirmed by conducting several investigations. In the late 2000s, Varma et al. (2008) investigated that foliar spray of *T. viride* reduced EB severity caused by *A. solani.* Other reports also evidenced that the antagonist's *Bacillus amyloliquefaciens, Pseudomonas fluorescens,* and *T. harzianum* were efficiently controlled EB incidence in tomato (El-Rafai et al. 2003; Camlica and Tozlu 2019).

The antagonistic potential of these MBCAs is attributed to several extracellular enzymes, PAL (phenylalanine ammonia-lyase), defense enzyme and oxidative enzymes (polyphenol oxidase, peroxidase and superoxide dismutase), several anti-fungal metabolites, presence of several enzymes and secondary metabolites ($\beta$-1,3-

glucanase) produced by these MBCAs (Montealegre et al. 2010; Chowdappa et al. 2013). However, the slow activity of the MBCAs based fungicides limits their application in EB management. In addition, the application of neem leaf extracts has also been used to control EB incidence in tomato (Raza et al. 2016). In particular, the active ingredient of neem leaf can be used for the formulation of next generation fungicides that will have broad application in IDM as well as to reduce the residue level and fungicide resistance problems.

## 4.6 Exploitation of Omics Approaches in Understanding Tomato × *Alternaria* Interactions and for EB Management

To reduce the losses in tomato due to EB, developing resistant varieties can be an economical and most effective management strategy (Panthee and Chen 2010; Adhikari et al. 2017). For the development of resistant varieties, investigators applied several genetic approaches. In particular, tomato plants show a high degree of similarity in gene sequence with other solanaceous crops (Kumar and Khurana 2014), making the investigation easy to understand the genetic programs based on interspecies knowledge transfer. The recent methodologies have established many efficient omics methods to untangle the molecular mechanisms of tomato plant response to *A. solani* to improve the detection and diagnosis of the pathogen (Fig. 4.2).

Historically, in an organism the genome is a whole set of chromosomes, which comprises all genes. The entire set of non-coding and coding RNAs is called a transcriptome, while the collected proteins derived from a genome are termed as proteome. Conversely, all metabolites present in the plant system are called the metabolome. However, the defense systems in plants against a particular pathogen cannot be studied uniquely through the genomic or transcriptomic methods, as they involve not only the expression of several defense-related genes, but also the incidence of post-translational modification or metabolites accumulation, affecting the final gene products expression.

The omics tools such as metabolomics and proteomics, enabling the proteins and metabolites interactions downstream of plant gene expression, may be practically pooled with genome and transcriptome. Although these approaches are complex, they can enhance our understanding of plant response mechanisms to fungal pathogen and other associated MBCAs, endophytes, and PGPR in a comprehensive way. Moreover, the methods of metagenomics enable the further understanding of the plant × associated microorganisms, offering an innovative prospect to sustain and manage the production of tomatoes at larger scale, based on microbiomes.

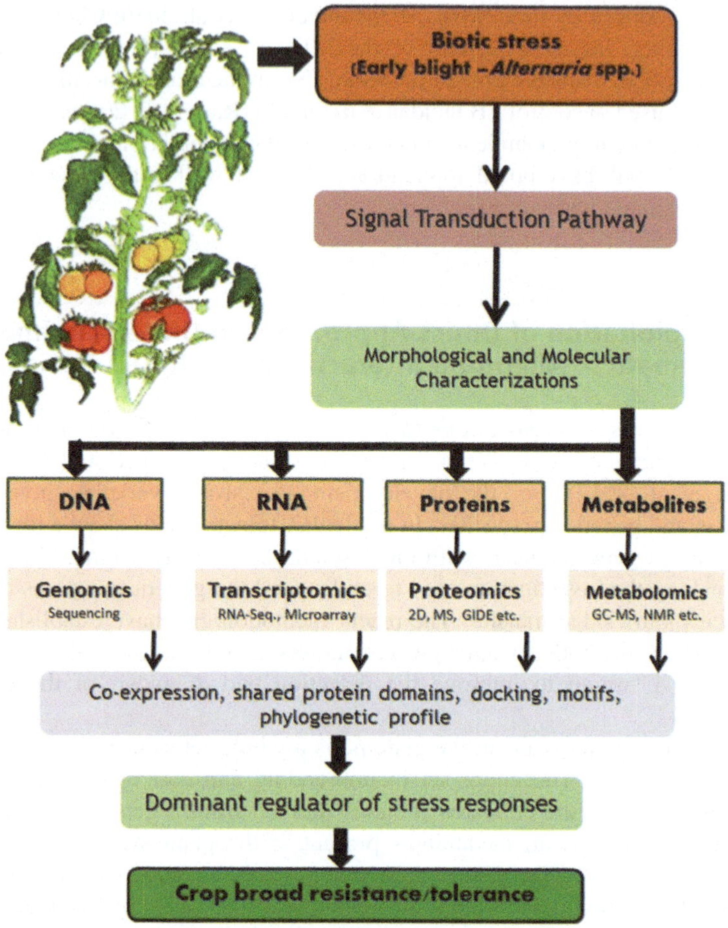


**Fig. 4.2** Schematic explanation of omics approaches used in future projects in the improvement of resistance/tolerant to EB of tomato (*PPI* protein–protein interaction, *DIGE* differential gel electrophoresis, *GC-MS* gas chromatography-mass spectrometry)

### 4.6.1 Search for the Resistant Cultivars against EB and Nature of Resistance

The investigations were carried out worldwide to search the resistant/tolerant cultivars of tomato against EB (Adhikari et al. 2017). However, till date there are few EB-resistant tomato genotypes available. Out of 401 tomato genotypes screened by Akhtar et al. (2019), only one genotype, i.e. "21,396" was found resistant against EB. In addition, some investigators found that several wild species (*Solanum pimpinellifolium*, *S. peruvianum*, *S. chilense*, and *S. habrochaites*) have been identified as potential sources of resistance against EB (Poysa and Tu 1996;

Thirthamallappa 2000; Foolad et al. 2000). Thus, these wild species can be exploited in the breeding program.

In addition, HRC-G90.158, HRC90.145, HRC90.159, (Poysa and Tu 1996), and IHR1816 F (Thirthamallappa 2000) have shown resistance toward EB. From India, Lohith et al. (2011) reported four genotypes, such as EC251717, EC251709, EC164295, and LE15 of tomato resistant against EB. In a recent report six genotypes of tomato, such as NCEBR-1, NCEBR-4, Arka Rakshak, Arka Alok, Arka Saurabh, and 8-3-3 have shown EB resistance (Amarnath et al. 2019a, b); however, these genotypes were resistant in lab conditions and need further screening in the field.

Unfortunately, in the germplasm of tomatoes, there are only a few studies describing wide explorations for promising resistance sources to an EB pathogen (Adhikari et al. 2017; Nasr Esfahani 2019). In addition to 401 genotypes, Akhtar et al. (2019) also screened inbred lines and 72 genotypes from ten species of wild *Solanum* and found that none of the inbred lines was immune, highly resistant, or resistant. However, some genotypes derived from *S. galapagense* (1), *S. peruvianum* (1), *S. pimpinellifolium* (5), *S. habrochaites* (5 introgression lines), *S. pennellii* (2 introgression lines), *S. lycopersicum* E-6203 × *S. pimpinellifolium* LA1589 (eight RILs) showed moderately resistant reaction. In tomato, the nature of resistance is reported as polygenic in nature. Consequently, some genes present in tomato may confer resistance to the leaf blight, whereas others may contribute stem or fruit rot resistance (Stancheva et al. 1991; Chaerani et al. 2007). However, Barksdale and Stoner (1977) reported that stem lesion resistance of EB was independent of EB resistance on the leaves.

In the past decades, in the genotypes C1943 and 71B2, the EB resistance genes were reported recessive and not allelic (Maiero et al. 1989). However, the F1 hybrids were intermediate when these two resistance genes were crossed with another susceptible genotype, indicating partial dominance or additive genetic control (Maiero et al. 1989). Besides, the recessive genes have also been identified in the genotypes 83,602,029 (Stancheva et al. 1991) and IHR1816 and IHR1939 (Thirthamallappa 2000) derived from *S. lycopersicum.* In addition to this, in *S. pimpinellifolium* and *S. habrochaites* the partial dominant inheritance has been reported (Martin and Hepperly 1987). Another tomato genotype, i.e. 87B187 derived from PI390662 (*S. habrochaites*), shared common resistance genes with the genotype NCEBR-2 (Maiero et al. 1990a, b), even though this genotype was developed via *S. lycopersicum* source, C1943. Moreover, Thirthamallappa (2000) investigated independent genes in the genotypes IHR1816 and IHR1939, which were derived from *S. pimpinellifolium* and *S. habrochaites*, respectively.

### 4.6.2 Identification of Quantitative Trait Loci (QTLs) for Resistance to EB

The quantitative trait nature of EB makes selection more problematic as compared to the qualitative traits. In the tomato breeding programs, QTL analysis and development of molecular markers has been carried out in order to cognize the genetic

**Table 4.1** Quantitative traits loci (QTLs) detected for EB resistance in tomato

| QTLs | Chromosome | Interval | Phenotypic[a] variation explained (%) |
|---|---|---|---|
| *$BC_1$* | | | |
| *EBR1.1* | 1 | TG559—TG208A | 21.9 |
| *EBR2.1* | 2 | TG337—CT59 | 15.3 |
| *EBR5.2* | 5 | CT202—TG318 | 8.4 |
| *EBR6.1* | 6 | TG279—CT107B | 7.6 |
| *EBR8.1* | 8 | TG176—CT92 | 7.3 |
| *EBR9.1* | 9 | RLRR-130—CLRR-950 | 13.6 |
| *EBR9.2* | 9 | SS14-520.3—TG429 | 16.2 |
| *EBR9.3* | 9 | SS19-530—CT143 | 15.9 |
| *EBR10.1* | 10 | TG241—TG403 | 20.2 |
| *EBR11.1* | 11 | CT168—TG508 | 13.3 |
| *EBR12.2* | 12 | SS14-520.1—SS1-530.1 | 13.4 |
| *$BC_1S_1$* (self-pollinated progeny of $BC_1$) | | | |
| *EBR1.1* | 1 | TG559—TG208A | 11.9 |
| *EBR2.1* | 2 | TG337—CT59 | 15.9 |
| *EBR3.1* | 3 | TG411—TG214 | 9.1 |
| *EBR5.1* | 5 | TG441—CT242 | 7.9 |
| *EBR5.2* | 5 | XLRR-370—SAS5-250.3 | 11.2 |
| *EBR8.1* | 8 | CD40—TG176 | 10.3 |
| *EBR8.2* | 8 | TG330—TG294 | 21.0 |
| *EBR9.1* | 9 | CLRR-950—SAS5-250.1 | 25.0 |
| *EBR10.1* | 10 | TG241—TG403 | 16.3 |
| *EBR11.1* | 11 | TG508—TG651 | 11.5 |
| *EBR11.2* | 11 | CT55—CD17 | 9.9 |
| *EBR11.3* | 11 | SAS11-760.2—TG393 | 11.5 |
| *EBR12.1* | 12 | TG68—CT79 | 8.2 |

[a]Based on simple interval mapping in $BC_1$ and $BC_1S_1$ populations of an interspecific cross between *L. hirsutum* (PI126445; EB resistant) and *L. esculentum* (NC84173; EB susceptible), Source: Foolad et al. (2002)

control of EB resistance and to enable its introgression in tomatoes. Foolad et al. (2002) identified ten QTLs for EB through the crossing of resistant (PI126445) genotype derived from *S. habrochaites* and susceptible genotype (NC84173) of tomato, and each QTL explained total phenotypic variation in the range of 8.4–25.9%, while the collective effect was more than 57%. A list of QTLs identified by Foolad et al. (2002) for EB resistance in tomato is presented in Table 4.1. Later on, by selective genotyping, Zhang et al. (2003) identified QTLs conferring EB resistance in a *L. esculentum* × *L. hirsutum* cross. In addition, they also detected seven QTLs for EB resistance in a trait marker analysis (Zhang et al. 2003). However, the success in incorporating resistance in tomato is limited because most of the breeding lines such as NCEBR-4 (Gardner and Shoemaker 1999), NCEBR1, NCEBR-2 (Gardner 1988), and HRC90.303, HRC91.341 (Poysa and Tu 1996) were late maturing, relatively low yielding, and indeterminate. These accessions were derived from *L. hirsutum*. In 2007, Chaerani et al. (2007) identified three resistant QTLs to stem lesions from $F_2$ and $F_3$ populations derived from a cross between *S. peruvianum* LA2157 (resistant) and *S. lycopersicum* cv. Solentos (susceptible), and that explained 35% of the phenotypic variance. These QTLs can be used for the development of markers against EB in tomatoes.

### 4.6.3 Genomics Studies of Host and Pathogen

In the recent era of molecular biology, gene sequencing-based approaches remain economical, and both the pyro-sequencing and traditional Sanger dideoxy nucleotide have demonstrated their usefulness for confirmatory sequencing (Pareek et al. 2011). The EB pathogen, *A. alternata or A. solani* has become perfect for dividing the complexity of necrotrophic fungal pathogens and a wide range of pathogenicity of various crops. The pathogen may survive in diverse ecological stresses that promote or inhibit the infections on their host plants such as tomato (Ahlem et al. 2012).

Recently, based on conserved DNA sequences the genus *Alternaria* has been modernized (Ozkilinc et al. 2017; Woundenberg et al. 2014). It was confirmed that some species of *Alternaria*, i.e. *A. grandis* and *A. protenta* closely related to *A. solani* (Duarte et al. 2014), can also incite EB in tomato and potato (Ayad et al. 2017; Bessadat et al. 2016). To understand the *A. alternata or A. solani*–plant interactions in-depth at whole genome level, the whole genome sequence of *A. alternata* isolated from onion was studied. Its total genome size was 33.12 Mb with 50.9% GC content and 11,701 predicted coding sequences (Bihon et al. 2016). In addtion, *A. alternata* isolates from sorghum had 27 scaffolds, and the total genome size was 33.5 Mb (Nguyen et al. 2016). However, the partial sequence for *A. alternata* isolated from tomato is available (Gherbawy et al. 2018). Although, in the past, genomes of many *Alternaria* species (Hu et al. 2012), including *A. solani*, have been sequenced (Dang et al. 2015; Woudenberg et al. 2015), but due to analysis based on short-read sequencing, most of these genome assemblies were highly fragmented. Still, in discovering new genes, this information can be useful to clarify the classification and taxonomy of *Alternaria* species, and they enable comparative genomics.

Therefore, to produce fungal genomes having high-quality assemblies, use of long reads derived from PacBio-SMRT (Pacific Biosciences-single-molecule real-time) sequencing tools is a most prevalent method (Faino et al. 2015). This has been recently explained for the pathogen *A. alternata* (Nguyen et al. 2016). In particular, the assembly of a contiguous genome for the study of plant pathogenic fungi is essential because the genes coding the disease development effector proteins are often existing in fast-evolving that are challenging to assemble (Thomma et al. 2016). Likewise, understanding about related chromosomes and the gene organization helps in the gene cluster identification that has a major role in the secondary metabolite production, and together the characterization of potential provisionally expendable chromosomes helps in studying the pathogenicity of *Alternaria* spp. (Thomma et al. 2016).

Recently, Wolters et al. (2018) sequenced the *A. solani* causing EB in tomato and potato of genome size 33.1 Mb comprises about 99% of the total length of chromosomes. They identified that *A. solani* has ten chromosomes. Similar results were reported in an earlier study, in which genome sizes of *A. solani* was in the range of 32.6–32.9 Mb (Dang et al. 2015; Woudenberg et al. 2015). Besides *A. solani* genome sequenced by Wolters et al. (2018) showed a major advancement than that of the earlier *A. solani* genome assemblies, which consisted of over 100 separate contigs. Their genome sequencing analysis provides a concrete basis for the performance of comparative genomics, which will help to understand the molecular basis of pathogenicity of *A. solani* and other *Alternaria* species.

As far as the host is concerned, the first full genome sequence of tomato was carried out (Tomato Genome Consortium 2012), which describes 35,000 genes on 12 chromosomes. Later on, Li et al. (2018) sequenced genome of 360 varieties of tomato followed by Bolger et al. (2014) who sequenced the genome of *S. pennellii,* a stress-tolerant tomato wild species. The tomato plant contains 83 SlWRKY genes, which have several roles in the defense responses to both biotic and abiotic stresses (Bai et al. 2018). In the tomato plant, most of the WRKYs genes act as positive regulators of host responses to biotic stresses, whereas a lesser number of genes act as negative regulators.

The releases of sequences of whole genome of *A. alternata* and *A. solani* and their hosts will help in tackling the candidate genes responsible for virulence of *Alternaria* species and the potential target genes in the tomato plant associated with resistance against it. The genome sequences of tomatoes are very useful in understanding the plant defense system against *Alternaria* species. The sequencing of the genome of both host and pathogen will be also useful for the tomato breeders in developing resistant hybrids through the selection of defense-related genes in host crop or modification in virulent genes of the pathogen. As long as both *Alternaria* sp. and tomato genomes have been sequenced, the gene expression analysis through whole genome sequencing will tackle the critical factors in the pathogenesis of *Alternaria* spp. and mechanisms of EB resistance in tomato.

### 4.6.4 Transcriptomics

The comparative gene expression analyses can be utilized to mine the guiding information through transcriptomic technologies to generate data on biotic stress modulations of gene expression in tomato plants. In the modern era of molecular biology, RNAseq-based approaches are being used to study the transcriptomics in both model and non-model plants or pathogens (Warren et al. 2007). Remarkably, the transcriptome analysis of an organism helps to determine the pathogenesis-related proteins to be efficient to various biotic stress conditions (Ali et al. 2018). For instance, after the infection by a pathogen, plants produce pathogenicity-related (PR) proteins and chitinase in response to chitin, which is a major component of the cell wall of fungi (Adhikari et al. 2017).

The antifungal influence of chitinases and several hydrolytic enzymes has been determined against several foliar fungal pathogens, including *A. solani*. In addition, the genes accountable for the production of PR proteins have considerable enhanced resistance against pathogens causing EB and other pathogens in several arable crops (McNeece et al. 2019; Upadhyay et al. 2014a, b). When *Alternaria* infects the tomato plants, it suppresses both photosynthesis and metabolic processes such as glycolysis, electron transport chain, etc. At the same time, the defense-related genes, for instance, that encode chitinase, PR protein (PR2 and 3), and $\beta$-1, 3-glucanase showed a higher level in the highly EB-resistant species of tomato (Moghaddama et al. 2019). In addition, the expression of many secondary metabolites and defense-related genes in tomato plants were also upregulated when attacked by *Alternaria*.

In addition to the PR proteins, WRKY proteins also have a major role in the plant defense against the pathogens (Yang et al. 2018). In this regard, Moghaddam et al. (2019) reported that the expression pattern of antifungal genes 7 PR and 5 SlWRKYs genes in tomato increased 1–50-fold, when infected by *A. alternata,* and were upregulated among the resistant tomato varieties. In addition, the differential expression patterns of genes SlWRKY1 and SlWRKY11 were consistent with the expression pattern of genes PR7 and PDF1.2, which suggest that these transcription factors have a possible role in the enhancement of expression of PR genes in response to *A. alternata* infection.

Apart from EB, tomato plants also showed improved resistance to late blight caused by *Phytophthora infestans* (Cui et al. 2019) and *S. arcanum* to EB, *A. solani* (Shinde et al. 2018) due to the overexpression of WRKY1 gene. In an investigation, SlWRKY39 gene present in tomato was significantly upregulated in response to *Pseudomonas syringae* infection (Bai et al. 2018). In the same line, in response to *Botrytis cinerea* and *A. brassicicola,* the expression pattern of AtWRKY70 gene was altered, and changes in activity of AtWRKY70 genes might increase the susceptibility to *B. cinerea, Erysiphe cichoracearum,* gall formation by *Linaria vulgaris*, and *Macrophomina phaseolina* (Ulker et al. 2007; Lawaju et al. 2018; Pandey et al. 2016; Zorića et al. 2019).

In addition, the variable expression pattern of WRKY and PR defense-related genes is controlled not only by salicylic acid and jasmonic acid mediated signal events, but in between the resistant and susceptible genotypes of tomato infected by *Alternaria* species, the level of gene expression also varied (Pathak et al. 2017; Yang et al. 2015). Conversely, some plants showed resistance at seedling stages while becoming moderately resistant/susceptible at mature stage as has been reported for EB in potatoes incited by *A. alternata* (Nasr Esfahani et al. 2017). The experiment of Moghaddama et al. (2019) revealed that the tomato variety Esfahan local inoculated with pathogen showed an enhanced expression of defense-related genes and significant resistance at both young and mature stages, while the tomato variety Rio Grande showed resistance only at maturity stage. They also reported that in the inoculated EB-resistant tomato variety (H.a.s 2274) the expression of PR7 was upregulated at transplanting stage, and a strong expression in the inoculated resistant genotypes (Esfahan local, H.a.s 2274 and Rio Grande) was reported at the maturing stage (Moghaddama et al. 2019). Therefore, from the above findings, it is suggested that these are the key genes activating the defense response in host plant to the pathogen.

Besides, in tomato plants PR7 gene encoding 69 endopeptidase (Moghaddama et al. 2019), has been reported as proteases induced by the pathogen (Jorda and Vera 2000), and the fungal activities of PR7 gene is shown by another investigation (Golshani et al. 2015). The PR7 defense gene has also been found to be expressed during in several others interaction of the pathogen with hosts, comprising *Pseudomonas syringae* (Jorda and Vera 2000) and *Phytophthora infestans* (Tian et al. 2007) infections. The enzymatic activity of PR2 and PR3 proteins (β-1, 3-glucanase and chitinase) in the enhancement of defenses in tomato against EB has also been studied (Moghaddama et al. 2019), which revealed that both enzymes had a significant contribution to the protection of tomato from EB. Some studies revealed that the release of glucanase and chitinase in the form of hydrolytic products of induced PR genes disturbs the virulence of fungal pathogens and endorses the plant immunity responses (Kumar et al. 2018; Pusztahelyi 2018). Further, an investigation reported that among 32 genes present in the resistant genotype of tomato (EC-520061), 20 genes were upregulated against EB whereas in case of the CO-3, a susceptible genotype, no significant upregulation in fold change was examined (Upadhyay et al. 2016). Thus, these studies showed that these enzymes and genes significantly impact the EB resistance in tomatoes.

These results approve the crosstalk existence at the tomato plant retorts to *Alternaria* spp., involving several hormone signaling pathways, which alter the rate of photosynthesis, transport of proteins and their synthesis, thereby emphasizing the complexity of cellular signaling networks in tomato plants (AbuQamar et al. 2016). In addition, the incorporation of genomics and transcriptomics data of tomato or EB pathogen, along with proteomics will detect the biomarkers for EB pathogen. These omics data sets (transcriptomics and proteomics data) can build a vigorous model of functional features of biological pathways linking the transcripts and proteins.

### 4.6.5 Proteomics

In a host plant, the outcome of the incompatible and compatible host-pathogen interaction is determined through proteome analysis and associated metabolites. Independently, proteomic and metabolic profiling, or in the permutation with transcriptome data, provides additional understanding about the mechanisms of host defense response at the molecular level (Sharma et al. 2007; Tenenboim and Brotman 2016; Kumar et al. 2014). As far as EB of tomato is concerned, fewer studies regarding the proteomics analysis of tomato plants infected with EB have been carried out. However, literature is available for the other hosts such as *Brassica* and other crops infected by *Alternaria* species. The level of 48 proteins was significantly affected at several points in the tolerant lines of *Brassica* spp. when infected by *A. brassicae,* which suggested that the role of ROS (reactive oxygen supply) mediated auxin signaling in pathosystem of *Alternaria* sp. (Sharma et al. 2007).

Likewise, the level of 210 proteins in the *Mentha arvensis* leaves affected/altered during infection by *A. alternata* identified by matrix assisted laser desorption or ionization time of flight-mass spectrometry of them 29% of the proteins was defense-related proteins (Sinha and Chattopadhyay 2011). In another pathosystem of tomato, it was found that there was alternation in 186 proteins in wild-type mature green fruits infected by *Botrytis cinerea,* which were unaltered in wild-type red ripe fruits (RR). However, less defense-related proteins were altered in mature wild-type green tomato fruits than in RR tomato fruits (Shah et al. 2012). Therefore, further investigations are required to study the proteome analysis of tomato plants infected with EB to understand changes in the protein level. However, as far as tomato-EB interaction is concerned, the proteomic study can be compounded by the existence of pathogen proteins, which can be determined through the accessible full genome sequence of tomato and *Alternaria* spp.

### 4.6.6 Metabolomics

Each plant or pathogen contained metabolites, and these are organic compounds classify under the end product of plant metabolism or gene expression. Secondary metabolites present in plants have several roles in defense against pests and pathogens, and any changes in these metabolites affect the plant defense to the pathogens (Yuan et al. 2017). Conversely, secondary metabolites react with particular stress conditions, either biotic or abiotic, for example, ROS scavengers, pathogens, coenzymes, regulatory molecules, and antioxidants. Metabolomic profiling is carried out through NMR (nuclear magnetic resonance spectrometry) or MS (mass spectrometry), such as GC (gas chromatography)-MS and LC (liquid chromatography)-MS (Gathungu et al. 2014; Sharma et al. 2018).

Like other necrotrophic fungi infecting arable crops, the genus *Alternaria* often produces various phytotoxins and secondary metabolites, as "killing" weapons to the host cells from a wide range of plant species (Encinas-Basurto et al. 2017). Approximately, the 70 phytotoxins sill has been recognized that is produced by the several species of *Alternaria*, some are host-specific, and some are non-host specific (Johann et al. 2012; Escrivá et al. 2017). The major toxins produced by *Alternaria* include alternariol monomethyl ether (AME), alternariol (AOH), and altertoxin I and altertoxin II (Jarolim et al. 2017) which have several side effects in humans as well as in plants (Wenderoth et al. 2019). However, these phytotoxins have a phytotoxic minor impact on the host plant. Still, majorly they support in the colonization process of the pathogen inside the host by compensating the response of plant hypersensitive (Touhami et al. 2018). After colonization, they inhibit the enzymatic reactions within the host tissue or lead to death or necrosis of plant cells. In more resistant tomato genotypes, a correlation between the reduction in the production of AOH and a hogA mutant of *A. alternata* was taken as a sign for the role of *Alternaria* toxin AOH as a supporting factor in the virulence and colonization (Wojciechowska et al. 2014). However, AOH supports the colonization of the fungus (Wenderoth et al. 2019).

In addition, during interaction of wild tomato × *A. solani,* a significant modulation in secondary metabolites have been identified. In this regard, Shinde et al.

(2017) reported that secondary metabolites (phytoalexins, phenylpropanoids, lignin accumulation) synthesized in a resistant wilt tomato species, namely *S. arcanum* through steroidal-glycoalkaloid and phenylpropanoid pathways has significant role in protection against EB. The WRKY and MYB in WRKY1 genes had major role in secondary metabolites synthesis pathways, and in resistant plant, the lignin biosynthesis that was regulated by transcription factors was upregulated (Shinde et al. 2017).

During the infection, both host and pathogen release metabolites mediate the resistance response in the host. Some secondary metabolites, such as 3-methyl-2-butenal, dimethyl disulfide, 1-butanol, hexanol, and 2-methyl-1-butanol acetate responsible for resistance in tomato fruits were synthesized on tomato only when infected by *A. alternata* (Johanson and Thurston 1990). The primary, secondary metabolites correlated to tomato-EB resistance include a higher level of flavonol, tannin, and phenolic compounds in both stems and leaves, as has been reported in EB-resistant cultivars (Bhatia et al. 1972).

The production of these metabolites is associated with several mechanisms. The peroxidase (PO) present in the host plant plays an important role in the production of reactive oxygen (RO), these RO are directly or indirectly toxic to the fungal pathogen infecting plants (Hammond-Kosack and Jones 1996). The phenylalanine lyase (PAL) is also an important enzyme in the secondary molecules synthesis (Mauch-Mani and Slusarenko 1996), which help in the activation of the expression pattern of a variety of pathogenesis-related genes. Moreover, in response to the *A. solani,* the polyphenol oxidase (PPOs) is systemically upregulated, examined in the upper nodes of leaves, but absent in the lower nodes of tomato leaves (Thipyapong and Steffens 1997). This induction pattern of PR genes in tomato leaves accords with the observation of transient resistance of young leaves of tomato to *A. solani* infection (Johanson and Thurston 1990). The oxidation of phenols to quinones is catalyzed by PPOs and sensitive molecules that encourage the death of pathogen cell and blocks to the secondary infection in the host plant (Thipyapong and Steffens 1997).

The expression pattern of PR-1B increased when salicylic acid was applied on tomato roots to prevent the infection from EB (Spletzer and Enyedi 1999). After leaf treatment of tomato with arachidonic acid, the PR-1-like protein level increased (Coquoz et al. 1995), and the sequential expression of ST-ACS4 and ST-ACS5, ACC synthase genes also reported in potato plants (Schlagnhaufer et al. 1997). In a recent report, the total phenol contents of tomato were significantly increased as a response to *A. solani* infection (Attiaa et al. 2020). The remarkable metabolic changes in tomato upon infection with *Alternaria* spp. cause metabolic perturbations, both in the plant and the fungal pathogen. A recent study reports that, at transplanting stage, the activity of PAL increased 5-fold and TPC 4-fold, when the resistant tomato plant was inoculated by EB pathogen, while 2–3 fold increased in TPC activity and 3-fold in POD was reported at maturity stage (Alizadeh-Moghaddam et al. 2020).

Thus, the above described results suggest the resistant genotypes of tomato can be differentiated from susceptible genotypes through using both genetic and enzymatic diversity to EB.

## 4.7 Conclusions and Future Remarks

The descriptive information on the EB of tomato unveils a wealth of information which is regarded, for instance, pathogen epidemiology, integrated disease management, and role of omics in EB management. However, investigations in some important areas need further attention. There are few EB-resistant varieties of tomato and QTLs are available, and it will need to be investigated in the future research program. Although few QTLs are available for EB of tomato, however, to avoid integration of large parts of the donor genome along with the resistance gene, fine mapping is needed before these can be used in a marker-assisted breeding program. Also, before QTLs are deployed in a tomato breeding program, their pleiotropic effects on other traits should be investigated in future research projects.

In the era of molecular approaches, there is no doubt that the exploitation of omics in the potential disease mitigation is delivering toward understanding the mechanism at the molecular level of the tomato plant resistance to the *Alternaria* sp. The breeders are making potential efforts to link the resistant genes with traits to improve the resistance of tomato cultivars and understand the mechanisms of disease resistance. Therefore, to make the sustainable production of tomato, scientists must adopt innovative technologies to develop the high yield and EB-resistant varieties of tomato.

Omics enables the researchers to identify, isolate the desired genes and traits. It helps to interpret the complex interaction among genes and helps in creating tools to enhance crop productivity. This article provides a comprehensive overview of "omics" technologies and its application in agriculture to combat major problems of crops especially related to field pathogens, for example, EB of tomato. Through omics technologies, the consistency and predictability of plant genetic engineering and breeding will be significantly improved by reducing the time and expense for producing EB-resistant tomato crops. There is an urgent need to create an environment where modern tools like omics can be conveniently used and comprehensively regarded as important keys to combat other diseases of tomato crops, including EB. These can still be used with conventional tools of disease diagnostics and management, thus bridging the knowledge gaps and enabling us with a better understanding of plant disease management under conditions like climate change.

## References

AbuQamar SF, Moustafa K, Tran L-SP (2016) 'Omics' and plant responses to *Botrytis cinerea*. Front Plant Sci 7:1658

Adhikari P, Oh Y, Panthee DR (2017) Current status of early blight resistance in tomato: an update. Int J Mol Sci 18:1–22

Agrios GN (2005) Plant pathology, 5th edn. Academic Press, Oxford

Ahlem H, Mohammed E, Badoc A, Ahmed L (2012) Effect of pH, temperature and water activity on the inhibition of *Botrytis cinerea* by *Bacillus amyloliquefaciens* isolates. Afr J Biotechnol 11:2210–2217

Akhtar KP, Ullah N, Saleem MY, Iqbal Q, Asghar M, Khan AR (2019) Evaluation of tomato genotypes for early blight disease resistance caused by *Alternaria solani* in Pakistan. J Plant Pathol 101:1159–1170

Ali S, Ganai BA, Kamili AN, Bhat AA, Mir ZA, Bhat JA et al (2018) Pathogenesis-related proteins and peptides as promising tools for engineering plants with multiple stress tolerance. Microbiol Res 212:29–37

Alizadeh-Moghaddam G, Rezayatmand Z, Nasr-Esfahani M, Khozaei M (2020) Bio-genetic analysis of resistance in tomato to early blight disease, *Alternaria alternata*. Phytochemistry 179:112486

Amarnath M, Ravishankar KV, Satish D, Sriram S, Sadashiva AT (2019a) Screening of tomato genotypes against early blight pathogen *Alternaria solani* using detached leaf assay method. Pest Manage Horticult Ecosyst 25:93–101

Amarnath M, Ravishankar KV, Satish D, Sriram S, Sadashiva AT (2019b) Screening of tomato genotypes against early blight pathogen *Alternaria solani* using detached leaf assay method. Pest Manage Hortic Ecosyst 25:93–101

Arunkumar KT (2006) Studies on *Alternaria solani* (Ellis and Martin) Jones and Grout causing early blight of tomato. M.Sc. thesis, University of Agricultural Science, Bangalore

Attiaa MS, El-Sayyad GE, Abd Elkodousd M (2020) The effective antagonistic potential of plant growth-promoting rhizobacteria against *Alternaria solani*-causing early blight disease in tomato plant. Sci Hortic 266:109289

Ayad D, Leclerc S, Hamon B, Kedad A, Bouznad Z, Simoneau P (2017) First report of early blight caused by *Alternaria protenta* on potato in Algeria. Plant Dis 101:836

Bai Y, Sunarti S, Kissoudis C, Visser RGF, van der Linden CG (2018) The role of tomato WRKY genes in plant responses to combined abiotic and biotic stresses. Front Plant Sci 9:801

Barksdale TH, Stoner AK (1977) A study of the inheritance of tomato early blight resistance. Plant Dis Rep 61:63–65

Bessadat N, Hamon B, Henni DE, Simoneau P (2016) First report of tomato early blight caused by *Alternaria grandis* in Algeria. Plant Dis 100:533

Bessadat N, Berruyer R, Hamon B, Bataille-Simoneau N, Benichou S, Kihal M, Henni DE, Simoneau P (2017) *Alternaria* species associated with early blight epidemics on tomato and other Solanaceae crops in northwestern Algeria. Eur J Plant Pathol 148:181–197

Bhatia IS, Uppal DS, Bajaj KL (1972) Study of phenolic contents of resistant and susceptible varieties of tomato (*Lycopersicum esculentum*) in relation to early blight disease. Indian Phytopathol 25:231–235

Bihon W, Cloete M, Gerrano AS, Oelofse D, Adebola P (2016) Draft genome sequence of *Alternaria alternata* isolated from onion leaves in South Africa. Genome Announc 4:e01022–e01016

Bolger A et al (2014) The genome of the stress-tolerant wild tomato species *Solanum pennellii*. Nat Genet 46:1034–1038

Camlica E, Tozlu E (2019) Biological control of *Alternaria solani* in tomato. Fresen Environ Bull 28(10):7092–7100

Cerkauskas R (2005) Early blight, AVRDC, the world vegetable centre, www.avrdc.org

Chaerani R, Voorrips RE (2006) Tomato early blight (*Alternaria solani*): the pathogen, genetics, and breeding for resistance. J Gen Plant Pathol 72:335–347

Chaerani R, Smulders MJM, Van Der Linden CG, Vosman B, Stam P, Voorrips RE (2007) QTL identification for early blight resistance (*Alternaria solani*) in a *Solanum lycopersicum* × *S arcanum* cross. Theor Appl Genet 114:439–450

Chowdappa P, Kumar SM, Lakshmi MJ, Upreti KK (2013) Growth stimulation and induction of systemic resistance in tomato against early and late blight by *Bacillus subtilis* OTPB1 or *Trichoderma harzianum* OTPB3. Biol Control 65:109–117

Coquoz JL, Buchala AJ, Meuwly PH, Métraux JP (1995) Arachidonic acid induces local but not systemic synthesis of salicylic acid and confers systemic resistance in potato plants to *Phytophthora infestans* and *Alternaria solani*. Phytopathology 85:1219–1224

Cotty JP (1987) Modulation of sporulation of *Alternaria tagetica* by carbon dioxide. Mycologia 74:508–513

Cui J, Jiang N, Meng J, Yang G, Liu W, Zhou X, Ma N (2019) LncRNA33732- respiratory burst oxidase module associated with WRKY1 in tomato- *Phytophthora infestans* interactions. Plant J 97:933–946

Damalas CA, Koutroubas DS (2015) Farmers' exposure to pesticides: toxicity types and ways of prevention. Toxics 4:1–10

Dang HX, Pryor B, Peever T, Lawrence CB (2015) The *Alternaria* genomes database: a comprehensive resource for a fungal genus comprised of saprophytes, plant pathogens, and allergenic species. BMC Genomics 16:239

Data VV, Mayee CD (1981) Epidemiology of early blight of tomato foliage to early blight caused by *Alternaria solani*. Indian Phytopathol 33:83–86

Datar VV, Mayee CD (1982) Conidial dispersal of *Alternaria solani* in tomato. Indian Phytopathol 35:69–70

Duarte HSS, Zambolim L, Rodrigues FA, Paul PA, Pádua JG, Ribeiro Júnior JI, Júnior AFN, Antonio F, Rosado AW (2014) Field resistance of potato cultivars to foliar early blight and its relationship with foliage maturity and tuber skin types. Trop Plant Pathol 39:294–306

El-Rafai IM, Asswah SMW, Awdalla OA (2003) Biocontrol of some tomato diseases using some antagonistic microorganisms. Pak J Biol Sci 6:399–406

Encinas-Basurto D, Valenzuela-Quintanar MI, Sánchez-Estrada T-HM, Rodríguez-Félix A, Troncoso-Rojas R (2017) Alterations in volatile metabolites profile of fresh tomatoes in response to *Alternaria alternata* (Fr) Keissl infection. Chilean J Agric Res 77:194

Engindeniz S, Ozturk GC (2013) An economic comparison of pesticide application for processing and table tomatoes, a case study for Turkey. J Plant Prot Res 53:230–237

Escrivá L, O'ueslate S, Font G, Manyes L (2017) *Alternaria* mycotoxins in food and feed. An overview. J Food Quality 2017:1569748

European Commission (2012) Amending Annex I to Regulation (EC) No 669/2009 implementing Regulation (EC) No 882/2004 of the European Parliament and of the Council as regards the increased level of official controls on imports of certain feed and food of non-animal origin. Off J Eur Union 54:45–50

Faino L, Seidl MF, Datema E, van den Berg GCM, Janssen A, Wittenberg AHJ, Thomma BPHJ (2015) Single-molecule real-time sequencing combined with optical mapping yields completely finished fungal genome. M Bio 6:e00936–e00915

FAO (2018) Tomato production in 2018, crops/regions/world list/production quantity (pick lists). UN Food and Agriculture Organization, Corporate Statistical Database (FAOSTAT), Rome

Farooq S, Jat RR, Gupta A, Singh R, Majeed M, Nabi SU et al (2019) Evaluation of different fungicides against *Alternaria solani* (Ellis & Martin) Sorauer cause of early blight of tomato under laboratory conditions. Pharma Innov J 8:140–142

Foolad MR, Ntahimpera N, Christ BJ, Lin GY (2000) Comparison between field, greenhouse, and detached-leaflet evaluations of tomato germplasm for early blight resistance. Plant Dis 84:967–972

Foolad MR, Zhang LP, Khan AA, Niño-Liu D, Lin GY (2002) Identification of QTLs for early blight (*Alternaria solani*) resistance in tomato using backcross populations of a *Lycopersicon esculentum* X *Lycopersicon hirsutum* cross. Theor Appl Genet 104:945–958

Fourtouni A, Menetas Y, Christias C (1998) Effect of UV-B radiation on growth, pigmentation and spore production in phytopatho-genic fungus *Alternaria solani*. Can J Bot 76:2093–2099

Gardner RG (1988) NC EBR-1 and NC EBR-2 early blight resistant tomato breeding lines. Hort Sci 23:779–781

Gardner RG, Shoemaker PB (1999) "Mountain supreme" early blight resistant hybrid tomato and its parents, NC EBR-3 and NC EBR-4. Hort Sci 34:745–746

Gathungu RM, Bird SS, Sheldon DP, Kautz R, Vouros P, Matson WR et al (2014) Identification of metabolites from LC-EC profiling: GC-MS and re- fractionation provide essential information orthogonal to LCMS/micro NMR. Anal Biochem 454:23–32

Gherbawy Y, Hussein MA, Runge F, Spring O (2018) Molecular characterization of *Alternaria alternata* population isolated from upper Egyptian tomato fruits. J Phytopathol 166:709–721
Golshani F, Fakheri BA, Behshad E, Mohammadpour Vashvaei R (2015) PRs proteins and their mechanism in plants. BFAIJ 7:477–495
Gondal AS, Ijaz M, Riaz K, Khan AR (1993) Compendium of tomato diseases. American Phytopathological Society, St. Paul, pp 28–29
Hammond-Kosack KE, Jones JDG (1996) Resistance gene-dependent plant defense responses. Plant Cell 8:1773–1791
Hu J, Chen C, Peever T, Dang H, Lawrence C, Mitchell T (2012) Genomic characterization of the conditionally dispensable chromosome in *Alternaria arborescens* provides evidence for horizontal gene transfer. BMC Genomics 13:171
Iqbal T, Altaf S, Dar JA (2019) Status of early blight [*Alternaria solani* (Ellis and Martin) jones and grout] disease of tomato in Kashmir. J Pharmacog Phytoch 8:2152–2154
Islam RMD, Mondal C, Hossain I, Meah MB (2013) Organic management: an alternative to control late blight of potato and tomato caused by *Phytophthora infestans*. Int J Theor Appl Sci 5:32–42
Jarolim K, Del Favero G, Pahlke G, Dostal V, Zimmermann K, Heiss E et al (2017) Activation of the Nrf2-ARE pathway by the *Alternaria alternata* mycotoxins altertoxin I and II. Arch Toxicol 91:203–216
Johanson A, Thurston H (1990) The effect of cultivar maturity on the resistance of potatoes to early blight caused by *Alternaria solani*. Am Potato J 67:615–623
Johann S, Rosa LH, Rosa CA, Perez P, Cisalpino PS, Zani CL, Cota BB (2012) Antifungal activity of altenusin isolated from the endophytic fungus *Alternaria* sp. against the pathogenic fungus *Paracoccidioides brasiliensis*. Rev Iberoam Micol 29:205–209
Jorda L, Vera P (2000) Local and systemic induction of two defense-related Subtilisin-like protease promoters in transgenic *Arabidopsis* plants, Lucifer in induction of PR gene expression. Plant Physiol 124:1049–1057
Khulbe RD, Sati MC (1987) Seed transmission of *Alternaria solani* in tomato. Indian Phytopathol 40:106–107
Kumar R, Khurana A (2014) Functional genomics of tomato: opportunities and challenges in post-genome NGS era. J Biosci 39:917–929
Kumar S, Srivastava K (2013) Screening of tomato genotypes against early blight *Alternaria solani* under field condition. Bioscan 8:189–193
Kumar V, Gupta RC, Singh PC, Pandey KK, Kumar R, Rai AS et al (2007) Management of early blight disease of tomato cv. Ksahi Amrit through fungicides, bioagents and cultural practices in India. Veg Sci 34:206–207
Kumar V, Haldar S, Pandey KK, Singh RP, Singh AK, Singh PC (2008) Cultural, morphological, pathogenic and molecular variability amongst tomato isolates of *Alternaria solani* in India. World J Microbiol Biotechnol 24:1003–1009
Kumar A, Bimolata W, Kannan M, Kirti PB, Qureshi IA, Ghazi IA (2014) Comparative proteomics reveals differential induction of both biotic and abiotic stress response associated proteins in rice during *Xanthomonas oryzae* pv. *oryzae* infection. Funct Integr Genomics 15(4):425–437
Kumar M, Brar A, Yadav M, Chawade A, Vivekanand V, Pareek N (2018) Chitinases-potential candidates for enhanced plant resistance towards fungal pathogens. Agri 8:88. https://doi.org/10.3390/agriculture8070088
Lawaju BR, Lawrence KS, Lawrence GW, Klink VP (2018) Harpin-inducible defense signaling components impair infection by the ascomycete *Macrophomina phaseolina*. Plant Physiol Biochem 129:331–348
Li T, Zhang XY, Huang Y, Xu ZS, Wang F, Xiong AS (2018) An R2R3-MYB transcription factor, SlMYB28, involved in the regulation of TYLCV infection in tomato. Sci Hortic 237:192–200
Loganathan M, Venkataravanappa V, Saha S, Rai AB, Tripathi S, Rai RK, Chowdappa P (2016) Morphological, pathogenic and molecular characterizations of *Alternaria* species causing early blight of tomato in northern India. Proc Nat Acad Sci, Ind Sec B: Biol Sci 86:25–330

Lohith MR, Chandrashekar Reddy K, Venkata Ramana C, Venkata Rao P, Ravendra Reddy K, Lokanadha Reddy D (2011) Screening of tomato genotypes against early blight (*Alternaria solani*) by detached leaf method. Acta Hortic 914:465–468

Maiero M, Ng TJ, Barksdale TH (1989) Combining ability estimates for early blight resistance in tomato. J Am Soc Hort Sci 114:118–121

Maiero M, Ng TJ, Barksdale TH (1990a) Genetic resistance to early blight in tomato breeding lines. Hort Sci 25:344–346

Maiero M, Ng TJ, Barksdale TH (1990b) Inheritance of collar rot resistance in the tomato breeding lines C1943 and NC EBR-2. Phytopathology 80:1365–1368

Majumder S, Shakil NA, Kumar J, Banerjee T, Sinha P, Singh BB, Garg P (2016) Eco-friendly PEG-based controlled release nano-formulations of mancozeb: synthesis and bioefficacy evaluation against phytopathogenic fungi *Alternaria Solani* and *Sclerotium Rolfsii*. J Environ Sci Health B 51:873–887

Martin FW, Hepperly P (1987) Sources of resistance to early blight, *Alternaria solani*, and transfer to tomato, *Lycopersicon esculentum*. J Agric Univ Puerto Rico 71:85–95

Mauch-Mani B, Slusarenko AJ (1996) Production of salicylic acid precursors is a major function of phenylalanine ammonium lyase enzymes in the resistance of *Arabidopsis* to *Peronospora parasitica*. Plant Cell 8:203–212

Mauricio-Castillo JA, Salas-Muñoz S, Reveles-Torres LR, Salas-Luevano MA, Salazar-Badillo FB (2020) Could *Alternaria solani* IA300 be a plant growth-promoting fungus? Eur J Plant Pathol 157:413–419

McNeece BT, Sharma K, Lawrence GW, Lawrence KS, Klink VP (2019) The mitogen activated protein kinase (MAPK) gene family functions as a cohort during the *Glycine max* defense response to *Heterodera glycines*. Plant Physiol Biochem 137:25–41

Mizubuti GSE, Junior VL, Forbes GA (2007) Management of late blight with alternative products. Pest Technol 1:106–116

Moghaddama GA, Rezayatmanda Z, Esfahania MN, Khozaeic M (2019) Genetic defense analysis of tomatoes in response to early blight disease *Alternaria alternata*. Plant Physiol Biochem 142:500–509

Momel TM, Pemezny KL (2006) Florida plant disease management guide: tomato. Florida Cooperation Extensive Service, Institute of Food and Agriculture Sciences, Florida

Montealegre J, Valderrama L, Sánchez S, Herrera R, Besoain X, Pérez LM (2010) Biological control of *Rhizoctonia solani* in tomatoes with *Trichoderma harzianum* mutants. Electron J Biotechnol 13:2

Nasr Esfahani M (2019) An IPM plan for early blight disease of potato *Alternaria solani sorauer* and *A. alternata* (Fries.) Keissler. Arch Phytopathol Plant Protect. https://doi.org/10.1080/03235408.2018.1489600

Nasr Esfahani M, Alizadeh Moghaddam G, Karimkhah MA (2017) The relation of leaf micromorphological components with early blight resistant potatoes varieties. Iran Plant Prot Sci 39:51–64

Nguyen HD, Lewis CT, Lévesque CA, Gräfenhan T (2016) Draft genome sequence of *Alternaria alternata* ATCC 34957. Genome Announc 4:e01554–e01515

Ozkilinc H, Sarpkaya K, Kurt S, Can C, Polatbilek H, Yasar A, Sevinc U, Uysal A, Konukoglu F (2017) Pathogenicity, morpho-species and mating types of *Alternaria* spp. causing *Alternaria* blight in *Pistacia* spp. in Turkey. Phytoparasitica 45:719–728

Pandey D, Kunnumal Rajendran SRC, Gaur M, Sajeesh PK, Kumar A (2016) Plant defense signaling and responses against necrotrophic fungal pathogens. J Plant Growth Regul 35:1159–1174

Panthee DR, Chen F (2010) Genomics of fungal disease resistance in tomato. Curr Genomics 11:30–39

Pareek CS, Smoczynski R, Tretyn A (2011) Sequencing technologies and genome sequencing. J Appl Genet 52:413–435

Pasche JS, Wharam CM, Gudmestad NC (2004) Shift in sensitivity of *Alternaria solani* in response to QoI fungicides. Plant Dis 88:181–187

Pathak RK, Baunthiyal M, Shukla R, Pandey D, Taj G, Kumar K (2017) In silico identification of mimicking molecules as defense inducers triggering jasmonic acid mediated immunity against Alternaria blight disease in *Brassica* species. Front Plant Sci 8:609

Poysa V, Tu JC (1996) Response of cultivars and breeding lines of *Lycopersicon* spp. to *Alternaria solani*. Can Plant Dis Surv 76:5–8

Prasad B, Dutt BL (1974) Inducing sporulation of *Alternaria solani* II. Effect of light. Mycopathol Mycol Appl 49:141–146

Pusztahelyi T (2018) Chitin and chitin-related compounds in plant–fungal interactions. Mycology 9:189–201

Rao AP, Subrahamanyam K (1976) A new record of *Alternaria* on coconut leaf [India]. FAO, Rome

Raza W, Ghazanfar MU, Iftikhar Y, Ahmed KS, Haider N, Rasheed MH (2016) Management of early blight of tomato through the use of plant extracts. Int J Zool Stud 1:1–4

Sahi HPS (1990) Epidemiology and management of Alternaria leaf spots of tomato in Himachal Pradesh. Ph.D thesis, Dr. Y.S. Parmar University of Horticulture and Forestry, Solan, p 97

Sain SK, Pandey AK (2016) Biological spectrum of *Trichoderma harzianum* Rifai isolates to control fungal diseases of tomato (*Solanum lycopersicon* L.). Arch Phytopath Plant Prot 49:507–521

Schlagnhaufer CD, Arteca RN, Pell EJ (1997) Sequential expression of two 1-aminocyclopropane-1-carboxylate synthase genes in response to biotic and abiotic stresses in potato (*Solanum tuberosum* L.) leaves. Plant Mol Biol 35:683–688

Shah P, Powell ALT, Orlando R, Bergmann C, Gutierrez-Sanchez G (2012) Proteomic analysis of ripening tomato fruit infected by *Botrytis cinerea*. J Proteome Res 11:2178–2192

Shahida P, Abdul G (1995) Seed borne mycoflora of tomato. Pakistan J Bot 27:201–208

Sharma N, Rahman MH, Strelkov S, Thiagarajah M, Bansal VK (2007) Proteome-level changes in two *Brassica napus* lines exhibiting differential responses to the fungal pathogen *Alternaria brassicae*. Plant Sci 172:95–110

Sharma K, Sarma S, Bohra A, Mitra A, Sharma NK, Kumar A (2018) Plant metabolomics: an emerging technology for crop improvement. In: Çelik O (ed) New visions in plant science. IntechOpen, London, pp 65–79

Shinde BA, Dholakia BB, Hussain K, Panda S, Meir S, Rogachev I, Aharoni A et al (2017) Dynamic metabolic reprogramming of steroidal glycol-alkaloid and phenylpropanoid biosynthesis may impart early blight resistance in wild tomato (*Solanum arcanum* Peralta). Plant Mol Biol 95:411–423

Shinde BA, Dholakia B, Hussain KH, Aharoni A, Giri AP, Kamble AC (2018) WRKY1 acts as a key component improving resistance against *Alternaria solani* in wild tomato, *Solanum arcanum* Peralta. Plant Biotechnol J 16:1502–1513

Singh RS (1995) Diseases of vegetable crops, 3rd edn. Oxford and IBH Publishers, New Delhi, pp 107–109

Singh PC, Singh D (2006) *In vitro* evaluation of fungicides against *Alternaria alternata*. Ann Plant Prot Sci 14:500–502

Singh RN, Kushal R, Chugh RK, Sangwan R (2020) Evaluation of plant extracts, fungicides and bio-agents against early blight disease of tomato incited by *Alternaria solani* under in vitro and field conditions. Int J Curr Microbiol App Sci 9:1914–1920

Sinha R, Chattopadhyay S (2011) Changes in the leaf proteome profile of *Mentha arvensis* in response to *Alternaria alternata* infection. J Proteomics 74:327–336

Spletzer ME, Enyedi AJ (1999) Salicylic acid induces resistance to *Alternaria solani* in hydroponically grown tomato. Phytopathology 89:722–727

Stancheva I, Lozanov I, Achkova Z (1991) Sources of resistance to *Alternaria solani* in the tomato in wild growing species of the genus *Lycopersicon*. Genetika-i-Selektsiya 24:126–130

Stangarlin JR, Kuhn OJ, Assi L, Schwan-Estrada KRF (2011) Control of plant diseases using extracts from medicinal plants and fungi. Formatex 2011:1033–1042

Tenenboim H, Brotman Y (2016) Omic relief for the biotically stressed: metabolomics of plant biotic interactions. Trends Plant Sci 21:781–791

The Tomato Genome Consortium (2012) The tomato genome sequence provides insights into fleshy fruit evolution. Nature 485:635–641

Thipyapong P, Steffens JC (1997) Tomato polyphenol oxidase: differential response of the polyphenol oxidase F promoter to injuries and wound signals. Plant Physiol 115:409–418

Thirthamallappa, Lohithaswa HC (2000) Genetics of resistance to early blight (*Alternaria solani* Sorauer) in tomato (*Lycopersicum esculentum* L.). Euphytica 113:187–193

Thomma BPHJ, Seidl MF, Shi-Kunne X, Cook DE, Bolton MD, van Kan JAL, Faino L (2016) Mind the gap; seven reasons to close fragmented genome assemblies. Fungal Genet Biol 90:24–30

Tian M, Win J, Song J, Van der Hoorn R, Van der Knaap E, Kamoun SA (2007) *Phytophthora infestans* cystatin-like protein targets a novel tomato papain-like apoplastic protease. Plant Physiol 143:364–377

Touhami N, Soukup ST, Schmidt-Heydt M, Kulling SE, Geisen R (2018) Citrinin as an accessory establishment factor of *P. expansum* for the colonization of apples. Int J Food Microbiol 266:224–233

Ulker B, Shahid M, Somssich IE (2007) The WRKY70 transcription factor of *Arabidopsis* influences both the plant senescence and defense signaling pathways. Planta 226:125–137

Upadhyay P, Rai A, Kumar R, Singh M, Sinha B (2014a) Differential expression of pathogenesis related protein genes in tomato during inoculation with *A Solani*. J Plant Pathol Microb 5:217

Upadhyay P, Rai A, Kumar R, Singh M, Sinha B (2014b) Differential expression of pathogenesis related protein genes in tomato during inoculation with *A solani*. J Plant Pathol Microbiol 5:217

Upadhyay P, Ganie SH, Rai A, Singh M, Sinha B (2016) Identification of transcription factors in tomato, potentially related to early blight resistance at invasion inhost tissue, using microarray expression profiling. S Afr J Bot 106:165–173

van derWaals JE, Korsten L, Slippers B (2004) Genetic diversity among *Alternaria solani* isolates from potatoes in South Africa. Plant Dis 88:959–964

Varma PK, Gandhi SK, Singh S (2008) Biological control of *Alternaria solani*, the causal agent of early blight of tomato. J Biol Contr 22:67–72

Warren RL, Sutton GG, Jones SJM, Robert RA (2007) Assembling millions of short DNA sequences using SSAKE. Bioinformatics 23:500–501

Wenderoth M, Garganese F, Schmidt-Heydt F, Soukup ST, Ippolito A, Sanzani SM, Fischer R (2019) Alternariol as virulence and colonization factor of *Alternaria alternata* during plant infection. Mol Microbiol 112:131–146

Wojciechowska E, Weinert CH, Egert B, Trierweiler B, Schmidt-Heydt M, Horneburg B et al (2014) Chlorogenic acid, a metabolite identified by untargeted metabolome analysis in resistant tomatoes, inhibits the colonization by *Alternaria alternata* by inhibiting alternariol biosynthesis. Eur J Plant Pathol 139:735–747

Wolters PJ, Faino L, van den Bosch TBM, Evenhuis B, Visser RGF, Seidl MS et al (2018) Gapless genome assembly of the potato and tomato early blight pathogen *Alternaria solani*. Resour Announce 31:7

Woudenberg JHC, Seidl MF, Groenewald JZ, de Vries M, Stielow JB, Thomma BP, Crous PW (2015) *Alternaria* section *Alternaria*: species, *formae speciales* or pathotypes? Stud Mycol 82:1–21

Woundenberg JHC, Truter M, Groenewald JZ et al (2014) Large-spored *Alternaria* pathogens in section Porri disentangled. Stud Mycol 79:1–47

Yang X, Hu H, Yu D, Sun Z, He X, Zhang J, Chen Q (2015) Candidate resistant genes of sand pear (*Pyrus pyrifolia* Nakai) to *Alternaria alternata* revealed by transcriptome sequencing. PLoS One 10:e0135046

Yang X, Li H, Yang Y, Wang Y, Mo Y, Zhang R, Zhang Y, Ma J, Wei C, Zhang X (2018) Identification and expression analyses of WRKY genes reveal their involvement in growth and abiotic stress response in watermelon (*Citrullus lanatus*). PLoS One 16:e0191308

Yuan Y et al (2017) Metabolomic analyses of banana during postharvest senescence by $^{1}$H-high resolution-NMR. Food Chem 218:406–412

Zhang LP, Lin GY, Nino-Liu D, Foolad MR (2003) Mapping QTLs conferring early blight (*Alternaria solani*) resistance in a *Lycopersicon esculentum* × *L hirsutum* cross by selective genotyping. Mol Breed 12:3–19

Zorića AS, Morina F, Toševskic I, Tostie T, Jovićd J, Krstićd O, Veljović-Jovanović S (2019) Resource allocation in response to herbivore and gall formation in *Linaria vulgaris*. Plant Physiol Biochem 135:224–232

# Microbial-Mediated Remediation of Environmental Contaminants by Integrated Multi OMIC's Approaches

**5**

Eswar Rao Tatta, Madangchanok Imchen, Girish Kumar Rasineni, and Ranjith Kumavath

**Abstract**

The global environment offers everything like good health, good food, to the human population. Therefore, optimization for efficient bioremediation requires augmentation of complex input to minimize the environmental contaminants. These pollutants hold toxic molecules that influence the ecosystem and existing living things, the forthcoming effects could be devastating to human populations and the environment. Moreover, bioremediation is an option to enhance the efficacy of the natural biodegradation process by using living microbes, which can degrade the toxic compound into less or non-toxic forms. This chapter focused on OMIC's approaches bringing paradigm swift in understanding microbial sources and their application in the bioremediation for cleaning the environment.

**Keywords**

OMIC's approaches · Bioremediation · Genomics · Transcriptomics · Proteomics · Metabolomics · Polluted environment

E. R. Tatta · M. Imchen · R. Kumavath (✉)
Department of Genomic Science, School of Biological Sciences, Central University of Kerala, Kasaragod, Kerala, India

G. K. Rasineni
Tanet MedCorp Pvt. Ltd., Hyderabad, India



A. Kumar et al. (eds.), *Omics Technologies for Sustainable Agriculture and Global Food Security Volume 1*, https://doi.org/10.1007/978-981-16-0831-5_5

## 5.1 Introduction

A healthy environment is proficient enough to provide both wealth and health to humans. Since the earliest times the human population, wildlife, and environment are equally distributed and they could not disturb each other. A healthy environment is interdependent on the activities of living creatures and the rest physical world. Today, the environmental health is besmirched directly or indirectly with anthropogenic activities. Release of toxic components above normal levels in the form of liquid or solid into the environment through industrial practice and daily human activities adversely affects not only environment but also people and animal health (Krishna et al. 2017). The use of natural resources at a higher rate than nature's restoring capacity results in extreme environmental pollution, majorly affecting soil, groundwater, rivers, and atmosphere (Gouma 2009). Industries such as Textile, Pharmaceutical, Oil and Gas, and other heavy manufacturing industries contaminating the air, water, and soil through several toxic chemicals such as solvents, polychlorinated biphenyls, petroleum hydrocarbons, radioisotopes, polycyclic aromatic hydrocarbons, salts, and heavy metals (Shah 2014; Gouma 2009). In order to prevent environmental pollution due to these toxic chemicals, they need to be not only controlled but also need to degrade them to non-toxic levels. The processing the toxic pollutants from the environment such air, water and soil is termed as remediation. Environmental remediation can be done using physical, chemical, and biological processes or techniques. Bioremediation is an emerging technology used to complete management of diverse groups of environmental pollutants and other physical and chemical treatment methods (Pardeep Singh et al. 2020).

Bioremediation is defined as the process whereby toxic waste can be biologically degraded under controlled conditions to an innocuous state or below the respective toxic concentration limits. Bioremediation employs the living organisms, most notably microorganisms, to degrade the pollutants and convert them into less toxic or non-toxic form (Zouboulis and Moussas 2011). The microorganisms survive at diverse environmental circumstances through their tensile metabolic activities (Abatenh et al. 2017). Microorganisms are natural decomposers as they have unique ability to utilize diverse types of organic substances as a source of energy and convert toxic ones into harmless by-products (Rawat and Rangarajan 2019). Microbial bioremediation is an ecofriendly and cost-competitive strategy for eliminating xenobiotic and or anthropogenic compounds from the polluted environments (Desai et al. 2010). However, implementation of the bioremediation strategies requires a detailed understanding of factors governing the growth, metabolism, dynamics, and functions of the microbial communities. The recent advances in molecular biology and genetic engineering have opened new avenues to study bioremediation in detail. Omics technologies allow for an in-depth view of the microbial community and its surroundings at the molecular level to better understand their cellular and molecular dynamics in contaminated environments. It will help to make bioremediation more efficient and effective over a wide range of pollutants. Omics approaches have been utilized individually to study the microbial system in the context of bioremediation, providing information about biodegradation pathways and the involved organisms in terms of changes in genome, transcriptome, proteome, metabolome, etc. in a given time and conditions (Rawat and Rangarajan 2019).

### 5.1.1 OMIC's Approaches to Environmental Bioremediation

Breakthrough innovative discoveries in high-throughput technologies such as genomics, transcriptomics, proteomics, and metabolomics, along with bioinformatics tools, have led to a better understanding of different environmental approaches bioremediation. Earlier bioremediation research focused mainly on chemical kinetics, intermediate products, and final product identification and quantification of organic pollutants. The studies on the molecular mechanisms behind the contaminant transformation processes received less attention largely due to the technical difficulties. The advancement in modern molecular biology, system biology, and availability of whole genome sequence data, fosters new techniques including genomics, transcriptomics, proteomics, and metabolomics, which might potentially be applied in the bioremediation of organic chemicals in the environment (Malla et al. 2018). The OMIC's technology has been employed for analysis of available whole genome sequencing data of different microorganisms from different environmental communities. Different tools are essential to evaluate the entire genome sequencing, protein profiles, and metabolites by the combination of different omic platforms (Schneider and Orchard 2011). Several approved OMIC's platforms (Table 5.1) available to understand the microbial systems biology and measure or determine the dynamic changes within cell overtime (Zhang et al. 2010). OMIC's technologies opened a new era in system biology and a possibility of being a pioneering and proficient research avenue for studying natural systems (Fig. 5.1).

The high-throughput technology could summarize the gene, proteins, small metabolites, and their metabolic pathway by coordinating with bioinformatics tools. OMIC's approach along with bioinformatics tools provide significant and essential insights into specific microorganism and their ability for bioremediation (Desai et al. 2010). Bioinformatics a new branch of science that analyzes experimentally generated biological data (Schneider and Orchard 2011; Pérez-Llano et al. 2018; Mayer 2011). Understanding and quantitative discovery of biochemical pathways through computational analysis enables understanding and detecting of alterations in the biological systems (Desai et al. 2010). The identification of large annotated pathway sequences of genes, proteins, and metabolites together is possible using these bioinformatic tools (Robertson 2005; Dangi et al. 2019). The usefulness of genome arrangement information and associated OMIC's approaches linking microbial population's functions in polluted sites and their bioremediation capacity is an immense challenge (Kumavath and Pratap 2012).

**Table 5.1** The list of OMIC tools for the study of the bioremediation

| Acronym | Full name | Omic domain | Website |
|---|---|---|---|
| MIGS | Minimum Information About a Genome Sequence | Genomic | http://gensc.org |
| SAM | Sequence Alignment/Map | Genomic | http://samtools.sourceforge.net |
| PML | Polymorphism Markup Language | Genomic | http://www.openpml.org |
| PaGE-OM | Phenotype and Genotype Experiment Object Model | Genomic | http://www.pageom.org |
| Gen2Phen | Gen2Phen | Genomic | http://www.gen2phen.org |
| ensembl | ensembl | Genomic | http://www.ensembl.org |
| GO | Gene Ontology | Genomic | http://www.geneontology.org |
| MINSEQE | Minimum Information about a high-throughput Nucleotide SEQuencing Experiment | Transcriptomics | http://www.mged.org/minseqe |
| MIAME, | Minimum Information About a Microarray Experiment | Transcriptomics | http://www.mged.org |
| MAGE | MicroArray Gene Expression | Transcriptomics | http://www.mged.org |
| MAQC | MicroArray Quality Control | Transcriptomics | http://www.fda.gov/nctr/science/centers/toxicoinformatics/maqc |
| ERCC | External RNA Control Consortium | Transcriptomics | http://www.cstl.nist.gov/biotech/Cell&TissueMeasurements/GeneExpression/ERCC.htm |
| NIST | National Institute for Standards Technology | Transcriptomics | http://www.cstl.nist.gov/biotech/ |
| MGED O | Microarray Gene Expression Data Ontology | Transcriptomics | http://www.mged.org |
| GEO | Gene Expression Omnibus | Transcriptomics | http://www.ncbi.nlm.nih.gov/geo |
| CR | Comparative RNA | Transcriptomics | http://www.rna.ccbb.utexas.edu |
| RNAmods | RNA modification database: | Transcriptomics | http://library.med.utah.edu/RNAmods |
| MIAPE | Minimum Information About a Proteomics Experiment | Proteomic | http://www.psidev.info/index.php?q=node/91 |
| MIMIx | Minimum Information about a Molecular Interaction Experiment | Proteomic | http://www.psidev.info/index.php?q=node/277 |
| PSI | Proteomics Standards Initiative | Proteomic | http://www.psidev.info/ |

(continued)

**Table 5.1** (continued)

| Acronym | Full name | Omic domain | Website |
|---|---|---|---|
| PSI-MI | Proteomics Standards Initiative Molecular Interactions | Proteomic | http://www.psidev.info/mif |
| Pseq | Protein sequences | Proteomic | http://www.uniprot.org |
| PIR | Protein information Resources | Proteomic | http://pir.georgetown.edu |
| PD | Protein databank | Proteomic | http://www.rcsb.org/pdb/home/home.do |
| CIMR | Core Information for Metabolomics Reporting | Metabolomics | http://msi-workgroups.sourceforge.net |
| MeMo | Metabolic Modelling | Metabolomics | http://dbkgroup.org/memo |
| ArMet | Architecture for Metabolomics | Metabolomics | http://www.armet.org |
| IntEnz | Integrated relational Enzyme | Metabolomics | http://www.ebi.ac.uk/intenz |

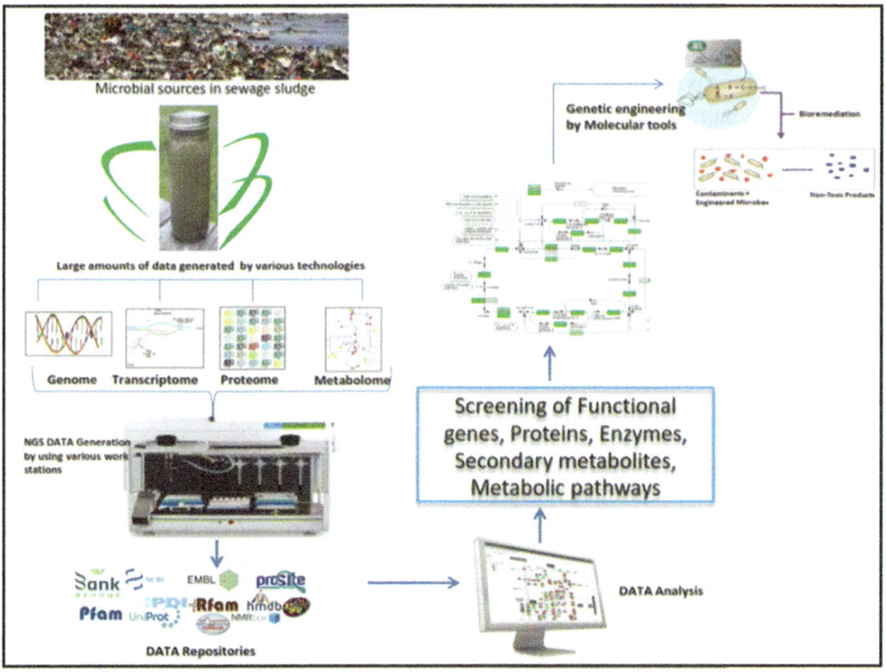


**Fig. 5.1** The schematic representation of the soil-contaminate and its OMIC's tools for bioremediation

### 5.1.2 Tools and Strategies for Bioremediation

The conventional remediation of waste pollutants is generally carried out by ignition and landfill. On the other hand, bioremediation or biodegradation takes advantage of microbial metabolic processes to detoxify and degrade the toxic chemicals in contaminated environments.

- Chemical processes such as treatment of chemical oxidation, electrochemical degradation, and volatilization again produce the flotation, toxic gases (Mustapha and Lens 2018).
- Physical processes such as tipping or recycling, incineration, landfilling, adsorption, and membrane filtration provide the concentrates or reduce the volume of waste compounds and sedimentations (Otero-Blanca et al. 2018).
- Biological methods such as bioaccumulation, bio-augmentation, bio-venting, bio-stimulation, biosorption the final products like precipitation, biotransformation or biomineralization or natural biogeochemical cycle (Moreira et al. 2018).

A combination of physicochemical and biological remediation processes like using a specific bacterial population and or by its metabolic products can neutralize or detoxify the toxic substances (Ju and Zhang 2015). It is easy to understand the population dynamics and diversity in microbial communities at precise sites of the polluted area through this advanced techniques. In the recent past, scientists started working on discovering and developing of novel enzymes and biosensors approaches for bioremediation of wastes from the environment. Among all remediation technologies, the sophisticated, efficient method removes dangerous chemicals from the contaminated/spoiled water or soils through the microbial treatment (Moreira et al. 2018). Genetically engineered microorganisms (GEMs) generated by the biotechnology tools are efficient in bioremediation, especially at heavily polluted sites (Rayu et al. 2012).

## 5.2 Genomic Tools in Environmental Bioremediation

Molecular techniques like genome mining and bioinformatics are valuable tools to construct potential genomics data of beneficial microorganisms for bioremediation. This contribution of genomics paves the way to understand the metabolisms and microorganisms interactions with pollutants in both single species and microbial communities. Such approaches aid in the utilization of microorganisms for efficient, and environmentally sound remediation strategies (Plewniak et al. 2018). For example, benzoate, catechol, arsenic, cobalt, cadmium, degraded by brevibacterium epidermidis EZ-K02 (Ziganshina et al. 2018); Nitroacetate and nitriloacetate degraded by *Mycobacterium oleivorans* (Miller et al. 2016); Nitrate and uranium degraded by *Anaeromyxobacter* sp. Fw109-5 (Hwang et al. 2015). The whole genome sequence analysis is important in understanding the microbial population, including its role in biodegradation processes.

Next-Generation Sequencing (NGS) ignited a real revolution in environmental sciences. It triggered the spread of its novel, cutting-edge disciplines, e.g. metagenomics and metatranscriptomics in to bioremediation and biodegradation, leading it to enter into an *omics* era (Ma and Zhai 2012). At least 95 percent of microbes are uncultivable in the laboratory applying standard culture conditions hampering their investigation for a long time during bioremediation studies. However, by using *comparative metagenomic* approach, millions of individual microbes can be identified and quantified based on their PCR-amplified 16S or 18S rDNA segments in a single run. Long-read sequencing of two hyper variable regions enables the species-level taxonomic characterization of bacteria, archaea, fungi, protozoa, algae, etc. (Bihari 2013). In addition, sequencing methods targeting small subunit (SSU) rRNA hypervariable regions have allowed the identification of signature microbial species that serve as bioindicators for sewage contamination in these environments (Tan et al. 2015). NGS-based molecular ecology can be best framed into on-site bio-augmentation or bio-stimulation regimes. During regular sampling of the test site and chemical and physical, a comprehensive microbiological datasets can be gained to create multidimensional matrices, where coherent dynamics and association networks can be determined (Eiler et al. 2012). In this way, the niches microbial key players and relevant interactions can be identified (Bihari 2013).

### 5.2.1 Transcriptomic Tools in Environmental Bioremediation

The regulation of gene expression is one of the key processes for adapting to environmental conditions changes and thus for survival. Transcriptomics describes this process in a genome-wide range (Singh and Nagaraj 2006). The transcriptome is the set of all RNA transcripts, including coding and non-coding, in an individual or a population of cells. Microarrays are an extremely powerful platform in transcriptomics that enable determination of the mRNA expression level of practically every gene of an organism (Singh and Nagaraj 2006). Various microbial transcriptomics methods reveal the capabilities of biodegradation of various toxic compounds by adapting to their stress conditions. A comprehensive 50-mer-based oligonucleotide microarray was developed for effective monitoring of biodegrading populations based on known genes and pathways involved in biodegradation and metal resistance (Rhee et al. 2004). This type of DNA microarray was effectively used develop soil bioremediation methods and analyze naphthalene-amended enrichment. A global gene expression analysis revealed the co-regulation of several thus-far-unknown genes during the degradation of alkylbenzenes (Kuhner et al. 2005). DNA microarrays have also been used to determine bacterial species, in quantitative applications of stress gene analysis of microbial genomes and in genome-wide transcriptional profiles (Muffler et al. 2002; Singh and Nagaraj 2006). Transcriptomic approaches were used for the profiling of bacterial expression levels with comparison to various stress conditions. The studies on transcriptomic differential expression genes (DEG) revealed the mechanism behind microbial degradation of several toxic compounds such as *Aspergillus niger, Trichoderma harzianum, Talaromyces purpurogenus,* and *Aspergillus flavus* in degradation of hydrocarbons; *Acinetobacter venetianus* RAG-1 Degradation of Alkanol (Asemoloye et al. 2018; Kothari et al. 2016).

### 5.2.2 Proteomic Tools in Environmental Bioremediation

Proteomics an emerging and powerful discipline aimed at the study of the whole proteome or the sum of all proteins from an organism, tissue, cell or biofluid resulting in an information rich landscape of expressed proteins and their modulations under specific conditions (Husi and Albalat 2014). The advent of proteomics leads to an extensive examination of global changes in the composition or abundance of proteins and the identification of key proteins involved in the microorganisms in a given physiological state. A number of reports have described sets of proteins that are up or downregulated in response to the presence of specific pollutants (Hivrale et al. 2016). In proteomic studies, chromatographic and mass spectroscopy (MS) associated techniques play an important role in the protein identification, expression, posttranslational modifications, and including structure and functions of proteins (Seibert et al. 2005; Wang et al. 2016). Liquid chromatography and mass spectroscopy (LC-MS) other proteomic techniques including two-dimensional polyacrylamide gel electrophoresis (2-D PAGE) and MALDI-TOF (Matrix-assisted laser desorption/ionization time-of-flight), SELDI-TOF (surface-enhanced laser desorption/ionization time-of-flight), protein microarrays—help in systematic mapping and identification of whole-cell proteome and hundreds of proteins profiling (Singh 2006; Bianco and Perrotta 2015; Wang et al. 2016). A proteomics method applied to observe the ability of remediation by various microbes. For instance, *Sphingomonas* sp. GY2B degrades to Phenanthrene (Izrael-Živković et al. 2018), Penicillium oxalicum degrades to polycyclic aromatic hydrocarbons (Liu et al. 2017). Six major proteins were identified as significantly induced and over expressed on 2-DE when Mycobacterium sp. Starin PYR-1 was exposed to phenanthrene, dibenzothiophene, and pyrene (Hivrale et al. 2016). Through advances in proteomics, post-translational modification which involves in protein functionality, protein–protein interaction have been studied extensively. Prospective applications of modern proteomics techniques in microbial ecology identify of novel functional genes, identify new enzymatic and metabolic pathways and identify the novel proteomes in the biodegrading organisms. Besides this, it is helpful to monitor dynamic and sustainability of other environmental factors (Hivrale et al. 2016).

## 5.3 Metabolomic Tools in Environmental Bioremediation

Metabolomics, a comprehensive analysis of metabolites in a biological specimen, is an emerging technology that holds promise to inform the practice of bioremediation.. Metabolomics has emerged as a new field in biology with the promise to speed up the functional analysis of genes with unknown function (Villas-Bôas et al. 2005). In addition to genomics, transcriptomics, and proteomics, cutting-edge research is now expanding toward the analysis of microbial cellular metabolites. The very important analytical methods to the analysis of metabolite are High-Performance Liquid Chromatography (HPLC), direct injection mass spectrometry (DIMS), gas chromatography (GC), Fourier transform-infrared (FT-IR) spectroscopy, and NMR (Chakraborty and Das 2017). Application of metabolome-based approaches to the environmental samples has made it possible to develop models that can envisage microbial activities under different bioremediation strategies. Metabolomics allows us to better understand the microbial communities dynamic operations and their functional contributions to the environments in which they live (Malla et al. 2018). Metabolic engineering (ME) technique has been developed in order to improve cellular properties or introduce new ones through the use of recombinant DNA technology (Stephanopoulos et al. 1998). This approach is an alternative to classical mutagenesis for the improvement of industrial microorganisms. Metabolic Engineering (ME) tools and principles are atmost relevant for bioremediation due to the shortcomings of natural microbial population to degrade recalcitrant xenobiotics such as chlorotoluens, PCBs PAH, and others (Urgun-Demirtas et al. 2006). Accumulation of these compounds in the environment represents a potential pollution problem, since many of them are highly toxic, mutagenic, and/or carcinogenic (Villas-Boas and Bruheim 2007).

The use of metabolomics trappings could also significantly extend and enhance bioremediation approaches by providing a better sign of the biodegradation progression. However, microbial cellular physiology and metabolic activities were frequently fluctuated along with environmental changes (Wang et al. 2016). The metabolomics aims to expand the understanding of the microbiomes role and could mineralize xenobiotic compounds too (Bharagava et al. 2019). Villas-Boas et al. and Robertson (2005) have reviewed the environmental applications of metabolomics, and the diversity of work is captivating. Metabolites play a key role in connecting the numerous pathways that function within or outside of the living cell. Metabolic footprinting analysis can be a very useful approach to assessing xenobiotics mineralization process in the environment. Techniques developed for metabolic footprinting analysis aim to analyze low molecular mass compounds which is a highly recommended approach to ensure that a polluting compound is being totally degraded or that the biodegradation process results in the accumulation of hazard recalcitrant catabolic products. The use of metabolomics tools can significantly extend and enhance the power of existing bioremediation approaches by providing a better overview of the biodegradation process.

## 5.4 Enzymes and Biomarkers Involved in Environmental Bioremediation

Among natural bio-agents, enzymes have an immense potentiality to effect alteration and detoxification of toxic substances, to practice cell-free bioremediation (Rao et al. 2010). Hence, the biocatalysts are powerful tools to detoxify toxic substances and restore polluted environments. However, the biomarkers act as biosensors for the detection of pollutants and noticeable contaminants in an environment (Table 5.2). The biocatalysts and biomarkers naturally occur in prokaryote and eukaryotic cells (Dangi et al. 2019). The revolution of the OMIC's techniques lead to a paradigm shift in bioremediation research, there is an insistent call to develop a novel or improved biomarkers for rapid assessment and detection of pollutants at polluted environmental sites (Sharma et al. 2018). The Enzymatic proteins have several beneficial characteristics, for instance, they have broad specificity, microbial = 1 metabolites cannot inhibit them, they can be used under extreme environment (Rao et al. 2017). They can act on large classes of organic and inorganic pollutant mixtures and can completely convert them into an inoffensive end product. Furthermore, enzymes have advantages over microbial remediation in effectiveness and rapidity (Moller and Jansson 1998; Jansson et al. 2000; Rao et al. 2010). Reactive molecules production and their use as biomarkers for the environmental bioremediation are presented in Table 5.3. The biomarkers have monitored the efficiency of bioremediation using biosensors in various harsh environments.

The microbial enzymes have versatile applications in remediation technology and there have been focused on developing methods for biosensors. The major advantage of bacteria tag with the *lacZY* genes and luciferase genes (*luxAB*), was encoding lactose permease and b-galactosidase (Flemming et al. 1994), can be detected and counted as a blue colonies and luminescent colonies formation, respectively, in oil-contaminated soils. Furthermore, the advances in genetic-engineering techniques have opened up new opportunities towards providing the novel biocatalysts and biomarkers from (GEMs) genetically engineered microorganisms to function as designer biocatalysts and biomarkers, that may be able to remediate the polluted environment (Dua et al. 2002; Rao et al. 2017; Dangi et al. 2019).

## 5.5 Concluding and Future Perspectives

OMIC's, an amalgamation of different emerging technologies can help decipher the relationship between microbes in a polluted environment and their potential role in bioremediation. The emergence of such innovative high-throughput OMIC's technologies (Fig. 5.1) in identifying and characterizing of the whole environmental microbial population has opened up a new landscape. This has enlightened us with a myriad of concealed microbes with crucial biogeochemical functions. The highly active enzymes of novel origin have a promising future in detoxification and degradation of pollutants (Siggins et al. 2012; Bharagava et al. 2019). Integration of such OMIC's approaches can provide a deeper understanding of the pathways for

**Table 5.2** The list of Enzymes involved for bioremediation process and their mode of actions towards environmental bioremediation

| Enzyme | EC code | Source of organism | Mode of action | Polluted area | References |
|---|---|---|---|---|---|
| Laccase | 1.10.3.2 | *Bacillus vallismortis* | Degradation of dye | Pulp and paper industry | Chauhan et al. 2017 |
| Manganese peroxidase | 1.11.1.13 | *Salmonella enterica* | Maillard reactions products | Fermentation industries | Kumar and Chandra 2018 |
| Lignin peroxidases | 1.11.1.14 | *Phanerochaete chrysosporium* | Lignin-degrading | Wood processing | Coconi-Linares et al. 2015 |
| Phenol oxidase | 1.10.3.1 | *Azospirillum brasilense* | Degradation phenolic substance | Industrial chemicals | Nikitina et al. 2010 |
| Carboxylesterases | 3.1.1.1 | *Archaeoglobus fulgidus* | Biocide detoxification | Pharmaceutical industry | Sood et al. 2016 |
| Dioxygenase | 1.13.11 | *Synechocystis* | Carotenoid cleavage | Pharmaceutical industry | Marasco and Schmidt-Dannert 2008 |
| Peroxidase | 1.11.1.7 | *Horseradish* | Phenols, chlorophenols, anilines | Dewatering of slime | Duran and Esposito 2000 |
| Permease | 2.7.1.202 | *Acinetobacter venetianus* | Dodecanol breakdown | Oil industries | Kothari et al. 2016 |
| Catechol dioxygenases | 1.13.11.1 | *Pseudomonas pseudoalcaligenes* | Cleavage of the aromatic ring | Pesticides and refrigerants | Broderick 1999 |
| Haloalkane dehalogenases | 3.8.1.5 | *Mycobacterium sp.* | α/β-hydrolases | Contaminated ground water | Nagata et al. 2015 |
| Phosphotriesterases | 3.1.8.1 | *Streptomyces setonii* | Hydrolysis | Toxic insecticides | Santillan et al. 2016 |
| Cyanide dioxygenase | 4.2.1.66 | Rhizopus oryzae | Oxidative degradation | Chemical industrial | Cabuk et al. 2006 |

**Table 5.3** Reactive molecules production used as biomarkers for the environmental bioremediation

| Gene | Biomarker | Organism/bacteria | Bioremediation | Reference |
|---|---|---|---|---|
| cytochrome P450 1A | CYP1A1 | Human | Neurotoxic substances | Sarkar 2006 |
| *luxAB* | luciferase | Firefly | Gasoline | Moller and Jansson 1998 |
| Metallothioneins | MTs | Multiple bacterial sp | Oxidative stress | Paniagua-Michel and Olmos-Soto 2016 |
| *luxAB* and *lacZY* | lac-lux | *Pseudomonas fluorescens,* Pseudomonas cepacia | 2,4-dichlorophenoxy Acetic acid | Masson et al. 1993 |
| *nptII* | Antibiotic resistance gene | Antibiotic resistant bacterial sp. | Antibiotic selection | |
| *Mer* | Heavy metal resistance gene | Multiresistant *Enterobacteriaceae* | Heavy metal resistance selection | Andrade et al. 2018 |
| *lacZY* | Chromogenic marker genes | Pseudomonas fluorescens | Gasoline | Jaderlund et al. 2008 |
| *lacZY* and *luxAB* | lux-lac | *Pseudomonas aeruginosa* | Antibiotic selection | Flemming et al. 1994 |
| *Gfp* | Green fluorescent protein | *Methylobacterium* | Chlorophenol | Tresse et al. 1998 |
| Cytochrome P450 | CYP1A | Multiple bacterial sp | Oxidative | Kelly and Kelly 2013 |

microbe based bioremediation and could even provide a novel ecofriendly alternative to eliminate unwanted xenobiotic, pharmaceutical wastes, pesticides, insecticides, polycyclic aromatic hydrocarbons from contaminated sites (Malla et al. 2018). The analysis of high-throughput data from public repositories is the current challenge that can provide strong predictive based models through deep machine learning and artificial intelligence that can in turn be implemented for bioremediation and sustainable development. To make the best use of public repositories, the scientific community could be encouraged to provide descriptive metadata.

Bioremediation is one of the most promising methods for sustainable development. The integration of OMIC's in the field of bioremediation has opened up an efficient research avenue with a promising future. No wonder, the developments in OMIC's filed have progressed exponentially (Schneider and Orchard 2011) in the last decade. However, it largely remains unimplemented for the public. Nonetheless, various regulatory badies are being taken positive steps for the implementation of OMIC's technologies in decision making.

**Acknowledgments** Authors thanks to SERB-EMEQ/051/2014 for partial financial assistance and the research facilities supported by the Central University of Kerala,

**Competing Interests** The authors have declared no competing interests.

## References

Abatenh E, Gizaw B, Tsegaye Z, Wassie M (2017) Application of microorganisms in bioremediation-review. J Environ Microbiol 1(1):02–09

Andrade LN, Siqueira TE, Martinez R, Darini ALC (2018) Multidrug-resistant CTX-M-(15, 9, 2)- and KPC-2-producing Enterobacter hormaechei and Enterobacter asburiae isolates possessed a set of acquired heavy metal tolerance genes including a chromosomal sil operon. Front Microbiol 9:539

Asemoloye MD, Ahmad R, Jonathan SG (2018) Transcriptomic responses of catalase, peroxidase and laccase encoding genes and enzymatic activities of oil spill inhabiting rhizospheric fungal strains. Environ Pollut 235:55–64

Bharagava RN, Purchase D, Saxena G, Mulla SI (2019) Applications of metagenomics in microbial bioremediation of pollutants: from genomics to environmental cleanup. In: Das S, Dash H (eds) Microbial diversity in the genomic era, 1st edn. Academic Press, Cambridge, MA, pp 459–477

Bianco L, Perrotta G (2015) Methodologies and perspectives of proteomics applied to filamentous fungi: from sample preparation to secretome analysis. Int J Mol Sci 16(3):5803–5829

Bihari Z (2013) Current trends in bioremediation and biodegradation: next-generation sequencing. J Bioremed Biodegr 4:e138

Broderick JB (1999) Catechol dioxygenases. Essays Biochem 34:173–189

Cabuk A, Unal AT, Kolankaya N (2006) Biodegradation of cyanide by a white rot fungus, Trametes versicolor. Biotechnol Lett 28(16):1313–1317

Chakraborty J, Das S (2017) Application of spectroscopic techniques for monitoring microbial diversity and bioremediation. Appl Spectrosc Rev 52(1):1–38

Chauhan PS, Goradia B, Saxena A (2017) Bacterial laccase: recent update on production, properties, and industrial applications. 3 Biotech 7(5):323

Coconi-Linares N, Ortiz-Vázquez E, Fernández F, Loske AM, Gómez-Lim MA (2015) Recombinant expression of four oxidoreductases in Phanerochaete chrysosporium improves degradation of phenolic and non-phenolic substrates. J Biotechnol 209:76–84

Dangi AK, Sharma B, Hill RT, Shukla P (2019) Bioremediation through microbes: systems biology and metabolic engineering approach. Crit Rev Biotechnol 39(1):79–98

Desai C, Pathak H, Madamwar D (2010) Advances in molecular and "-omics" technologies to gauge microbial communities and bioremediation at xenobiotic/anthropogen contaminated sites. Bioresour Technol 101(6):1558–1569

Dua M, Singh A, Sethunathan N, Johri A (2002) Biotechnology and bioremediation: successes and limitations. Appl Microbiol Biotechnol 59(2–3):143–152

Duran N, Esposito E (2000) Potential applications of oxidative enzymes and phenoloxidase-like compounds in wastewater and soil treatment: a review. Appl Catal B Environ 28(2):83–99

Eiler A, Heinrich F, Bertilsson S (2012) Coherent dynamics and association networks among lake bacterioplankton taxa. ISME J 6(2):330–342

Flemming CA, Leung KT, Lee H, Trevors JT, Greer CW (1994) Survival of lux-lac-marked biosurfactant-producing Pseudomonas aeruginosa UG2L in soil monitored by nonselective plating and PCR. Appl Environ Microbiol 60(5):1606–1613

Gouma S (2009) Biodegradation of mixtures of pesticides by bacteria and white rot fungi. Ph.D. thesis, School of Health Cranfield University, 416

Hivrale AU, Rane NR, Pawar PK, Gowindwar SP (2016) Application of genomics and proteomics in bioremediation. In: Rathoure AK, Dhatwalia VK (eds) Toxicity and waste management using bioremediation. IGI Global, Hershey, PA, pp 97–112

Husi H, Albalat A (2014) Proteomics. In: Handbook of pharmacogenomics and stratified medicine. Academic Press, Cambridge, MA, pp 147–179

Hwang C, Copeland A, Lucas S, Lapidus A, Barry K et al (2015) Complete genome sequence of Anaeromyxobacter sp. Fw109-5, an anaerobic, metal-reducing bacterium isolated from a contaminated subsurface environment. Genome Announc 3(1):e01449–e01414

Izrael-Živković L, Rikalović M, Gojgić-Cvijović G, Kazazić S, Vrvić M et al (2018) Cadmium specific proteomic responses of a highly resistant Pseudomonas aeruginosa san ai. RSC Adv 8 (19):10549–10560

Jaderlund L, Hellman M, Sundh I, Bailey MJ, Jansson JK (2008) Use of a novel nonantibiotic triple marker gene cassette to monitor high survival of Pseudomonas fluorescens SBW25 on winter wheat in the field. FEMS Microbiol Ecol 63(2):156–168

Jansson JK, Björklöf K, Elvang AM, Jørgensen KS (2000) Biomarkers for monitoring efficacy of bioremediation by microbial inoculants. Environ Pollut 107(2):217–223

Ju F, Zhang T (2015) Experimental design and bioinformatics analysis for the application of metagenomics in environmental sciences and biotechnology. Environ Sci Technol 49 (21):12628–12640

Kelly SL, Kelly DE (2013) Microbial cytochromes P450: biodiversity and biotechnology. Where do cytochromes P450 come from, what do they do and what can they do for us? Philos Trans R Soc B Biol Sci 368(1612):20120476

Kothari A, Charrier M, Wu YW, Malfatti S, Zhou CE et al (2016) Transcriptomic analysis of the highly efficient oil-degrading bacterium Acinetobacter venetianus RAG-1 reveals genes important in dodecane uptake and utilization. FEMS Microbiol Lett 363(20):fnw224

Krishna IM, Manickam V, Shah A, Davergave N (2017) Environmental management: science and engineering for industry. Butterworth-Heinemann, Oxford, p 664

Kuhner S, Wohlbrand L, Fritz I, Wruck W, Hultschig C et al (2005) Substrate-dependent regulation of anaerobic degradation pathways for toluene and ethylbenzene in a denitrifying bacterium, strain EbN1. J Bacteriol 187(4):1493–1503

Kumar V, Chandra R (2018) Characterization of manganese peroxidase and laccase producing bacteria capable for degradation of sucrose glutamic acid-Maillard reaction products at different nutritional and environmental conditions. World J Microbiol Biotechnol 34(2):32

Kumavath RN, Pratap D (2012) Omics in bioremediation: a consolidated overview. Indian J Biotechnol 6(8,9):242–253

Liu S, Guo C, Dang Z, Liang X (2017) Comparative proteomics reveal the mechanism of Tween80 enhanced phenanthrene biodegradation by Sphingomonas sp. GY2B. Ecotoxicol Environ Saf 137:256–264

Ma J, Zhai G (2012) Microbial bioremediation in omics era: opportunities and challenges. J Bioremed Biodegr 3:e120

Malla MA, Dubey A, Yadav S, Kumar A, Hashem A et al (2018) Understanding and designing the strategies for the microbe-mediated remediation of environmental contaminants using omics approaches. Front Microbiol 9:1132

Marasco EK, Schmidt-Dannert C (2008) Identification of bacterial carotenoid cleavage dioxygenase homologs that cleave the interphenyl α, β double bond of stilbene derivatives via a monooxygenase reaction. ChemBioChem 9(9):1450–1461

Masson L, Comeau Y, Brousseau R, Samson R, Greer C (1993) Construction and application of chromosomally integrated lac-lux gene markers to monitor the fate of a 2, 4-dichlorophenoxyacetic acid-degrading bacterium in contaminated soils. Microb Releases 1:209–216

Mayer B (ed) (2011) Bioinformatics for omics data: methods and protocols. Humana Press, New York (No. 57: 004 BIO)

Miller NT, Fuller D, Couger MB, Bagazinski M, Boyne P et al (2016) Draft genome sequence of Pseudomonas moraviensis strain devor implicates metabolic versatility and bioremediation potential. Genom Data 9:154–159
Moller A, Jansson JK (1998) Detection of firefly luciferase-tagged bacteria in environmental samples. In: Bioluminescence methods and protocols. Humana Press, New York, pp 269–283
Moreira IS, Amorim CL, Murphy CD, Castro PM (2018) Strategies for biodegradation of fluorinated compounds. In: Approaches in bioremediation. Springer, Cham, pp 239–280
Muffler A, Bettermann S, Haushalter M, Hörlein A, Neveling U et al (2002) Genome-wide transcription profiling of Corynebacterium glutamicum after heat shock and during growth on acetate and glucose. J Biotechnol 98(2–3):255–268
Mustapha HI, Lens PN (2018) Constructed wetlands to treat petroleum wastewater. In: Approaches in bioremediation. Springer, Cham, pp 199–237
Nagata Y, Ohtsubo Y, Tsuda M (2015) Properties and biotechnological applications of natural and engineered haloalkane dehalogenases. Appl Microbiol Biotechnol 99(23):9865–9881
Nikitina VE, Vetchinkina EP, Ponomareva EG, Gogoleva YV (2010) Phenol oxidase activity in bacteria of the genus Azospirillum. Microbiology 79(3):327–333
Otero-Blanca A, Folch-Mallol JL, Lira-Ruan V, Carbente MDRS, Batista-García RA (2018) Phytoremediation and fungi: an underexplored binomial. In: Approaches in bioremediation. Springer, Cham, pp 79–95
Paniagua-Michel J, Olmos-Soto J (2016) Modern approaches into biochemical and molecular biomarkers: key roles in environmental biotechnology. J Biotechnol Biomater 6(216):2
Pérez-Llano Y, Martínez-Ávila L, Batista-García RA (2018) Omics approaches: impact on bioremediation techniques. In: Approaches in bioremediation. Springer, Cham, pp 43–59
Plewniak F, Crognale S, Rossetti S, Bertin PN (2018) A genomic outlook on bioremediation: the case of arsenic removal. Front Microbiol 9:820
Rao MA, Scelza R, Scotti R, Gianfreda L (2010) Role of enzymes in the remediation of polluted environments. J Soil Sci Plant Nutr 10(3):333–353
Rao TE, Imchen M, Kumavath R (2017) Marine enzymes: production and applications for human health. Adv Food Nutr Res 80:149–163
Rawat M, Rangarajan S (2019) Omics approaches for elucidating molecular mechanisms of microbial bioremediation. In: Smart bioremediation technologies. Microbial enzymes. Elsevier, Amsterdam, pp 191–203
Rayu S, Karpouzas DG, Singh BK (2012) Emerging technologies in bioremediation: constraints and opportunities. Biodegradation 23(6):917–926
Rhee S-K, Liu X, Wu L, Chong SC, Wan X et al (2004) Detection of genes involved in biodegradation and biotransformation in microbial communities by using 50-Mer oligonucleotide microarrays. Appl Environ Microbiol 70(7):4303–4317
Robertson DG (2005) Metabonomics in toxicology: a review. Toxicol Sci 85(2):809–822
Santillan JY, Dettorre LA, Lewkowicz ES, Iribarren AM (2016) New and highly active microbial phosphotriesterase sources. FEMS Microbiol Lett 363(24):fnw276
Sarkar A (2006) Biomarkers of marine pollution and bioremediation. Ecotoxicology 15(4):331–332
Schneider MV, Orchard S (2011) Omics technologies, data and bioinformatics principles. In: Bioinformatics for omics data. Humana Press, New York, pp 3–30
Seibert V, Ebert MP, Buschmann T (2005) Advances in clinical cancer proteomics: SELDI-ToF-mass spectrometry and biomarker discovery. Brief Funct Genomics 4(1):16–26
Shah MP (2014) Environmental bioremediation: a low cost nature's natural biotechnology for environmental clean-up. J Pet Environ Biotechnol 5(4):1
Sharma B, Dangi AK, Shukla P (2018) Contemporary enzyme based technologies for bioremediation: a review. J Environ Manag 210:10–22
Siggins A, Gunnigle E, Abram F (2012) Exploring mixed microbial community functioning: recent advances in metaproteomics. FEMS Microbiol Ecol 80(2):265–280
Singh OV (2006) Proteomics and metabolomics: the molecular make-up of toxic aromatic pollutant bioremediation. Proteomics 6(20):5481–5492

Singh OV, Nagaraj NS (2006) Transcriptomics, proteomics and interactomics: unique approaches to track the insights of bioremediation. Brief Funct Genomics 4(4):355–362

Singh P, Singh V, Singh R, Borthakur A, Madhav S et al (2020) Bioremediation: a sustainable approach for management of environmental contaminants. In: Abatement of environmental pollutants trends and strategies, pp 1–23

Sood S, Sharma A, Sharma N, Kanwar SS (2016) Carboxylesterases: sources, characterization and broader applications. Insights Enzyme Res 1:1

Stephanopoulos GN, Aristidou AA, Nielsen J (1998) Metabolic engineering: principles and methodologies. Academic Press, San Diego

Tan B, Ng C, Nshimyimana JP, Loh LL, Gin KY-H et al (2015) Next-generation sequencing (NGS) for assessment of microbial water quality: current progress, challenges, and future opportunities. Front Microbiol 6:1027

Tresse O, Errampalli D, Kostrzynska M, Leung KT, Lee H et al (1998) Detection of nptII (kanamycin resistance) genes in genomes of transgenic plants by the marker-rescue transformation. Mol Gen Genet MGG 257(6):606–613

Urgun-Demirtas M, Stark B, Pagilla K (2006) Use of genetically engineered microorganisms (GEMs) for the bioremediation of contaminants. Crit Rev Biotechnol 26:145–164

Villas-Boas SG, Bruheim P (2007) The potential of metabolomics tools in bioremediation studies. Omics 11(3):305–313

Villas-Bôas SG, Rasmussen S, Lane GA (2005) Metabolomics or metabolit profiles? Trends Biotechnol 23(8):385–386

Wang DZ, Kong LF, Li YY, Xie ZX (2016) Environmental microbial community proteomics: status, challenges and perspectives. Int J Mol Sci 17(8):1275

Zhang W, Li F, Nie L (2010) Integrating multiple 'omics' analysis for microbial biology: application and methodologies. Microbiology 156(2):287–301

Ziganshina EE, Mohammed WS, Doijad SP, Shagimardanova EI, Gogoleva NE et al (2018) Draft genome sequence of Brevibacterium epidermidis EZ-K02 isolated from nitrocellulose-contaminated wastewater environments. Data Brief 17:119–123

Zouboulis AI, Moussas PA (2011) Groundwater and soil pollution: bioremediation. In: Encyclopedia of environmental health pages. Elsevier, Amsterdam, pp 1037–1044

# Harnessing the Potential of Modern Omics Tools in Plant Tissue Culture

6

Ranjana Gautam, Rajesh Kumar Meena, Gulab Khan Rohela, Naveen Kumar Singh, and Pawan Shukla

**Abstract**

Advancement in plant biology using high-throughput molecular tools has offered several exciting opportunities to gain new insight towards mechanisms controlling growth and developmental processes. In the past few years, omics technology has shown tremendous potential in the field of plant tissue culture. This chapter focuses on the adoption of omics-based techniques such as genomics, transcriptomics and proteomics to stimulate research interest in model plants and agriculturally important crops through tissue culture. It provides information about new candidate genes which encode for signalling and regulatory proteins that control early embryogenesis and morphogenesis. Omics has given an alternative mean for tissue culture to characterize the genomic data for marker development related to somaclonal variants. The large amount of data generated using these tools provide a quantum increase in knowledge related to differentiation, redifferentiation, somaclonal variation and somatic embryogenesis and lay foundation for analysing tissue culture samples at cellular and molecular levels.

Ranjana Gautam and Rajesh Kumar Meena contributed equally with all other contributors.

R. Gautam · R. K. Meena
Department of Plant Sciences, School of Life Sciences, University of Hyderabad, Hyderabad, Telangana, India

G. K. Rohela · P. Shukla (✉)
Central Sericultural Research and Training Institute, Pampore, Jammu and Kashmir, India

N. K. Singh
Department of Life Science, Central University of South Bihar, Gaya, Bihar, India





A. Kumar et al. (eds.), *Omics Technologies for Sustainable Agriculture and Global Food Security Volume 1*, https://doi.org/10.1007/978-981-16-0831-5_6

**Keywords**

Plant tissue culture (PTC) · Genomics · Proteomics · Transcriptomics · Somatic embryogenesis · Micropropagation

## 6.1 Introduction

The potential contributions of plant tissue culture (PTC) in the field of food security and crop improvement in the last few years are highly impressive. It has emerged as an indispensable tool for *in vitro* culture of plant cells, tissues and organs under aseptic and controlled conditions to obtain genetically true clones. PTC was first attempted by Australian botanist Gottlieb Haberlandt in 1902. For his remarkable contribution in *in vitro* culture of plants, Haberlandt is regarded as the father of PTC. This technique has been explored in different areas including the genetic crop improvement and production of natural products, obtaining high-quality disease-free plants, micropropagation, generation of transgenic plants by introgression of novel genes and preservation of germplasm (Gautam et al. 2019; Ge et al. 2017; Shukla et al. 2014). Steps involved during plant tissue culture technique are shown in Fig. 6.1. During tissue culture a favourable environment like suitable temperature, growth regulators and nutrient rich medium is required for the growth and multiplication of tissues and organs. Plant cells have unique potential of totipotency and development plasticity to differentiate, redifferentiate and then regenerating into a mature plant. Establishment of tissue culture protocol is influenced by genotype, origin of the explants, *in vitro* environmental factors (such as hormones and nutrients), light quality, temperature and photoperiod. Plant tissue culture products are providing model systems for studying changes at morphogenesis, developmental genetic changes and gene regulation. Culture of tissues/cells sometimes results in physiological, genetic and epigenetic problems especially recalcitrant, vitrification and somaclonal variation that affect plant growth adversely (Cassells and Curry 2001). Though much has been studied but still several insights are lacking which are required to understand the molecular developmental program involved during tissue culture. Therefore, the knowledge of 'omics' will favourably help in understanding the complex regulatory mechanisms' during developmental processes (organogenesis, embryogenesis and totipotency) and identification of specific genes/proteins/metabolites which regulate genetic changes (at cell and tissue level) in plants. Hence, in the present scenario of modern biotechnology application of 'omics approach' in plant cell, tissue and organ culture has emerged as a preeminent tool of biological science. Omics era includes modern and powerful techniques like genomics, transcriptomics and proteomics (Fig. 6.2). These techniques have been complementing the classical techniques to illustrate an accurate picture of the entire fundamental cellular and the molecular processes in plants. In this chapter, current

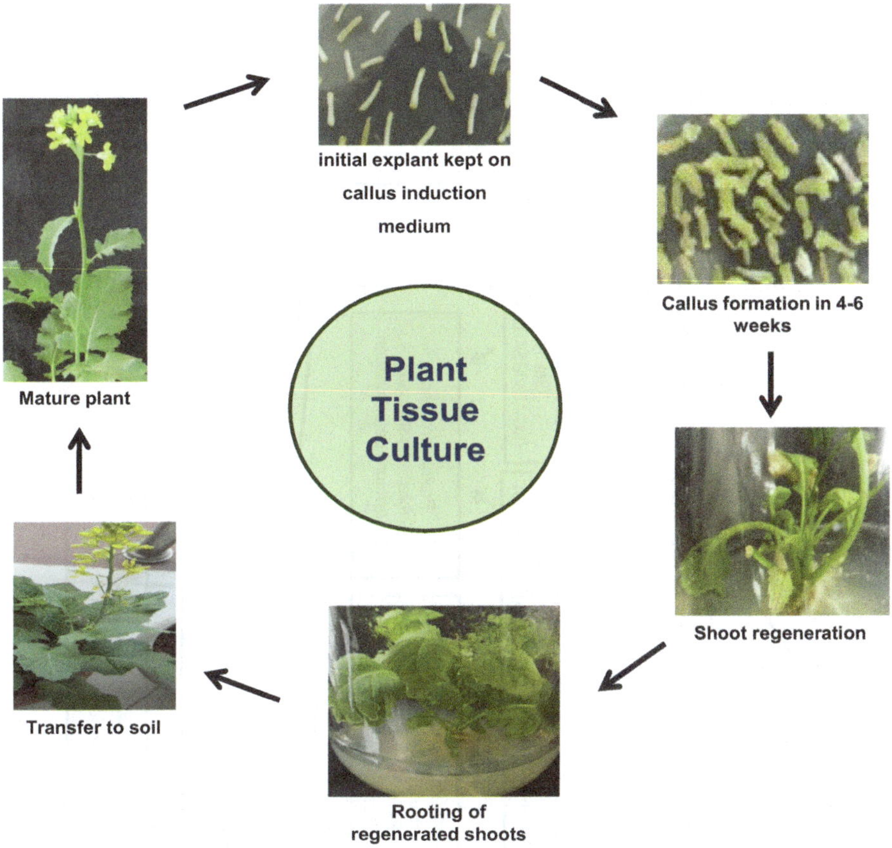


**Fig. 6.1** A representative diagram about different steps involved during plant tissue culture technique

status of the progress made towards plant tissue culture using 'omics-based technology' has been discussed.

## 6.2 Genomics and Progress in Plant Tissue Culture

The term genome includes the entire DNA present in an organism inclusive of its set of genes. Genomics aims to determine DNA sequences and performing genetic mapping of genes present in the genome. Genomic studies use high-throughput technologies like DNA sequencing, detection of mutant libraries and molecular markers based on DNA. These methods have the advantage that they help in substantial gene function and interaction analysis of gene products at cellular level (Domżalska et al. 2017; Zhou et al. 2013; Zhu et al. 2018). Recent sequencing programs of plant genomes facilitate in addressing the problems associated with

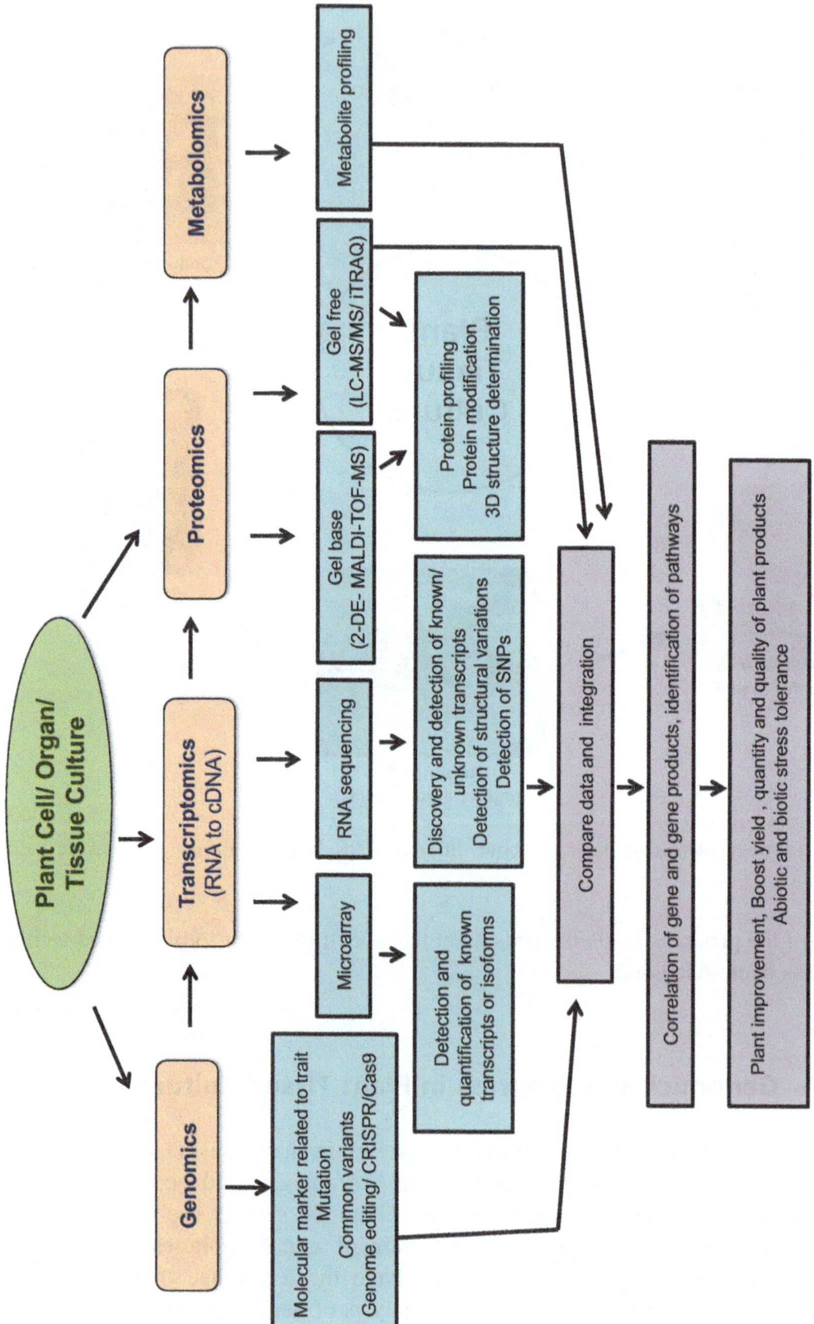


**Fig. 6.2** Application of different omic tools in plant tissue culture

epigenetic changes (somaclonal variation) and interpreting the functions of various genes at genetic level. The rapid accretion of high number of sequencing data and expressed sequence tags (ESTs) means that plenty of plant genes have been discovered so far, however still function of many genes needs to be established (Zhu et al. 2018). An *in vitro* culture condition induces dynamic changes at DNA level which cannot be observed but aid in facilitating the explants adaptation in culture conditions. These are associated with changes in chromosomal ploidy level, base substitution in DNA sequence and mutations (dos Santos et al. 2016).

## 6.3 Molecular Marker and QTL Mapping in Tissue Culture

Using molecular markers, the variations induced at morphological and physiological level between the source plant and original donor have been recorded well (Liu et al. 2016; Tan et al. 2013). Recently, Palama et al. (2010) using TMD (transposon methylation display) and metAFLP (modified methylation sensitive AFLP) reproducible markers have suggested that modifications at the nucleotide sequence and cytosine methylation are an important source of variation in *Arabidopsis* induced by tissue culture and transformation procedures. ISSR, SSR, RAPD have been widely used to identify the genetic relatedness among different genera and varieties of the plant species (Heringer et al. 2015; Nalousi et al. 2019). In muskmelon (*Cucumis melo* L.) by using the ISSR molecular markers the genetic homogeneity of the regenerants was assessed. It showed that regenerants generated by direct somatic embryogenesis (DSE) of muskmelon were 7% different as compared to mother (Kumaravel et al. 2017). RFLP analysis in an elite maize inbred has led to identification of chromosomal regions which are important for embryogenic initiation and regeneration during tissue culture (Tikendra et al. 2019). In flax (commonly known as *Linum usitatissimum* L.) use of RAPD, ISSR markers provides a cost-effective method to identify the microspore-derived plants as compared to plants raised from the tissues (somatic) of anther explants (Choi et al. 2019). In *Hordeum vulgare* L., 31 progeny lines were checked for the association of response of anther culture through molecular markers. Out of them three on chromosome 2H and 3H associated with regeneration rate, two on 2H and 4H with anther response and one on 4H linked with diploidization (Haoa et al. 2004). According to previous reports, by combining molecular marker and double haploids it is possible to do gene pyramiding, i.e. to pile up the resistance genes on top of one another (Wang et al. 2019). Some of the studies of plant tissue culture using molecular marker have been highlighted in Table 6.1.

QTL mapping has been conducted across various species like maize, cotton, rice and other crops to identify the callus formation and regeneration inducing genomic regions (Jung et al. 2008; Tikendra et al. 2019) which thus aid in improving the regeneration ability of tissues or cells by transferring the high regeneration ability possessing QTL genes into varieties with low response of regeneration. A *nitrite reductase* gene identified by QTL analysis was used to reduce the genetic hurdle in regeneration response of Koshihikari rice variety (Jung et al. 2008).

**Table 6.1** List of different types of molecular marker used in tissue culture responses in plants

| Marker type | Plant name | Phenotype (tissues) | Genotype (variety/ cultivar) | References |
|---|---|---|---|---|
| Inter-Simple Sequence Repeats (ISSR) | *Capsicum chinense* Jacq. | Direct organogenesis, direct and indirect somatic embryos and the embryogenic callus system | Registration No. 2367-chl-021-080110/c | Bello-Bello et al. (2014) |
| Amplified fragment length polymorphism (AFLP) and methylation sensitive amplified polymorphism (MSAP) | *Oryza sativa* L. | Calli | Japonica and indica | Wang et al. (2013) |
| Random amplified polymorphic DNA (RAPD) | Peach [*Prunus persica* (L.)] | Embryocallus | Batsch (Sunhigh) | Hashmi et al. (1997) |
| RAPD and ISSR | *Dendrobium chrysotoxum Lindl* | Micropropagation | Golden orchid | Tikendra et al. (2019) |
| ISSR | *Polianthes tuberosa* L. | Indirect shoot organogenesis | Pearl double | Nalousi et al. (2019) |
| Specific-locus amplified fragment sequencing (SLAF-seq) | *Oncidium* | Single nucleotide polymorphisms (SNPs) and insertion–deletions (InDels) identification in somaclonal variants | Milliongolds | Wang et al. (2019) |
| SSR marker | *Zea mays* L. | Double haploid | (HF1) and (11S6169) | Choi et al. (2019) |
| Methylation sensitive amplified polymorphism (MSAP) | Red Marsh | Somatic embryo | | Haoa et al. (2004) |
| RAPD and ISSR | *Musa acuminata* | Micropropagation | Nanjanagudu Rasabale (NR) | Lakshmanan et al. (2007) |
| AFLP markers | Potato (*Solanum tuberosum* L.) | Axillary-bud-proliferation, microtuberization and a novel somatic embryogenesis system | Desiree | Sharma et al. (2007) |
| RAPD and SSR | *Gossypium hirsutum* L. | Somatic embryogenesis | Coker 312 | Jin et al. (2008) |
| ISSR | *Brassica oleracea* L. | Somatic embryogenesis | Botrytis | Leroy et al. (2000) |

(continued)

**Table 6.1** (continued)

| Marker type | Plant name | Phenotype (tissues) | Genotype (variety/ cultivar) | References |
|---|---|---|---|---|
| RAPD and ISSR | *Linum usitatissimum* L. | Anther culture | AC McDuff, AC Emerson and their F1 hybrids | Chen et al. (1998) |
| RAPD, RFLP, Isozyme marker | *Brassica napus* | Double haploids | F1 hybrid obtained from cross of Darmor and Yudal | Foisset et al. (1996) |
| SSR | *Cucumis melo* L. | Ovary culture | F1 hybrid Jin Man Di | Malik et al. (2011) |
| Isozyme and microsatellite markers | *Pyrus communis* L. | Haploid and doubled haploid | 'Doyenné du Comice', 'Harrow Sweet' and 'Williams' | Bouvier et al. (2002) |

## 6.4 Mutagenesis and Genome Editing

Genetic mutagenesis has been the most efficiently used technique for characterizing the function of gene after sequencing of the genome which is just a part of the task. An effective way to characterize the mutants generated due to somaclonal variation is by using forward and reverse genetics. Among these insertional mutagenesis (T-DNA insertions and transposon or retrotransposons tagging) (Ram et al. 2019; Krysan et al. 1999), CRISPR-Cas9 (clustered regularly interspaced short palindromic repeats-associated Cas9 endonuclease) and TILLING (Targeting Induced Local Lesions IN Genomes) technology is presently widely used in crops to create mutant and knockouts which has resulted in boost of yield and improved agronomic traits (Viana et al. 2019; Xu et al. 2017). With CRISPR/Cas9 technology it has become more accessible to characterize the gene function by introducing precise mutations in the target gene(s). This technique has revolutionized the research due to its simplicity and specificity for studying the genome engineering and functional genomics of various crops.

CRISPR employs Cas9 endonuclease and a guide RNA complex (gRNA). The first 20 nucleotides of the target seed sequence of the guide RNA determine the site-specific target for the DNA recognition (Bouvier et al. 2002; Salvo et al. 2014). Protoplast, suspension culture, somatic embryogenesis have been widely used in producing transgenic plants harbouring the modifications of the target genes introduced by CRISPR/Cas9 technology (Gliwicka et al. 2013; Yakovlev et al. 2014; Zhang et al. 2017). This has proved to be valuable tool in the improvement of crops by considering the genetics and physiology of plants, study of developmental and biochemical pathways (Hashmi et al. 1997; Leroy et al. 2000).

The oilseed rape plants produced from hypocotyl by using CRISPR-Cas9 system with knocked out *alc* (*ALCATRAZ)* gene function produced siliques which could increase the level of shatter resistance, thus results in fewer seed loss during threshing (Bao et al. 2009). Mutagenesis of Phytoene desaturase (*PDS)* and PDR-type transporter (*PDR6)* genes in protoplast of *Nicotiana tabacum* was achieved by the application of CRISPR/Cas9. Transgenic plants with mutations in *PDS* and *PDR6* gene displayed etiolated leaves for the *pds* mutant and more branches for the *pdr6* mutant (Gao et al. 2015). Similarly, Klimek-Chodacka et al. (2018) targeted *flavanone-3-hydroxylase* (*F3H*) gene which is involved in anthocyanin biosynthesis pathway in purple-coloured callus and site directed editing resulted in the generation of *F3H* mutants that showed discoloration of callus. This helped in validating the functional significance of this gene in the anthocyanin biosynthesis along with serving as visual marker for examining successful edited events in carrot callus culture (Klimek-Chodacka et al. 2018). Combining CRISPR/Cas9 and microspore technology in wheat haploid cells, followed by regeneration of microspores into double haploid (DH) transgenic plants, demonstrated microspores as efficient explants by using *DsRed* gene and *TaLox2* and *TaUbiL1* (wheat) genes. This study investigated the factors that may affect the delivery of CRISPR/Cas9 into microspores and optimized microspores mutagenesis system in wheat genome to induce genetic modifications (Bhowmik et al. 2018). Ultimately, the studies involved the detection of the induced phenotypes by CRISPR/Cas9 technology within short span of time proves to be a revolutionary functional genomics tool in tissue culture by any other editing tool and powerful tool for genetic crop improvement.

## 6.5 Application of Transcriptomics for the Study of PTC

In recent years, global analysis of gene expression at the mRNA and proteomic level is providing a deeper insight into the *in vitro* plant regeneration in plants. Transcriptome consists of whole RNA (mRNA, tRNA, rRNA and noncoding RNA) molecules present in a cell. Transcriptome approach can be applied to an entire organism or to a specific cell. Although abundance of genomic data and marker studies are available for various developmental stages and agronomic traits of plant, still regulatory networks for crucial stages of plant development remain elusive. This tool allows the identification of genes which are switched on and off due to different stress responses in a cell. The fundamental interest behind this technique is to determine genes responsible for critical functions. Transcriptome sequencing represents the most straight forward approach that has been enforced to study the process of somatic embryogenesis and callus initiation in model systems, as well as in crops.

Plant cells have the ability to form pluripotent form of mass of cells known as callus from which new organs or whole plants are raised. Callus formation in plants is considered similar to the pluripotent stem cells of animals (iPS), whose induction is mediated by expression of a few transcription factors (Oct4, Nanog and Sox2) as plant and animal pluripotent stem cells show pluripotency characters. Callus formation from explants during *in vitro* condition is generally induced upon the media supplemented with hormonal combinations of auxin and cytokinin. The mechanism as to how these two plant hormones regulate the developmental fates is little known. Earlier work on the transcriptome profiling in different explants of *Arabidopsis* identified differentially expressed genes (DEGs) related with transcription, post-transcriptional regulation, hormone homeostasis and DNA-chromatin modification. This work helped to identify transcription factors like *CRF3* and *HB52*, which when overexpressed, formation of callus was obtained without supplementation of auxin hormone (Xu et al. 2012). A similar kind of approach was also performed in *Arabidopsis* seedlings (shoot and root explants) displaying the importance of transcription factor like *Lateral Organ Boundaries Domain* (*LBD*) in callus induction. Four genes *LBD16, LBD17, LBD18* and *LBD29* when overexpressed in the absence of callus induction medium were able to promote callus induction, whereas suppression of *LBD* gene inhibited callus induction. This showed that these genes are the regulators that mediate auxin signalling (Fan et al. 2012).

A transcriptome study of maize embryos, cultured on a medium containing 2,4,-dichlorophenoxyacetic acid for 0, 1, 2, 4, 6 and 8 days, identified genes related with signal transduction, iron ion binding, signal transmission. In this study *ZmBBM2*, an *AP2* transcription factor was instantly induced by auxin and its further characterization showed its role in callus induction and proliferation in maize (Du et al. 2019). Transcriptome analysis during the process of dedifferentiation and shoot regeneration through organogenesis was carried out in *Populus*, model woody species. It was found that genes encoding cell wall, mitochondria, organization of cell, ER and biogenetic genes were highly overexpressed during callus formation. Two F-box genes (involved in auxin signalling) were found to be differentially regulated during the process of callus induction and were closely related to transport inhibitor response 1 (TIR1) of *Arabidopsis* (Bao et al. 2009).

During tissue culture large transcriptomic studies have been generated and focussed on somatic embryogenesis (SE). SE is a process by which somatic embryos were formed from somatic cells either directly or indirectly due to the process of dedifferentiation with callus as an intermediate stage. Indirect process transits through different stages, viz. early pro-embryo, globular, heart-shape, torpedo shape and cotyledonary shape prior to regeneration. SE is usually bipolar and requires auxin as plant growth regulators but can also contain cytokinin in small amount. The regulation of embryogenesis in plants is identical to the developmental processes that occur through a series of intrinsic signals (hormones and transcription factors) playing a central role in giving up the information to new cells which are in the state of differentiation process (Long and Benfey 2006). Genes including late embryogenesis abundant (LEA) protein, WUSCHEL, MYB transcription factor and AGAMOUS have been reported previously for its key role in the formation of somatic embryos (Jamaluddin et al. 2017). Using transcriptome, numerous genes functions involved in somatic embryo development were identified in many plants such as *Arabidopsis* (Gliwicka et al. 2013), cotton (Jin et al. 2014), maize (Salvo et al. 2014), *Picea abies* (Dobrowolska et al. 2017) and papaya (Jamaluddin et al. 2017). Few examples of transcriptome application in different systems have been summarized in Table 6.2. Transcription factors involved in stress responses like WRKY and DREB families were identified as important ones in SE formation in *Arabidopsis* (Jin et al. 2014) whereas during ABA and NaCl treatment, SE development got induced and the level of gene transcripts related to stress responses in cotton is enhanced. In maize genes involved in hormone transport (Pinformed), transcription factors (Agamous, Boom and Leafy Cotyledon), receptor kinases (Clavata and Somatic Embryogenesis Receptor Kinases) were characterized to be the key genes responsible for onset of embryogenesis in immature embryo explants (Salvo et al. 2014). In recent times, complete picture of molecular events during *Arabidopsis* SE has been illustrated using Illumina HiSeq 2000 platform (Wickramasuriya and Dunwell 2015). This study described the gene expression in response to oxidative stress, osmotic stress and auxin signalling during SE and thereby provided the new insights in understanding the developmental mechanism underlying in SE formation.

In addition to this, the plants cultured *in vitro* encounter a range of deleterious responses which includes browning of explant. Explant browning leads to poor growth and can also cause death of explants. Browning damage use to be alleviated by supplements of media like active carbon, polyvinyl pyrrolidone, ascorbic acid, etc. This sometimes restricts the use of tissue culture in species and thus considerate the molecular mechanisms of the browning process are important. To understand the mechanism during explant browning, genome-wide transcriptome was carried out in Phalaenopsis explants. The results suggested that browning is due to the expression of genes that affect flavonoid biosynthesis and phenylpropanoid pathway as well as genes involved in ATPase activity and photosynthesis. These genes expression disturbs the energy metabolism, thus leading to impairment of development of plant tissues (Xu et al. 2015). Similarly, during transcriptomic analysis of *Medinilla formosana* explants obtained via ovary culture showed that genes involved in signal transduction and secondary metabolism were differentially expressed genes (DEGs). This study of morphological and anatomical observations also displayed that the browning of *M. formosana* was adaptive but not lethal (Wang et al. 2016).

The adoption of transcriptome has allowed researchers to look into the developmental mechanisms, signalling pathways governing the *in vitro* growth of cultured explants and tissues. These techniques are of greatest values not so much in tissue culture but rather in the identification of new transcripts on a genome-wide analysis to study their role underpinning the developmental mechanisms. The formation of tissues or callus is mediated by variable patterns of gene expression which results in changes in constituents of plant culture transcriptome, proteomes and metabolome.

**Table 6.2** Example of applications of transcriptomic technologies in plant tissue culture

| Species | Plant line | Sequencing platform | Gene name | Gene function | Phenotype | References |
|---|---|---|---|---|---|---|
| Maize | Inbred line A188 | Illumina HiSeq 2000 | Baby boom, leafy cotyledon and agamous | Involved in embryogenic pathway | Somatic embryogenesis | Salvo et al. (2014) |
| *Norway spruce* | | Illumina-based massive analysis | DNA-Cytosine Methyltransferase (DCM2) and Variant in Methylation (VIM1) | Catalysing methylation at CG context | Embryogenesis | Yakovlev et al. (2014) |
| | | | Histone methyltransferases | Histone modification | | |
| | | | Cysteine-rich polycomb-like protein 1 and Vernalization 1 | Epigenetic memory regulation | | |
| *Lilium pumilum* DC. Fisch. | | HiSeq 2000 platform | miR164 | Targets NAC (participating in the regulation of meristem differentiation, drought response) | Somatic embryogenesis | Zhang et al. (2017) |
| | | | miR159 | ABA-induced signalling | | |
| *Populus* | INRA 717-1 B4 | Microarray | F-box, Aux/IAA and ARF | Auxin signalling | Shoot organogenesis | Bao et al. (2009) |
| *Dimocarpus longan* Lour | Honghezi | Illumina paired-end sequencing | SWA1/2 | Gametogenesis in *Arabidopsis* | Embryogenic callus | Lai and Lin (2013) |
| | | | BEL1-LIKE | Cytokinin and auxin signalling during ovule development in *Arabidopsis* | | |
| | | | SWP (STRUWWE LPETER) | Duration of cell proliferation | | |
| *Arabidopsis thaliana* | Columbia-0 | Microarray | E3 genes (ASK3 and ASK18), E1 gene (UBA2) and E2 gene | Ubiquitin–proteasome pathways | Callus induction | Xu et al. (2012) |
| | | | IAA5 and IAA19 | Auxin signalling | | |
| | | | CKX5 and CKX3 | Cytokinin degradation | | |
| | | | ABI1 and PP2C-type phosphatises | Protein phosphorylation | | |
| | | | MYB112, MYB14, MYB63, MYB94 and At4g39160 | Cell development, hormone and environmental responses | | |

(continued)

**Table 6.2** (continued)

| Species | Plant line | Sequencing platform | Gene name | Gene function | Phenotype | References |
|---|---|---|---|---|---|---|
| Soybean (*Glycine max*) | Jack | Microarray | MADS box genes AGAMOUS-Like15 (GmAGL15 and GmAGL18) | Control of flowering time | Somatic embryogenesis | Zheng et al. (2013) |
| *Medinilla formosana* | Hayata | Illumina Hiseq 2000 platform | AUX1 | Auxin influx carrier protein | Browning | Wang et al. (2016) |
| | | | Aux/IAA, GH3 and small auxin-up RNAs (SAUR) | Primary auxin-related genes | | |
| | | | PFK-1 and HMGR | AMPK signalling pathway | | |
| *Arabidopsis thaliana* | Columbia | IG-CNS Illumina Hiseq2000 | BBM, AIL5 and SERK1 | Promote embryogenesis and organogenesis in the absence of exogenously applied growth regulators | SE, ground tissue and leaf primordial specification | Magnani et al. (2017) |
| *Carica papaya* L. | sekaki | Illumina HiSeq 2500 system | WUSCHEL, NAC, WRKY, MYB, Agamous-like MADS box protein and bHLH | Important role in somatic embryos in other crops | Embryogenic callus | Jamaluddin et al. (2017) |
| *Arabidopsis thaliana* | Col-0 | Microarray-based transcriptome analyses | AP2/EREBP, WRKY and NAC families | Biotic and abiotic stresses | Somatic embryogenesis | Gliwicka et al. (2013) |
| | | | WIND1 (wound induced dedifferentiation1) | Establishment and maintenance of the dedifferentiated status of somatic cells in the absence of exogenous growth hormones | | |
| | | | Monopteros (MP) | Auxin response factor | | |
| | | | Cytokinin response factors (CRFs) | Cytokinin regulatory genes | | |
| | | | ARF and AUX/IAA | Auxin signalling | | |
| | | | ERF | DNA binding | | |
| | | | WRKY | Biotic and abiotic stress, seed germination | | |

## 6.6 Proteomics in PTC

Rapid improvement in the genomic and transcriptomic technologies produces massive amount of nucleotide sequences in database. However, it is not sufficient to elucidate the biological functions of living cell due to limited correlation between genes and protein translation (Pandey and Mann 2000). Protein is the performer of gene expression, which regulates the biological function through several post-transcriptional and post-translational modification and/or protein–protein interaction in living system (Glisovic et al. 2008; Meena et al. 2018; Shukla et al. 2019). Therefore, the term proteomics designates the study of complete protein profile of a cell, tissue, organ and/or organism, often subjected to stress or developmental changes. Proteomics also accompanies the study of protein expression, modification and interaction with other proteins to regulate the several cellular and biological processes of living system (Kumar et al. 2014; Chin and Tan 2018). Moreover, due to technology improvement, several advanced tools have been developed for bioinformatics and computational science, which used to connect the proteomics with other omics technique and suggested the possible mode of signalling, regulatory and metabolic network of exclusive phenotype (Eldakak et al. 2013).

PTC involves several molecular events which participate in the transformation of explant/callus into fully developed plantlets. Various proteins and their interaction with other proteins regulate these molecular events. In the proteomic studies, a number of techniques, viz. nanoelectrospray ionization based liquid chromatography mass spectrometer (nESI LC-MS/MS), two-dimensional gel electrophoresis (2-DE) coupled with mass spectrometry (MS) and isobaric tags based relative and absolute quantitation (iTRAQ) were widely used to analyse the protein profile of cells, tissues and organs which are under *in vitro* culture studies (Ge et al. 2017; Jung et al. 2008). Proteomics is not only used to analyse the agronomically important crop like rice (Jung et al. 2008), wheat (Zhou et al. 2013), maize (Ge et al. 2017), cotton (Zhu et al. 2018), etc. but also implicated on trees fern *Araucaria angustifolia* (dos Santos et al. 2016) and *Cyathea delgadii* (Domżalska et al. 2017) tissue culture studies. Proteomics studies have been successfully used to identify the differentially expressed proteins (DEPs), which regulate the molecular events during different developmental stages and treatments in explant/callus during culture (Aguilar-Hernández and Loyola-Vargas 2018; Sharifi et al. 2012; Takáč et al. 2011; Meena et al. 2018).

Callus, mass of undifferentiated cells is an important phase of *in vitro* culture based plant regeneration and transformation (Tan et al. 2013; Liu et al. 2016). Proteomic study in rice (Yin et al. 2007, 2008), vanilla (Tan et al. 2013) and Lotus (Liu et al. 2016) on differentiation and callus formation identified the DEPs which regulate the process of callus formation. Tan et al. (2013) and Liu et al. (2016) have identified the most of the DEPs are involved in metabolism, stress and redox reaction, proteins related to cell and cell wall. These results suggested that DEPs might regulate the metabolism reprograming, ROS level and rapid cell division and proliferation during callus formation. Another proteomic study on *Vanilla planifolia* revealed that out of 15 significantly expressed proteins, mostly belong to amino acid-protein metabolism and photosynthesis during initial differentiation of shoots from protocorm callus (Palama et al. 2010). This investigation suggested that during induction of shoot differentiation, cell metabolism is stimulated at three levels: (1) photosynthetic, glycolytic compound synthesis, (2) sugar degradation and (3) synthesis of amino acid and proteins and their stabilization (Palama et al. 2010).

Somatic embryogenesis (SE) is a process of producing large amount of somatic embryo from somatic cells and it is an analogous to zygotic embryogenesis (Ji et al. 2011). Advanced proteomics-based approaches like 2-DE and iTRAQ provide an advanced tool to investigate the molecular mechanism of this morphogenetic route. Several proteomic studies have carried out on SE analysis in numerous plant species, viz. *Boesenbergia rotunda, Citrus sinensis Osbeck, Coffea arabica, Crocus sativus, Cucumis sativus, Cyclamen persicum, Daucus carota, Gossypium hirsutum, Medicago truncatula, Musa* spp., *Picea glauca, Vitis vinifera, Saccharum* spp. and *Zea mays* (Heringer et al. 2015; Kumaravel et al. 2017; Ge et al. 2017; Zhu et al. 2018) (Table 6.3). For example, Kumaravel et al. (2017) compared the protein profile of both non-embryogenic callus (NEC) and embryogenic callus (EC) of banana and found that controlled oxidative stress and plant growth regulators like adenylate isopentenyl transferase and indole-3-pyruvate monooxygenase are correlated with induction of SE in embryogenic callus. iTRAQ-based comparative proteomic analysis also provides evidences that signalling, glycolysis, plant hormone transduction, biosynthesis and metabolism of fatty acids related proteins were differentially expressed in cotton during SE compared to EC and NEC and regulate development of SE (Zhu et al. 2018). Further, research could be focussed on the functional validation of the identified proteins via proteomics and could open up the possibility to use these proteins as markers of different stages of plant tissue culture. This would help in uncovering and elucidating the mechanism as to how a single cell becomes a complete plant.

## 6.7 Metabolomics in Plant Tissue Culture

In the recent times, during the last two decades plant tissue culture has also gained its importance in metabolomics field to understand and trace out the metabolites of a specific metabolic pathway in plants (Kumar et al. 2017; Sharma et al. 2018). Unlike proteomics and genomics, in metabolomics the molecules or metabolites (<1.5 kDa) of specific metabolic pathway are studied with regard to their concentration, chemical and physical properties such as polarity, solubility, structural similarity and functional aspects (Kuehnbaum and Britz-McKibbin 2013). Molecular farming through tissue culture or cell suspension cultures-based production of commercially important secondary metabolites has gained rapid interest during recent times. Cell suspension cultures in combination with metabolomics can make and better understand the metabolic profile of a specific plant cell or organ (Oliveira et al. 2018). Plant tissue culture in combination with metabolomics can easily understand the metabolome of a plant species, i.e. identifying the complete set of molecules which includes signalling molecules, primary metabolites, plant growth hormones, intermediates, secondary metabolites and final products of metabolic pathways (De Luca and St Pierre 2000; Oliver et al. 1998).

Advanced computational tools based on metabolomics studies have also enabled to study the tissue specific metabolite diversity under control and under the influence

**Table 6.3** Example of applications of proteomic approaches in tissue culture

| S. No. | Species (common name) | Explant/tissue culture approach | Proteomic technique | Identified proteins | Regulated function | References |
|---|---|---|---|---|---|---|
| 1 | *Vanilla planifolia* | Nodal segment/callus | 2-DE | 23 | Defence and stress response, metabolism, protein synthesis | Tan et al. (2013) |
| 2 | *Oryza sativa* | Mature seed/callus | 2-DE | 60 | Carbohydrate metabolism, stress/defence | Yin et al. (2007) |
| 3 | *Oryza sativa* | Mature seed/callus | 2-DE | 157 | Cellular/metabolic process, amino acid biosynthesis, stress/defence | Yin et al. (2008) |
| 4 | *Citrus sinensis* | Embryogenic callus/SE | 2-DE | 24 | Glutathione (GSH) metabolism and anti-oxidative stress | Pan et al. (2009) |
| 5 | *Cyclamen persicum* | Somatic/zygotic embryo and non-embryogenic callus/SE | 2-DE-MALDI-TOF-TOF-MS | 35 | Metabolism/energy | Bian et al. (2010) |
| 6 | *Manihot esculenta* | Shoot apical meristem and immature leaf lobe/SE | LC-ESI-MS/MS | 383 | Carbohydrate/energy metabolism, protein biosynthesis, photosynthesis | Li et al. (2010) |
| 7 | *Vanilla planifolia* | Protocorm/callus | 2-DE-MALDI-TOF-TOF-MS | 15 | Amino acid metabolism, photosynthetic activity | Palama et al. (2010) |
| 8 | *Zea mays* (maize) | Immature embryo/callus | iTRAQ | 616 | Cell differentiation, division, communication and hormone biosynthesis | Ge et al. (2017) |
| 9 | *Saccharum spp.* (sugarcane) | Shoot apical meristems/SE | Nano UPLC-nano ESI-HDMS | 752 | Stress/defence, protein metabolism, cellular metabolism | Heringer et al. (2015) |
| 10 | *Vriesea reitzii* (Bromeliaceae) | Leaf basal region/nodular cluster cultures | 2-DE | 23 | Cellular/metabolic process, stress response | Corredor-Prado et al. (2019) |
| 11 | *Musa spp.* (banana) | Male flower buds/SE | 2-DE | 58 | Transport, stress/defence, growth and development | Kumaravel et al. (2017) |

of different stress (biotic and abiotic) factors (Li et al. 2016). This computational metabolomics studies will help in identifying specific tissues and suitable stress factors which can induce the production of commercial metabolites in large quantities among the plant species of several families. By using computational metabolomics in combination with tandem mass spectroscopy and information theory analysis, tissue specific metabolite profile was elucidated in *Nicotiana attenuata* (Li et al. 2016). *In vitro* culture based metabolomics studies could identify the specific metabolites like metabolites of glucosinolates synthesis in Brassicaceae family (Zang et al. 2009). In model plant *Arabidopsis* several 'omics' based studies such as tissue specific expression, storage of secondary metabolites in specific tissues, targeted degradation of proteins and proteins involved in alternate splicing mechanisms were successfully identified (Hirai et al. 2007; Rajniak et al. 2015; Sakurai et al. 2013). Further, metabolomics research could help in identifying specific metabolites which could enable us to identify different characteristics of a plant species such as shelf-life of fruits or grains, content of aroma compounds in aromatic plants under diversified conditions, nutrient contents of food crops, etc.

Several metabolomics studies were reported regarding the altered metabolic profile of transgenic callus or mutant lines in comparison to control or wild plants (Kumari et al. 2017). For instance, the overexpression of gene coding for *MdMYBA* transcription factor in mutant and transgenic lines was reported to produce high contents of anthocyanin pigments in flowers and fruits of mutants/compared to wild type (Ban et al. 2007; Kumar et al. 2017). There is a report of upregulation of MdMYB10 transcription factor which has reflected in high content of anthocyanin in the transformed callus and all the parts of transgenic apple plants compared to the wild cultivar type (Espley et al. 2007).

The molecular events that led the explant to undergo transformation for the formation of callus have been deduced recently in several plant species along with the comparative metabolic profiles of callus induced from the wild type and mutant lines (Fan et al. 2012; Ikeuchi et al. 2013; Sugiyama 2015). Similarly, metabolomics studies were conducted to identify the metabolic drift during deformation or sluggish growth of callus of mutant lines compared to wild types of different plant species. In tomato, cotyledonary callus of mutant lines (*pct1–2* and *shr*) showed differential expression compared to wild type plants (Kumari et al. 2017). Metabolites of glutamine and sucrose were reported to be differentially expressed in both the mutant lines compared to wild type. Trehalose carbohydrate was reported as overexpressed in the cotyledons of *pct1–2*, whereas sugars of talofuranose, β-D-glucopyranose, galactonate, tagatose, ribose, myo-inositol and galactoglycerol were overexpressed in cotyledons of *shr* mutant lines in comparison with wild type tomato plants during the transformation phase of cotyledon explants to callus (Kumari et al. 2017). Differential expression of sugars and amino acids was observed among the non-embryogenic and embryogenic callus of sugarcane (Mahmud et al. 2014).

Advancements made in *in vitro* based secondary metabolite production are highly helpful in producing plant-based healthcare products for the betterment of human life (Marchev et al. 2020). Advanced techniques of HPLC, NMR, GC-MS, MALTI-TOF-MS, ESI-MS, LC-MS, MS-MS were utilized in the *in vitro* culture based metabolomics studies (Table 6.4) of several plant species (Fischedick et al. 2015;

**Table 6.4** Plant metabolites identified by various spectroscopic techniques

| S. No. | Species (common name) | Explant/plant part | Spectroscopic technique | Identified metabolite | References |
|---|---|---|---|---|---|
| 1 | *Verbascum nigrum* L. | Hairy root culture | NMR | γ-Aminobutyric acid (GABA) | Georgiev et al. (2015) |
| 2 | *Verbascum eriophorum* Godr. | Hairy root culture | NMR | Verbascoside, Martynoside | Marchev et al. (2016) |
| 3 | *Arabidopsis thaliana* | Seed | LC-MS | Flavonoids' | Routaboul et al. (2012) |
| 4 | *Arabidopsis thaliana* | Leaf/seedling | LC-MS | Glucosinolates | Chan et al. (2011) |
| 5 | *Oryzae spp.* | Leaf | LC-MS | Phenolamides | Dong et al. 2015 |
| 6 | *Oryzae spp.* | Seed | LC-MS | Branched chain amino acids | Angelovici et al. (2013) |
| 7 | *Oryzae spp.* | Leaf | LC–Q-TOF-MS | Secondary metabolites | Matsuda et al. (2015) |
| 8 | *Oryzae spp.* | Seed | LC–Q-TOF-MS | Metabolome | Matsuda et al. (2012) |
| 9 | *Arabidopsis thaliana* | Leaf | GC–TOF-MS | Metabolome | Rowe et al. (2008) |
| 10 | *Arabidopsis thaliana* | Seedling | GC–TOF-MS | Metabolome | Lisec et al. (2009) |
| 11 | *Solanum lycopersicum* | Fruit | GC-MS | Primary metabolites | Schauer et al. (2008) |
| 12 | *Solanum tuberosum* | Tuber | GC-MS | Primary metabolites | Carreno-Quintero et al. (2012) |
| 13 | *Zea mays* | Leaf | GC-MS | Metabolome | Riedelsheimer et al. (2012) |
| 14 | *Triticum aestivum* | Leaf | GC-MS | Metabolome | Hill et al. (2013) |

Georgiev et al. 2015). Further, there is a need to utilize the advanced techniques of CRISPR/Cas9 for the metabolic engineering of medicinally important plant species through genome editing either by insertion, deletion, overexpression or suppression of specific structural genes coding for transcription factors/regulatory proteins resulting in expression profile of a specific metabolite in a desired way for providing the health and commercial benefits to mankind.

## 6.8 Future Perspectives

Omics represents as the most promising tool in the area of *in vitro* plant tissue culture giving perspective of the future. The application of 'omics' techniques has been used to understand the complexities of the developmental processes during micropropagation and somatic embryogenesis. Omics comprehensive nature offers an exclusively new pathway and research programs that should adapt accordingly. The combination of different omics has facilitated the group of gene identification which are involved in regulating the different steps of plant tissue culture. The integration of genomics, proteomics, metabolomics and transcriptomic studies data should be used to prepare mathematical models that can explain the problems associated with *in vitro* regeneration systems at different developmental stages. Further applications of these techniques will provide insights to understand the molecular mechanisms that control individual traits. It might help us to look deeper into the biology and physiology of the cultured explants, thus paving the way for sustainable agriculture by development of more robust plant tissue culture practices.

**Acknowledgements** The authors thank Anita Kumari, Research scholar, Department of biochemistry, UoH for fruitful discussion and helpful comments while writing the chapter.

**Conflict of Interest** The authors have no competing interest to declare.

## References

Aguilar-Hernández V, Loyola-Vargas VM (2018) Advanced proteomic approaches to elucidate somatic embryogenesis. Front Plant Sci 9:1658

Angelovici R, Lipka AE, Deason N, Gonzalez-Jorge S, Lin H, Cepela J, Buell R, Gore MA, DellaPenna D (2013) Genome-wide analysis of branched-chain amino acid levels in *Arabidopsis* seeds. Plant Cell 25:4827–4843

Ban Y, Honda C, Hatsuyama Y, Igarashi M, Bessho H, Moriguchi T (2007) Isolation and functional analysis of a MYB transcription factor gene that is a key regulator for the development of red coloration in apple skin. Plant Cell Physiol 48:958–970

Bao Y, Dharmawardhana P, Mockler TC, Strauss SH (2009) Genome scale transcriptome analysis of shoot organogenesis in populus. BMC Plant Biol 9:132

Bello-Bello JJ, Iglesias-Andreu LG, Avilés-Vinas SA, Gómez-Uc E, Canto-Flick A, Santana-Buzzy N (2014) Somaclonal variation in habanero pepper (*Capsicum chinense* Jacq.) as assessed ISSR molecular markers. Hort Sci 49:481–485

Bhowmik P, Ellison E, Polley B, Bollina V, Kulkarni M, Ghanbarnia K, Song H, Gao C, Voytas DF, Kagale S (2018) Targeted mutagenesis in wheat microspores using CRISPR/Cas9. Sci Rep 8:1–10

Bian F, Zheng C, Qu F, Gong X, You C (2010) Proteomic analysis of somatic embryogenesis in *Cyclamen persicum* Mill. Plant Mol Biol Rep 28:22–31

Bouvier L, Guerif P, Djulbic M, Durel C-E, Chevreau E, Lespinasse Y (2002) Chromosome doubling of pear haploid plants and homozygosity assessment using isozyme and microsatellite markers. Euphytica 123:255–262

Carreno-Quintero N, Acharjee A, Maliepaard C, Bachem CW, Mumm R, Bouwmeester H, Visser RG, Keurentjes JJ (2012) Untargeted metabolic quantitative trait loci analyses reveal a relationship between primary metabolism and potato tuber quality. Plant Physiol 158:1306–1318

Cassells AC, Curry RF (2001) Oxidative stress and physiological, epigenetic and genetic variability in plant tissue culture: implications for micropropagators and genetic engineers. Plant Cell Tiss Org Cult 64:145–157

Chan EK, Rowe HC, Corwin JA, Joseph B, Kliebenstein DJ (2011) Combining genome-wide association mapping and transcriptional networks to identify novel genes controlling glucosinolates in *Arabidopsis thaliana*. PLoS Biol 9:e1001125

Chen Y, Hausner G, Kenaschuk E, Procunier D, Dribnenki P, Penner G (1998) Identification of microspore-derived plants in anther culture of flax (*Linum usitatissimum* L.) using molecular markers. Plant Cell Rep 18:44–48

Chin CF, Tan HS (2018) The use of proteomic tools to address challenges faced in clonal propagation of tropical crops through somatic embryogenesis. Proteomes 6:21

Choi J-K, Sa KJ, Park DH, Lim SE, Ryu S-H, Park JY, Park KJ, Rhee H-I, Lee M, Lee JK (2019) Construction of genetic linkage map and identification of QTLs related to agronomic traits in DH population of maize (*Zea mays* L.) using SSR markers. Genes Genomics 41:667–678

Corredor-Prado JP, De Conti D, Júnior DR, Cangahuala-Inocente GC, Guerra MP, Dal Vesco LL, Pescador R (2019) Proteomic identification of differentially altered proteins during regeneration from nodular cluster cultures in *Vriesea reitzii* (Bromeliaceae). J Plant Growth Regul 38:586–599

De Luca V, St Pierre B (2000) The cell and developmental biology of alkaloid biosynthesis. Trends Plant Sci 5:168–173

Dobrowolska I, Businge E, Abreu IN, Moritz T, Egertsdotter U (2017) Metabolome and transcriptome profiling reveal new insights into somatic embryo germination in Norway spruce (*Picea abies*). Tree Physiol 37:1752–1766

Domżalska L, Kędracka-Krok S, Jankowska U, Grzyb M, Sobczak M, Rybczyński JJ, Mikuła A (2017) Proteomic analysis of stipe explants reveals differentially expressed proteins involved in early direct somatic embryogenesis of the tree fern *Cyathea delgadii* Sternb. Plant Sci 258:61–76

Dong X, Gao Y, Chen W, Wang W, Gong L, Liu X, Luo J (2015) Spatiotemporal distribution of phenolamides and the genetics of natural variation of hydroxycinnamoyl spermidine in rice. Mol Plant 8:111–121

dos Santos ALW, Elbl P, Navarro BV, de Oliveira LF, Salvato F, Balbuena TS, Floh EIS (2016) Quantitative proteomic analysis of *Araucaria angustifolia* (Bertol.) Kuntze cell lines with contrasting embryogenic potential. J Proteomics 130:180–189

Du X, Fang T, Liu Y, Huang L, Zang M, Wang G, Liu Y, Fu J (2019) Transcriptome profiling predicts new genes to promote maize callus formation and transformation. Front Plant Sci 10:1633

Eldakak M, Milad SI, Nawar AI, Rohila JS (2013) Proteomics: a biotechnology tool for crop improvement. Front Plant Sci 4:35

Espley RV, Hellens RP, Putterill J, Stevenson DE, Kutty-Amma S, Allan AC (2007) Red colouration in apple fruit is due to the activity of the MYB transcription factor, MdMYB10. Plant J 49:414–427

Fan M, Xu C, Xu K, Hu Y (2012) Lateral organ boundaries domain transcription factors direct callus formation in *Arabidopsis* regeneration. Cell Res 22:1169–1180

Fischedick JT, Johnson SR, Ketchum RE, Croteau RB, Lange BM (2015) NMR spectroscopic search module for Spektraris, an online resource for plant natural product identification–taxane diterpenoids from Taxus× media cell suspension cultures as a case study. Phytochemistry 113:87–95

Foisset N, Delourme R, Barret P, Hubert N, Landry B, Renard M (1996) Molecular-mapping analysis in *Brassica napus* using isozyme, RAPD and RFLP markers on a doubled-haploid progeny. Theor Appl Genet 93:1017–1025

Gao J, Wang G, Ma S, Xie X, Wu X, Zhang X, Wu Y, Zhao P, Xia Q (2015) CRISPR/Cas9-mediated targeted mutagenesis in *Nicotiana tabacum*. Plant Mol Biol 87:99–110

Gautam R, Shukla P, Kirti P (2019) Targeted expression of a cysteine protease (AdCP) in tapetum induces male sterility in Indian mustard, *Brassica juncea*. Funct Integr Genomics 19:703–714

Ge F, Hu H, Huang X, Zhang Y, Wang Y, Li Z, Zou C, Peng H, Li L, Gao S (2017) Metabolomic and proteomic analysis of maize embryonic callus induced from immature embryo. Sci Rep 7:1–16

Georgiev MI, Radziszewska A, Neumann M, Marchev A, Alipieva K, Ludwig-Müller J (2015) Metabolic alterations of *Verbascum nigrum* L. plants and SAArT transformed roots as revealed by NMR-based metabolomics. Plant Cell Tissue Organ Culture (PCTOC) 123:349–356

Glisovic T, Bachorik JL, Yong J, Dreyfuss G (2008) RNA-binding proteins and post-transcriptional gene regulation. FEBS Lett 582:1977–1986

Gliwicka M, Nowak K, Balazadeh S, Mueller-Roeber B, Gaj MD (2013) Extensive modulation of the transcription factor transcriptome during somatic embryogenesis in *Arabidopsis thaliana*. PLoS One 8:e69261

Haoa Y-J, Wen X-P, Deng X-X (2004) Genetic and epigenetic evaluations of citrus calluses recovered from slow-growth culture. J Plant Physiol 161:479–484

Hashmi G, Huettel R, Meyer R, Krusberg L, Hammerschlag F (1997) RAPD analysis of somaclonal variants derived from embryo callus cultures of peach. Plant Cell Rep 16:624–627

Heringer AS, Barroso T, Macedo AF, Santa-Catarina C, Souza GHMF, Floh EIS, de Souza-Filho GA, Silveira V (2015) Label-free quantitative proteomics of embryogenic and non-embryogenic callus during sugarcane somatic embryogenesis. PLoS One 10:e0127803

Hill CB, Taylor JD, Edwards J, Mather D, Bacic A, Langridge P, Roessner U (2013) Whole-genome mapping of agronomic and metabolic traits to identify novel quantitative trait loci in bread wheat grown in a water-limited environment. Plant Physiol 162:1266–1281

Hirai MY, Sugiyama K, Sawada Y, Tohge T, Obayashi T, Suzuki A, Araki R, Sakurai N, Suzuki H, Aoki K (2007) Omics-based identification of *Arabidopsis* Myb transcription factors regulating aliphatic glucosinolate biosynthesis. Proc Natl Acad Sci 104:6478–6483

Ikeuchi M, Sugimoto K, Iwase A (2013) Plant callus: mechanisms of induction and repression. Plant Cell 25:3159–3173

Jamaluddin ND, Noor NM, Goh H-H (2017) Genome-wide transcriptome profiling of *Carica papaya* L. embryogenic callus. Physiol Mol Biol Plan Theor 23:357–368

Ji A, Geng X, Zhang Y, Wu G (2011) Advances in somatic embryogenesis research of horticultural plants. Am J Plant Sci 2:727

Jin S, Mushke R, Zhu H, Tu L, Lin Z, Zhang Y, Zhang X (2008) Detection of somaclonal variation of cotton (*Gossypium hirsutum*) using cytogenetics, flow cytometry and molecular markers. Plant Cell Rep 27:1303–1316

Jin F, Hu L, Yuan D, Xu J, Gao W, He L, Yang X, Zhang X (2014) Comparative transcriptome analysis between somatic embryos (SE s) and zygotic embryos in cotton: evidence for stress response functions in SE development. Plant Biotechnol J 12:161–173

Jung Y-H, Jeong S-H, Kim SH, Singh R, Lee J-e, Cho Y-S, Agrawal GK, Rakwal R, Jwa N-S (2008) Systematic secretome analyses of rice leaf and seed callus suspension-cultured cells: workflow development and establishment of high-density two-dimensional gel reference maps. J Proteome Res 7:5187–5210

Klimek-Chodacka M, Oleszkiewicz T, Lowder LG, Qi Y, Baranski R (2018) Efficient CRISPR/Cas9-based genome editing in carrot cells. Plant Cell Rep 37:575–586
Krysan PJ, Young JC, Sussman MR (1999) T-DNA as an insertional mutagen in Arabidopsis. Plant Cell 11:2283–2290
Kuehnbaum NL, Britz-McKibbin P (2013) New advances in separation science for metabolomics: resolving chemical diversity in a post-genomic era. Chem Rev 113:2437–2468
Kumar A, Bimolata W, Kannan M, Kirti PB, Qureshi IA, Ghazi IA (2014) Comparative proteomics reveals differential induction of both biotic and abiotic stress response associated proteins in rice during *Xanthomonas oryzae* pv. oryzae infection. Funct Integr Genomics 15(4):425–437
Kumar R, Bohra A, Pandey AK, Pandey MK, Kumar A (2017) Metabolomics for plant improvement: status and prospects. Front Plant Sci 8:1302
Kumaravel M, Uma S, Backiyarani S, Saraswathi MS, Vaganan MM, Muthusamy M, Sajith KP (2017) Differential proteome analysis during early somatic embryogenesis in Musa spp. AAA cv. Grand Naine. Plant Cell Rep 36:163–178
Kumari A, Ray K, Sadhna S, Pandey AK, Sreelakshmi Y, Sharma R (2017) Metabolomic homeostasis shifts after callus formation and shoot regeneration in tomato. PLoS One 12: e0176978
Lai Z, Lin Y (2013) Analysis of the global transcriptome of longan (*Dimocarpus longan* Lour.) embryogenic callus using Illumina paired-end sequencing. BMC Genomics 14:561
Lakshmanan V, Reddampalli Venkataramareddy S, Neelwarne B (2007) Molecular analysis of genetic stability in long-term micropropagated shoots of banana using RAPD and ISSR markers. Electron J Biotechnol 10:106–113
Leroy X, Leon K, Charles G, Branchard M (2000) Cauliflower somatic embryogenesis and analysis of regenerant stability by ISSRs. Plant Cell Rep 19:1102–1107
Li K, Zhu W, Zeng K, Zhang Z, Ye J, Ou W, Rehman S, Heuer B, Chen S (2010) Proteome characterization of cassava (*Manihot esculenta* Crantz) somatic embryos, plantlets and tuberous roots. Proteome Sci 8:10
Li D, Heiling S, Baldwin IT, Gaquerel E (2016) Illuminating a plant's tissue-specific metabolic diversity using computational metabolomics and information theory. Proc Natl Acad Sci 113: E7610–E7618
Lisec J, Steinfath M, Meyer RC, Selbig J, Melchinger AE, Willmitzer L, Altmann T (2009) Identification of heterotic metabolite QTL in *Arabidopsis thaliana* RIL and IL populations. Plant J 59:777–788
Liu Y, Chaturvedi P, Fu J, Cai Q, Weckwerth W, Yang P (2016) Induction and quantitative proteomic analysis of cell dedifferentiation during callus formation of lotus (*Nelumbo nucifera* Gaertn. spp. baijianlian). J Proteomics 131:61–70
Long TA, Benfey PN (2006) Transcription factors and hormones: new insights into plant cell differentiation. Curr Opin Cell Biol 18:710–714
Magnani E, Jiménez-Gómez J, Soubigou-Taconnat L, Lepiniec L, Fiume E (2017) Profiling the onset of somatic embryogenesis in Arabidopsis. BMC Genomics 18:1–12
Mahmud I, Thapaliya M, Boroujerdi A, Chowdhury K (2014) NMR-based metabolomics study of the biochemical relationship between sugarcane callus tissues and their respective nutrient culture media. Anal Bioanal Chem 406:5997–6005
Malik A, Li C, Shuxia Z, Jin-feng C (2011) Efficiency of SSR markers for determining the origin of melon plantlets derived through unfertilized ovary culture. Hortic Sci 38:27–34
Marchev A, Yordanova Z, Alipieva K, Zahmanov G, Rusinova-Videva S, Kapchina-Toteva V, Simova S, Popova M, Georgiev MI (2016) Genetic transformation of rare *Verbascum eriophorum* Godr. plants and metabolic alterations revealed by NMR-based metabolomics. Biotechnol Lett 38:1621–1629
Marchev AS, Yordanova ZP, Georgiev MI (2020) Green (cell) factories for advanced production of plant secondary metabolites. Crit Rev Biotechnol 40:443–458

Matsuda F, Okazaki Y, Oikawa A, Kusano M, Nakabayashi R, Kikuchi J, Yonemaru JI, Ebana K, Yano M, Saito K (2012) Dissection of genotype–phenotype associations in rice grains using metabolome quantitative trait loci analysis. Plant J 70:624–636

Matsuda F, Nakabayashi R, Yang Z, Okazaki Y, Ji Y, Ebana K, Yano M, Saito K (2015) Metabolome-genome-wide association study dissects genetic architecture for generating natural variation in rice secondary metabolism. Plant J 81:13–23

Meena RK, Pullaiahgari D, Gudipalli P (2018) Proteomic analysis of heterotic seed germination in maize using F1 hybrid DHM 117 and its parental inbreds. Turk J Biol 42:352–370

Nalousi AM, Hatamzadeh A, Azadi P, Mohsenpour M, Lahiji HS (2019) A procedure for indirect shoot organogenesis of *Polianthes tuberosa* L. and analysis of genetic stability using ISSR markers in regenerated plants. Sci Hortic 244:315–321

Oliveira J, Hakimi O, Murgu M, Koblitz M, Ferreira M, Cameron L, Macedo A (2018) Tissue culture and metabolome investigation of a wild endangered medicinal plant using high definition mass spectrometry. Plant Cell Tissue Organ Culture (PCTOC) 134:153–162

Oliver SG, Winson MK, Kell DB, Baganz F (1998) Systematic functional analysis of the yeast genome. Trends Biotechnol 16:373–378

Palama TL, Menard P, Fock I, Choi YH, Bourdon E, Govinden-Soulange J, Bahut M, Payet B, Verpoorte R, Kodja H (2010) Shoot differentiation from protocorm callus cultures of *Vanilla planifolia* (Orchidaceae): proteomic and metabolic responses at early stage. BMC Plant Biol 10:82

Pan Z, Guan R, Zhu S, Deng X (2009) Proteomic analysis of somatic embryogenesis in Valencia sweet orange (*Citrus sinensis* Osbeck). Plant Cell Rep 28:281–289

Pandey A, Mann M (2000) Proteomics to study genes and genomes. Nature 405:837–846

Rajniak J, Barco B, Clay NK, Sattely ES (2015) A new cyanogenic metabolite in Arabidopsis required for inducible pathogen defence. Nature 525:376–379

Ram H, Soni P, Salvi P, Gandass N, Sharma A, Kaur A, Sharma TR (2019) Insertional mutagenesis approaches and their use in rice for functional genomics. Plant Theor 8:310

Riedelsheimer C, Czedik-Eysenberg A, Grieder C, Lisec J, Technow F, Sulpice R, Altmann T, Stitt M, Willmitzer L, Melchinger AE (2012) Genomic and metabolic prediction of complex heterotic traits in hybrid maize. Nat Genet 44:217

Routaboul J-M, Dubos C, Beck G, Marquis C, Bidzinski P, Loudet O, Lepiniec L (2012) Metabolite profiling and quantitative genetics of natural variation for flavonoids in Arabidopsis. J Exp Bot 63:3749–3764

Rowe HC, Hansen BG, Halkier BA, Kliebenstein DJ (2008) Biochemical networks and epistasis shape the *Arabidopsis thaliana* metabolome. Plant Cell 20:1199–1216

Sakurai T, Yamada Y, Sawada Y, Matsuda F, Akiyama K, Shinozaki K, Hirai MY, Saito K (2013) PRIMe update: innovative content for plant metabolomics and integration of gene expression and metabolite accumulation. Plant Cell Physiol 54:e5–e5

Salvo SA, Hirsch CN, Buell CR, Kaeppler SM, Kaeppler HF (2014) Whole transcriptome profiling of maize during early somatic embryogenesis reveals altered expression of stress factors and embryogenesis-related genes. PLoS One 9:e111407

Schauer N, Semel Y, Balbo I, Steinfath M, Repsilber D, Selbig J, Pleban T, Zamir D, Fernie AR (2008) Mode of inheritance of primary metabolic traits in tomato. Plant Cell 20:509–523

Sharifi G, Ebrahimzadeh H, Ghareyazie B, Gharechahi J, Vatankhah E (2012) Identification of differentially accumulated proteins associated with embryogenic and non-embryogenic calli in saffron (*Crocus sativus* L.). Proteome Sci 10:3

Sharma SK, Bryan GJ, Winfield MO, Millam S (2007) Stability of potato (*Solanum tuberosum* L.) plants regenerated via somatic embryos, axillary bud proliferated shoots, microtubers and true potato seeds: a comparative phenotypic, cytogenetic and molecular assessment. Planta 226:1449–1458

Sharma K, Sarma S, Bohra A, Mitra A, Sharma NK, Kumar A (2018) Plant metabolomics: An emerging technology for crop improvement. In: Çelik O (ed) New visions in plant science, 1st edn. IntechOpen, London, pp 65–79

Shukla P, Singh NK, Kumar D, Vijayan S, Ahmed I, Kirti PB (2014) Expression of a pathogen-induced cysteine protease (AdCP) in tapetum results in male sterility in transgenic tobacco. Funct Integr Genomics 14:307–317. https://doi.org/10.1007/s10142-014-0367-2

Shukla P, Gautam R, Singh NK, Ahmed I, Kirti PB (2019) A proteomic study of cysteine protease induced cell death in anthers of male sterile tobacco transgenic plants. Physiol Mol Biol Plants 25:1073–1082

Sugiyama M (2015) Historical review of research on plant cell dedifferentiation. J Plant Res 128:349–359

Takáč T, Pechan T, Šamaj J (2011) Differential proteomics of plant development. J Proteomics 74:577–588

Tan BC, Chin CF, Liddell S, Alderson P (2013) Proteomic analysis of callus development in Vanilla planifolia Andrews. Plant Mol Biol Rep 31:1220–1229

Tikendra L, Koijam AS, Nongdam P (2019) Molecular markers based genetic fidelity assessment of micropropagated *Dendrobium chrysotoxum* Lindl. Meta Gene 20:100562

Viana VE, Pegoraro C, Busanello C, AC O (2019) Mutagenesis in rice: the basis for breeding a new super plant. Front Plant Sci 10:1326

Wang X, Wu R, Lin X, Bai Y, Song C, Yu X, Xu C, Zhao N, Dong Y, Liu B (2013) Tissue culture-induced genetic and epigenetic alterations in rice pure-lines, F1 hybrids and polyploids. BMC Plant Biol 13:77

Wang Y, Wang Y, Li K, Song X, Chen J (2016) Characterization and comparative expression profiling of browning response in *Medinilla formosana* after cutting. Front Plant Sci 7:1897

Wang C, Tian M, Zhang Y, Ouyang T (2019) Molecular spectrum of somaclonal variation in PLB-regenerated Oncidium revealed by SLAF-seq. Plant Cell Tissue Organ Culture (PCTOC) 137:541–552

Wickramasuriya AM, Dunwell JM (2015) Global scale transcriptome analysis of *Arabidopsis* embryogenesis in vitro. BMC Genomics 16:301

Xu K, Liu J, Fan M, Xin W, Hu Y, Xu C (2012) A genome-wide transcriptome profiling reveals the early molecular events during callus initiation in *Arabidopsis* multiple organs. Genomics 100:116–124

Xu C, Zeng B, Huang J, Huang W, Liu Y (2015) Genome-wide transcriptome and expression profile analysis of Phalaenopsis during explant browning. PLoS One 10:e0123356

Xu Y, Li P, Zou C, Lu Y, Xie C, Zhang X, Prasanna BM, Oslen M (2017) Enhancing genetic gain in the era of molecular breeding. J Exp Bot 11:2641–2666

Yakovlev IA, Lee Y, Rotter B, Olsen JE, Skrøppa T, Johnsen Ø, Fossdal CG (2014) Temperature-dependent differential transcriptomes during formation of an epigenetic memory in Norway spruce embryogenesis. Tree Genet Genomes 10:355–366

Yin L, Tao Y, Zhao K, Shao J, Li X, Liu G, Liu S, Zhu L (2007) Proteomic and transcriptomic analysis of rice mature seed-derived callus differentiation. Proteomics 7:755–768

Yin L, Lan Y, Zhu L (2008) Analysis of the protein expression profiling during rice callus differentiation under different plant hormone conditions. Plant Mol Biol 68:597

Zang YX, Kim HU, Kim JA, Lim MH, Jin M, Lee SC, Kwon SJ, Lee SI, Hong JK, Park TH (2009) Genome-wide identification of glucosinolate synthesis genes in *Brassica rapa*. FEBS J 276:3559–3574

Zhang J, Xue B, Gai M, Song S, Jia N, Sun H (2017) Small RNA and transcriptome sequencing reveal a potential miRNA-mediated interaction network that functions during somatic embryogenesis in *Lilium pumilum* DC. Fisch Front Plant Sci 8:566

Zheng Q, Zheng Y, Perry SE (2013) AGAMOUS-Like15 promotes somatic embryogenesis in *Arabidopsis* and soybean in part by the control of ethylene biosynthesis and response. Plant Physiol 161:2113–2127

Zhou X, Wang K, Lv D, Wu C, Li J, Zhao P, Lin Z, Du L, Yan Y, Ye X (2013) Global analysis of differentially expressed genes and proteins in the wheat callus infected by *Agrobacterium tumefaciens*. PLoS One 8:e79390

Zhu H-G, Cheng W-H, Tian W-G, Li Y-J, Liu F, Xue F, Zhu Q-H, Sun Y-Q, Sun J (2018) iTRAQ-based comparative proteomic analysis provides insights into somatic embryogenesis in *Gossypium hirsutum* L. Plant Mol Biol 96:89–102

# Improving Nitrogen Use Efficiency of Legumes Under Changing Climate Through Omics Technologies

7

Rachapudi Venkata Sreeharsha, Shalini Mudalkar, Divya K. Unnikrishnan, S. Venkata Mohan, and Attipalli R. Reddy

**Abstract**

The assimilation rate of carbon per unit of nitrogen in the foliage is termed as nitrogen use efficiency (NUE) and this clearly depends on various factors including soil nitrogen availability, environmental conditions and climatic factors. In legumes, symbiotic nitrogen fixation occurs in root nodules which contain millions of nitrogen-fixing bacteroids. Root nodules possess leghaemoglobin as main constituent and the activity mainly depends on antioxidant levels and reactive oxygen species (ROS). Symbiotic N fixation influences a wide array of plant metabolic pathways including photosynthesis, protein metabolism in turn modulating the plant nitrogen use efficiency in response to different environmental conditions, viz. elevated $CO_2$, drought stress, elevated temperature, etc. Under elevated $CO_2$ conditions, several C3 plants experience photosynthetic acclimation due to the imbalance in C/N supply. Several legumes including pigeonpea, soybean were reported to enhance their nitrogen fixation capacity in response to increased carbon supply and overcame photosynthetic acclimation. Also, there was an increase in photosynthetic nitrogen use efficiency channelizing most of the fixed N to biosynthesis of photosynthetic enzymes. Abiotic stresses modulate the antioxidant system of root by increasing the ROS levels thus influencing the N fixation process and subsequently hampers plant metabolism and growth. Crops, shrubs and annuals are the most commonly explored species for their nitrogen-

R. V. Sreeharsha (✉) · S. Venkata Mohan
Department of Energy and Environmental Engineering, CSIR-Indian Institute of Chemical Technology (CSIR-IICT), Hyderabad, India

S. Mudalkar
Forest College and Research Institute, Hyderabad, Telangana, India

D. K. Unnikrishnan · A. R. Reddy
Department of Plant Sciences, University of Hyderabad, Hyderabad, India





A. Kumar et al. (eds.), *Omics Technologies for Sustainable Agriculture and Global Food Security Volume 1*, https://doi.org/10.1007/978-981-16-0831-5_7

fixing ability under various environmental conditions. However, perennial tree legumes are poorly characterized with respect to the mechanistic aspects of their nitrogen fixation ability and its downstream implications in plant nitrogen use efficiency and growth under different environmental conditions. Also, not much of data is generated in terms of molecular marker development and improvement of traits for NUE through omics technologies wherein the relationship between genes, proteins and metabolites is still obscure. Thus, studying legume nodule dynamics and nitrogen fixation along with in-depth 'omics' technologies is crucial to understand unexplored aspects of nitrogen fixation in legumes.

**Keywords**

Nitrogen use efficiency · Legume · Photosynthesis · $N_2$ fixation · Elevated $CO_2$ · Abiotic stress

## Abbreviations

| | |
|---|---|
| A | Assimilation rate |
| $A_{sat}$ | Light saturated photosynthesis |
| AMO | Ammonia monooxygenase |
| ANUE | Agronomic nitrogen use efficiency |
| ATP | Adenosine triphosphate |
| BNIs | Biological nitrification inhibitors |
| C | Carbon |
| $C_i$ | Internal carbon dioxide concentration |
| DCD | Dicyandiamide |
| DMPP | 3,4-dimethylpyrazole phosphate |
| FACE | Free air $CO_2$ enrichment |
| FBPase | Fructose-1,6-bisphosphatase |
| GS | Glutamine synthetase |
| HAO | Hydroxylamine oxidoreductase |
| K | Potassium |
| N | Nitrogen |
| $N_2O$ | Nitrous oxide |
| NADPH | Nicotinamide adenine dinucleotide phosphate |
| NADP-ME | NADP dependent malic enzyme; |
| NHI | Nitrogen harvest index |
| NNIs | Natural nitrification inhibitors |
| $NO_2$ | Nitrite |
| $NO_3$ | Nitrate |
| NR | Nitrate reductase |
| NUE | Nitrogen use efficiency |
| OTCs | Open top chambers |
| P | Phosphorus |

| | |
|---|---|
| Pn | Photosynthesis rate |
| PNUE | Photosynthetic nitrogen use efficiency |
| PSII | Photosystem 2 |
| R | Respiration |
| ROS | Reactive oxygen species |
| RUBISCO | Ribulose 1,5-bisphosphate carboxylase |
| RuBP | Ribulose 1,5-bisphosphate |
| SBPase | Sedoheptulose-1,7-bisphosphatase |
| UI | Utilization index |
| $V_{Cmax}$ | Maximum rate of carboxylation |
| WUE | Water use efficiency |

## 7.1 Introduction

The global crop production has seen a remarkable progress with the advent of green revolution and modern biotechnology. Though the genetic improvement programs in crops have been majorly responsible for the increased crop yields, the use of synthetic nitrogen (N), phosphorous (P) and potassium (K) fertilizers had also played immense role especially in developing countries (Duvick and Cassman 1999; Duvick 2005). The major nutrient responsible for plant development and growth is Nitrogen (N) thus influencing the crop biomass and yield. It is a predominant constituent of cellular organic compounds including proteins, nucleic acids, chlorophyll and alkaloids. N along with other minor elements improves root system for better absorption of water and nutrients under varied environmental conditions (Fageria and Baligar 2005; Garnett et al. 2009; Dechorgnat et al. 2010). There exists a strong correlation between leaf N content and $CO_2$ assimilation rates since major portion of leaf N is channelized for the synthesis of $CO_2$ assimilating enzymes such as Ribulose 1,5-bisphospahte carboxylase (RUBISCO), Phosphoenol pyruvate carboxylase (PEP carboxylase) which together accounts for nearly 70% of leaf protein (Parry et al. 2003; Sreeharsha et al. 2015). This kind of correlation was experimentally validated both in C3 and C4 plants as well as genetically variable cultivars of same species (Coque and Gallais 2006; Dawson et al. 2008; Chardon et al. 2010). The deficiency of nitrogen hampers vegetative and reproductive growth rates of the plant and induces premature senescence thus reducing the yields potentially. Flowering, fruit setting, protein and starch accumulation in seeds are also affected to considerable extent under poor N conditions (Grant et al. 2002). While, high N availability may prolong vegetative growth periods thus delaying crop maturity. Thus, the optimum N availability in the soil and efficiency of a plant for absorption and assimilation of N effectively will determine the productivity of a crop.

Inorganic fertilizers contribute to 46% of the total N loads of the world's cropland, whereas atmospheric N fixation from legumes and other N-fixing microorganisms add 20% of it, 12% comes by atmospheric deposition, precipitation, gases adsorption and 18% is lend by organic manures (farmyard manures, animal manures, green manures and crop residues) (Cassman et al. 2002; McAllister et al. 2012; Xu et al. 2012). However, crops utilise only about one-third of applied nitrogen, while the remaining is lost to the environment (McAllister et al. 2012; Xu et al. 2012). The nitrogen dynamics in soil-plant systems can be better understood by studying N cycling and other associated mechanisms. The cycling of the nitrogen is the consequence of various transformations undergone by N wherein it is used by living organisms, transformed upon the death and decomposition of organisms, and converted finally to its original oxidized state. Hence, the addition, transformation, utilization and release of N altogether form the main components of N cycling. The transformation of N in soil-plant systems happens through fixation, mineralization, nitrification and immobilization (Jansson and Persson 1982; Bolan et al. 1991; Stevenson and Cole 1999). Through N fixation, the atmospheric N is fixed into inorganic form while mineralization is the microbial and enzymatic conversion of organic forms of N into inorganic forms ($NH_4^+$ and $NO_3^-$) through ammonification. The oxidation of $NH_4^+$ to nitrite ($NO_2^-$) and subsequently to nitrate ($NO_3^-$) is termed as nitrification and it is an important soil process mediated by microorganisms through which nitrogen can be lost from terrestrial ecosystems. In addition to ammonification and nitrification other processes such as denitrification and ammonia volatilization are also important in N cycles. Ammonia volatilization is a process of conversion of $NH_4^+$ into $NH_3$ which is ultimately lost to environment. The process of nitrification and denitrification are tightly integrated wherein $NO_3^-$ is converted to gaseous $N_2$ under anaerobic conditions and both these processes are known to be responsible for the production of nitrous oxide ($N_2O$) (Fageria and Baligar 2005; Xu et al. 2012). Apart from these N cycles, certain levels of N in the natural ecosystems escape from the N pool and become unavailable for crop plants. The main reasons for this loss include leaching and surface runoff and these could be the potential reasons for N deficiency in crops.

In order to cope with the N loss through biological cycles, farmers especially in developing countries use excessive nitrogen fertilizer. Though this strategy is helpful to agriculture to certain extent, in a long run it impacts the environment negatively wherein eutrophication of water bodies and accumulation of nitrates occur in the underground water making it unsuitable for the consumption by humans. Also, the recent statistics of the worldwide agricultural productivity shows that the use of synthetic nitrogen fertilizers to improve crop plants may have reached a plateau wherein further increases in applied N may not result in yield improvements but will lead to serious environmental problems (Han et al. 2015). Approximately, 25% of the total input costs in crop production including seeds, fertilizers and pesticides is incurred for nitrogen fertilizers. To avoid the detrimental effects of excessive use of synthetic N fertilizers, it is crucial to improve the nitrogen use efficiency (NUE) of plants. Increasing NUE and decreasing N fertilizer usage can markedly contribute for conservation of air and water quality and also for economic sustainability.

## 7.2 Nitrogen Use Efficiency and Its Paradigms in Legumes

The nitrogen use efficiency is a broader phenomenon manifesting several other minor events such as agronomic efficiency, physiological efficiency, agro-physiological efficiency, apparent recovery efficiency and utilization efficiency. The NUE can be defined as the maximum grain yield produced per unit of N applied/absorbed/utilized by the plant at a given point of time (Fageria and Baligar 2005; Hirel et al. 2007; Dawson et al. 2008; Garnett et al. 2009; Han et al. 2015). It simply explains the plant efficiency to convert N inputs into outputs. So, NUE is the combination of the efficiency of absorption or uptake of N from the soil and efficiency of assimilation and remobilization as a function of grain yield. The other dimension of NUE is the agronomic nitrogen use efficiency (ANUE) which comprises both uptake and utilization efficiencies. The same can be applied to leaf N in a more specific way to determine the photosynthetic nitrogen use efficiency (PNUE) which can be defined as amount of C sequestered per N content present in unit leaf area (Sreeharsha et al. 2015). NUE can also be explained based on utilization index (UI) indicating total plant biomass produced multiplied by the ratio of the total plant biomass to total plant N (Han et al. 2015). The determination of NUE in crop plants is an important approach to evaluate the fate of applied chemical fertilizers and their role in improving crop yields. NUE and PNUE show different trends in legumes when compared to non-legume plants due to the innate capability of legume plants to fix atmospheric $N_2$. Symbiotic nitrogen fixation is a key biological process in which leguminous species form root nodules which can fix atmospheric nitrogen in symbiosis with rhizobial bacteria. Thus it is assumed that the nodule forming legumes can show superior NUE when compared to non-legumes particularly in nitrogen-poor soils (Rogers et al. 2006; Rogers et al. 2009; Cernusak et al. 2011; AbdElgawad et al. 2015). However, there are a large intraspecific and genotypic variations within the leguminous plants that exist in growth responses to various climatic factors depending upon their capacity to form nodules and the nodule mass ratio (ratio of nodule mass to total biomass) which correlates positively with nitrogen fixation. The NUE is majorly regulated by the nitrogen assimilatory and metabolizing enzymes including glutamine synthetase (GS), glutamate dehydrogenease (GDH) and nitrate reductase (NR). The advancements in the omics technologies including next generation sequencing (NGS), various platforms of proteomics and metabolomics help in the generation of massive genomic, proteomic and metabolomics data to better understand the dynamics of the nitrogen assimilatory enzymes and subsequently NUE under different environmental conditions.

## 7.3 Understanding the Influence of NUE on Plant Metabolism Through 'Omics' Technologies

Nitrogen use efficiency is a systemic phenomenon and almost every plant metabolism is connected and influenced by N levels and its assimilation. Among others, the major mechanisms that are tightly connected with NUE are photosynthesis,

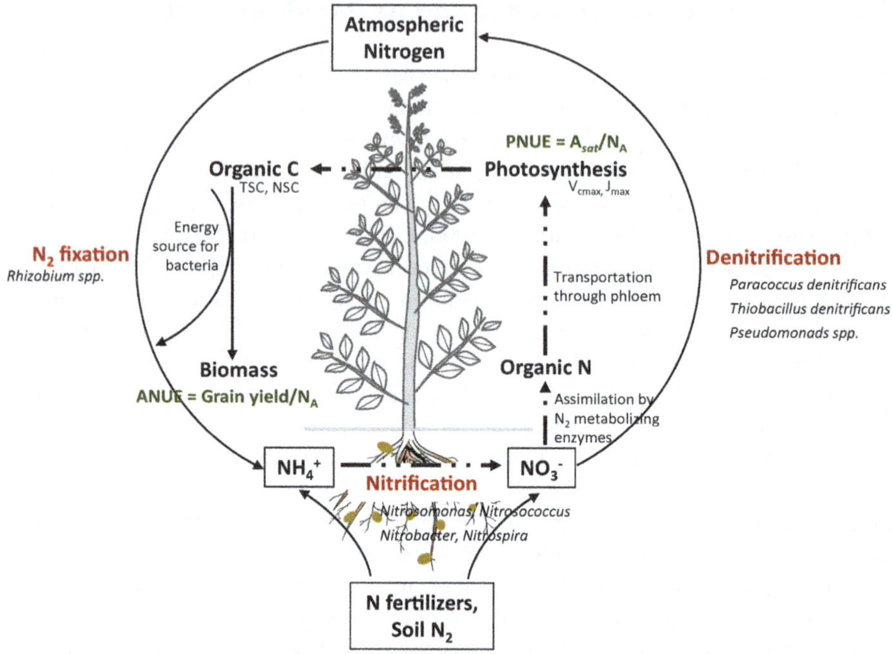


**Fig. 7.1** Schematic representation of NUE and the associated mechanisms along with the influencing factors in a model legume

respiration and grain yield. A multidisciplinary approach with various 'omics' technologies including genomics, transcriptomics, proteomics and metabolomics would help in dissecting specific plant pathways that are in association with NUE. This will also decipher identification of the genes corresponding to a particular trait along with the proteins and metabolites and also identify associated molecular markers. The NUE and PNUE and their interaction with other plant metabolism are depicted in Fig. 7.1.

## 7.3.1 Photosynthesis

The C sequestration through photosynthesis and NUE is interdependent on each other and tightly correlated since the leaf N levels invariably influence the photosynthetic process. An epistatic interaction between NUE and C assimilation will help in analysing their mutual association (Krapp and Truong 2005). Overexpression of cytosolic GS1 gene in tobacco from *Medicago sativa* demonstrates better nitrogen use efficiency thereby enhancing photosynthesis, growth and interestingly this was applied to many other crop plants also (Fuentes et al. 2001; Oliveira et al. 2002; Jing et al. 2004; Man et al. 2005). Long term growth under N deficient soils induces an imbalance in source-sink relationship associated with reduced nitrogen content and usually results in accumulation of leaf carbohydrates especially starch, which triggers a feedback mechanism that reduces photosynthetic capacity. This is due to reduction in Rubisco protein and its activity which is termed as RuBP carboxylation limitation (Spreitzer and Salvucci 2002; Parry et al. 2003; Sekhar et al. 2015). The Rubisco limitation can be assessed by $V_{Cmax}$ which can be measured by using A/Ci curves and these values are well correlated with PNUE. Approximately, 25–40% of the leaf nitrogen used for the synthesis of Rubisco in C3 plants, which leads to decreased nitrogen use efficiency. Thus, an increased PNUE can be seen when Rubisco-limited photosynthetic acclimation occurs within the leaf tissue because there is conservation and redistribution of nitrogen which was otherwise utilized for Rubisco synthesis. This holds true for many legumes including pigeonpea, soybean and other non-legumes including mulberry and Jatropha where there was a decrease in PNUE as the growth proceeds (Sekhar et al. 2015; Kumar et al. 2017). The other factor which can limit the photosynthesis is RuBP regeneration capacity which depends on the expression and activity of sedoheptulose-1,7-bisphosphatase (SBPase), fructose-1,6-bisphosphatase (FBPase). In addition, RuBp regeneration can be hampered by inefficient function of photosystem-II linked with decreased electron transport capacity leading to reduced amounts of ATP and NADPH, which are vital for activation of key enzymes in RuBP regeneration process (Sekhar et al. 2014).

There are many traits affecting NUE like root structure in case of uptake efficiency and photosynthesis in case of assimilation efficiency. NUE can be improved by improving photosynthesis for given concentration of leaf N or by reducing N content per unit leaf area without changing photosynthesis (Foulkes et al. 2009). Nitrogen uptake efficiency of plant also depends upon the growth capacity of the plant along with external factors and the feedback is regulated by both N and C signalling in shoot (Lemaire and Millard 1999; Gastal et al. 2015). N uptake is positively regulated by photosynthetic assimilate transported from leaf to root through phloem when levels of N are low (Forde 2002). When levels of N are saturated within respective organs, a negative feedback is sent from shoot to root. It is also found that N can also be sequestrated into stem along with leaf which can lead to more storage of N thus delaying the absorption of N from soil. So, the increase in growth rate of plant increases photosynthesis and leaf area leading to positive C signal which can enhance the uptake of N and also lead to N sequestration to various organs. It is also observed that decreased plant growth rate causes a drop in C signalling which increases the organic N- compounds recirculating in phloem and represses the N absorption by root. Hence the N uptake is co-regulated by both soil N and plant growth rate which in turn regulated by photosynthesis.

The NUE is different in $C_4$ and $C_3$ plants since the former plants show different photosynthetic mechanism from $C_3$ plants and have advantage under N limiting conditions. Further, legumes have an advantage over non-legume plants wherein they require less use of N fertilizer since they can fix nitrogen through symbiotic relationship between N-fixing diazotrophs (Vance and Heichel 1991; Cernusak et al. 2011). Legumes are known to have more leaf N content and this results in more photosynthesis and subsequent higher growth rates (Adams et al. 2018). In cereals and dicots the leaf N was not significant measure of $A_{sat}$ whereas in case of legumes $A_{sat}$ and leaf N are interlinked (Feng et al. 2009; Adams et al. 2018). Light is often co-regulated by leaf N and also affects the allocation of N to various photosynthetic components. Reports have shown that NADP-ME (NADP dependent Malic enzyme) containing plant species tend to have higher PNUE relative to other C4 grasses except for one NADP-ME lineage (Aristidoideae) that had PNUE similar to C3 counterparts (Taub and Lerdau 2000; Ghannoum et al. 2005; Taylor et al. 2010). In grasses, the higher PNUE in NADP-ME was driven by a faster Rubisco enzyme this in turn was associated with differential allocation of nitrogen, chlorophyll and PSII between the mesophyll and bundle sheath cells (Edwards and Barber 1976; Hatch and Osmond 1976; Ghannoum et al. 2005).

### 7.3.2 Respiration

The response of respiration to nitrogen and their correlation is important as it influence the plant, the ecosystem and global carbon budgets (King et al. 2006; Houghton 2007). The respiration of whole plant differs not only with the amount of nitrogen per plant but also across the plant sizes (Reich et al. 2006). This in turn is in proportion to biomass and N levels of various tissues including leaves, stems and roots which differ not only chemically but also structurally and metabolically (Reich et al. 2008). Various studies have shown that the plant respiration is influenced by both the enzymes, co-factors which contain N and the substrate which is usually carbohydrates irrespective of the organ type (Amthor 1994; Amthor 2000; Cannell and Thornley 2000; Amthor and Baldocchi 2001; Atkin and Tjoelker 2003; Gifford 2003; Bouma 2005; Lambers et al. 2005). However, it is still unclear whether the respiration–nitrogen relationship is same or different for each organ. One of the many hypothesis regarding R–N is that the slope would be similar for plant parts such as roots, leaves and stems but at the same time respiratory cost related to N partitioning or processes like turnover of protein, phloem loading, uptake of nutrient and assimilation of nitrogen will differ consistently among different plant organs (Reich et al. 2006). Previously, it was believed that the leaf tissue might have higher respiration since it contained most of N and is metabolically active. However, later studies showed that leaf had lower respiration since the total N in leaves majorly involve in photosynthesis and less in respiratory related components (Reich et al. 2006, 2008). But in shoots and roots they are involved in the storage of non-structural carbohydrates, its conversion, nutrient uptake as well as transport. Hence the respiration and N relationship among the organs is different. Next generation sequencing through genome wide association studies, quantitative trait loci mapping and mining of molecular markers could help in understanding the interplay between respiration and nitrogen use efficiency.

### 7.3.3 Grain Yield

Plant life cycle is divided into vegetative/pre-flowering phase and reproductive/post-flowering phase. The requirement of N also changes accordingly during these phases wherein during vegetative phase the N is allocated and sequestrated among leaves, shoot and roots which are potential sink tissues during that phase. While in reproductive phase the leaves act as source for both N and C and seeds and fruits are the major sink tissues. So, the N uptake and NUE also change according to the phase of the plant life cycle. According to the definition, the assimilation and remobilization of plant nitrogen to produce grains/yield are one of the components of NUE. Thus, the plants with better efficiency to remobilize plant N to seeds will be considered to have better NUE. In case of N deficiency the yield and grain number was found to be less in various crops including maize, wheat and rice (Uhart and Andrade 1995; Mae 1997; Sadras and Slafer 2012). There were a handful of studies which were benefited from the advanced omics technologies and gene engineering wherein the NUE was modulated and ultimately resulted in better growth and productivity. For instance, the overexpression of OsENOD93-1 gene which is known to be responsive to nitrogen in rice resulted in an increase in the grain yield (Bi et al. 2009). Similarly, a transcription factor encoding gene Dof1 from maize when overexpressed in Arabidopsis and grown under low nitrogen levels led to an increased nitrogen uptake as well as increased levels of amino acid (Yanagisawa et al. 2004). Enhanced biomass and seed yield were observed in rape seed upon overexpression of alanine aminotransferase (AlaAT) gene from Barley (Good et al. 2007) The N deficiency leads to lower growth rate which in turn affects flowering and consequently the production of grain. The grain filling in plants in turn depends upon the remobilization of C and N from vegetative phase and also post-flowering photosynthesis and root N absorption (Gastal et al. 2015). Quantitative Trait Loci (QTLs) for NUE and carbon assimilation were found to be co-localizing in maize thus representing an enhanced utilisation of 'C' facilitated by N for filling the grains (Gallais and Coque 2005). Delay in leaf senescence can prolong photosynthesis and hence produce more C which in turn will help in better N uptake. If the plant is better at N uptake and has better growth rate the grain yield will also be optimum in which case, the legumes will have strategic advantage. Cereals like wheat and rice require heavy N fertilization to have high leaf N contents and persistent photosynthesis (Makino 2011). At the same time nitrogen harvest index (NHI) is also important for crops that reflect the protein content of grain which is an important nutritional quality.

## 7.4 Nitrogen Dynamics Under Changing Climatic Scenario

Climate change is one of the major factors that influence the plant productivity on a global scale. Plants respond differently to these climatic changes depending on the species, cultivar type and growth conditions and likewise NUE also. Elevated $CO_2$ and drought stress are two scenarios which can interact with each other and influence the plant growth, physiology and yield. Under elevated $CO_2$ conditions, the shoot

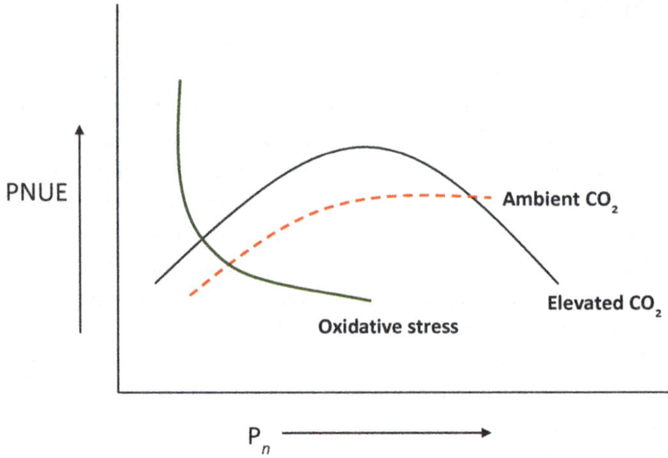


**Fig. 7.2** Relationship between nitrogen use efficiency and photosynthesis under ambient, elevated $CO_2$ and drought stress conditions in a model legume crop

traits contribute for dynamics of NUE while under drought and oxidative stress conditions, root or below ground tissues majorly contribute for altered NUE. The effect of $CO_2$ concentrations and drought stress on PNUE as a function of P*n* is depicted in Fig. 7.2.

### 7.4.1 Elevated $CO_2$

Natural and anthropogenic mediated $CO_2$ emissions can lead to change in the climate wherein there is an alteration in the budget of earth's energy. Elevated atmospheric $CO_2$ has significant effect on physiology, growth and yield of major food crops as well as tree species. Elevated $CO_2$ is known not only to stimulate photosynthesis (A) but also the growth of most plants (Ainsworth and Rogers 2007; Rogers et al. 2009; Sreeharsha and Reddy 2015). To understand the plant responses to elevated $CO_2$ and associated C and N dynamics, Free Air $CO_2$ Enrichment (FACE) and Open Top Chambers (OTCs) are highly useful. Various studies have shown the differential responses of $C_3$, $C_4$ plants as well as legume, non-legumes plants under elevated $CO_2$ conditions (Sekhar et al. 2015; Kumar et al. 2017; Sreeharsha et al. 2019). Generally, elevated $CO_2$ enhances photosynthesis of plants due to excessive C supply. However, most of the crop and tree species showed photosynthetic acclimation (saturation and subsequent reduction in photosynthesis) after certain period of growth due to N limitation. This could be due to feedback regulation of Rubisco through increased sugars and saturated sink capacity. However, some other plants showed persistent stimulation of $P_n$ despite an increase in foliar starch under elevated $CO_2$ conditions. For instance, mulberry grown under elevated $CO_2$ showed accumulation of starch and total sugars with respect to area and mass whilst showed an enhancement of $P_n$ by 40% (Sekhar et al. 2014, 2015). So, there is no consensus phenomenon of high starch accumulation and photosynthetic acclimation. Also, it was believed that a decline in Nm under elevated $CO_2$ was associated with photosynthetic down regulation. However, the recent studies showed the stimulation of photosynthesis nevertheless a reduction in Nm which infers the responses of plants to elevated $CO_2$ varies significantly from species to species. The N demand can be met by supplying $N_2$ fertilizers externally to balance the excessive C supply. Hence, when growth occurs under elevated $CO_2$ there is decrease in soil nitrogen availability and there is increased C pool in the biomass leading to lower leaf nitrogen (mass and area basis) which results in decrease of net photosynthetic capacity and grain yield.

The phenomenon is somewhat different in legume crops when grown under elevated $CO_2$. There exists mutualism between the plant and the bacteroid in the root nodules wherein the carbon supply which provides energy is received by the bacteroid from the host in the form of sucrose and the bacteroid returns $NH_4^+$ to the host cells. Therefore, the increased carbon source in the legumes during elevated $CO_2$ may thus be diverted to other available sinks such as nodules for their growth and productivity (Rogers et al. 2009; Cernusak et al. 2011). In legumes, the responses of nodulation to elevated $CO_2$ such as nodule size, number and specific nitrogenase activity provide valuable information about the NUE. In several model legume crops, the increased carbon source is invested in nodule development thereby enhancing nodule number and nodule mass and subsequently increased atmospheric nitrogen is being assimilated through the nitrogen-fixing bacteria. Also, there is a positive correlation between nodule mass ratio and $P_n$ values substantiating the fact that more the number of nodules, there is increased mass which indirectly promoted photosynthesis by overcoming the nitrogen limitation and hence the plant escapes the photosynthetic acclimation. Due to their increased nodule size, nodule number and total N contents, several legume crops including *Trifolium repens, Lupinus albus, Pisum sativum, Cajanus cajan* and *Glycine max* showed increased seed yields under high $CO_2$. (Zanetti et al. 1996; Rogers et al. 2009; Butterly et al. 2016).

One of the primary components of NUE is the efficiency of nitrogen uptake. Plants can uptake N in the form of both nitrates as well as ammonium. Interestingly, there was a strong correlation between inhibition of nitrification and the ratio of uptake of root ammonium to nitrate. Higher ammonium uptake can lead to higher NUE of plant and this could be of significance in the context of adaptation to climate change. Total protein and N content in plants generally reduce under elevated $CO_2$, especially when the nitrogen source is nitrate, its assimilation is slower under this condition thus leading to harmful effects on food quality. Growth under elevated $CO_2$ decreases stomatal conductance and transpiration rates which potentially decrease N uptake from the soil thus influencing the NUE. Elevated $CO_2$ also inhibits the photorespiration-dependent nitrate assimilation in shoot of many species. Thus, it is highly relevant to focus on developing plants with higher NUE to cope with the limiting N sources under future elevated $CO_2$ conditions. Otherwise, the future agriculture demands excessive use of N fertilizers which in turn have deleterious effects on the environment.

### 7.4.2 Drought Stress

Drought and high temperature are two major environmental factors which limits the growth and productivity by depressing carbon assimilation and nitrogen uptake. Drought causes oxidative stress and decreases photosynthetic rates and interrupts photosynthetic electron transport in plants. Root and nodule development are also extremely sensitive to drought and decrease along with gas exchange parameters. The imbalance in carbon sequestration and electron transport rates produces reactive oxygen species (ROS) including methylglyoxalase (MG). Subsequently, ROS will cause oxidative damage to photosynthetic apparatus and affect NUE. Plants evolved various mechanisms to mitigate ROS which include Ascorbate-Glutathione pathway, glyoxalase pathway and aldo-keto reductase mediated mechanisms (Mudalkar et al. 2016; Mudalkar et al. 2017). Legumes can mediate improved mitigation of ROS in drought stressed conditions when grown under elevated $CO_2$ (Sreeharsha et al. 2019). It was evident that optimum ROS concentrations play an important role in nitrogen fixation process by activating the expression of nitrogenase enzyme (Serraj et al. 1999; Streeter 2003). However, under oxidative stress excessive ROS will be generated becoming detrimental to biomolecules and hamper the nodule functionality. The efficiency of N acquisition and utilization of plant under oxidative stress conditions could be related to the capacity of the plant to buffer excess ROS by enzymatic and non-enzymatic antioxidative system thus minimizing the damage to the root nodules and other tissues. In turn, RUBISCO carboxylation capacity and kinetics are interrelated to plant performance under stress conditions since, higher carboxylation efficiencies result in less photorespiration and subsequent ROS generation. So it is evident that plants with higher RUBISCO kinetics can show higher PNUE especially under stress conditions.

High temperature stress leads to reduced photosynthesis which is caused due to the disruptions in the structure and function of chloroplasts and hence reduced chlorophyll content in leaves thereby decreasing the energy available for assimilation of nitrate which could be negatively affecting the status of plant nitrogen. Further, plant nitrogen uptake under these stress can be hampered by decreasing the activity of enzymes involved in nitrogen metabolism (Prasad et al. 2008). The assimilation of intracellular ammonium into organic compounds occurs majorly due to the enzymes NR and GS. Combined heat and drought stress decreased the activity of NR but increased the proteolytic enzyme activity suggesting a rapid mobilization of nitrogen compounds to the grain (Prasad et al. 2008; Zinta et al. 2014). The reduced nitrogen uptake of plants under heat and drought stress could be attributed to the lower availability of nitrogen in the soil under drought and/or a decreased capacity of root to uptake nitrogen and reduction in nitrogen demand by the plant. Reports have shown that drought priming can enhance the uptake of the nitrogen to the plants exposed to sole or combined drought and heat stress. The overall effect of drought on plant NUE showed that maximum fertilizer use efficiency can be obtained with low N rates applied under well-watered conditions.

Each step of plant NUE including N uptake, transport, assimilation, and remobilization is regulated by both genetic and environmental factors. The interactions between genotype, soil N levels, moisture content and soil type define NUE. The analysis of different plant genotypes under different environmental conditions for their yield, total N uptake or NUE will demonstrate the genetic component of NUE and gene, environment interaction for regulation of NUE. Clearly, the most important environmental factor affecting NUE is water availability. In water scarce conditions, plant faces difficulty in nutrient extraction from soil and the yield is constrained by moisture availability not N availability. This sort of imbalance between water supply and nitrogen fertilization under drought stress finally reduces NUE. So there is an indirect negative correlation between drought and NUE. In turn, a balanced N status in the plant can alleviate the effects of drought stress by preventing cell membrane damage and enhancing osmoregulation. Sufficient N nutrition resulted in higher water use efficiency (WUE) as most of the leaf N is used to synthesize components of the photosynthetic apparatus in particular RuBisCo thus playing a major role in carbon assimilation. Root morphology and physiology are closely associated with soil resource acquisition and growth and development of aboveground tissues (Lynch 2013). The nutrient acquisition is effected by altered root architecture under stress conditions and for efficient N acquisition, the plant has to develop deeper roots with vigorous lateral root growth and strong responses of lateral roots to localized N supply. Alternatively, improving the drought tolerance or selecting drought tolerant varieties can also enhance the crop yield without application of additional N. Recent advanced technologies in drought assessment of crop plants through hydraulic conductivity will help in selecting elite accessions that can withstand the drought stress and mediate improved NUE (Reddy et al. 2019)

## 7.5 Improving NUE of Crop Plants

Improving NUE of crop plants is crucial in obtaining targeted yields and to face the future climate change scenarios with respect to the agricultural production and protection of environment. Of late, several studies were successful in enhancing the NUE of several model crop plants through breeding as well as advanced biotechnological tools. Different approaches implemented so far in crop plants to enhance NUE are summarized in Table 7.1. Under N stress condition, overexpression of alflafa glutamine synthestase (GS1) in tobacco plants led to enhanced NUE. Similarly, increase in biomass was observed when pea GS1 was overexpressed under N stress and non-stressed conditions (Oliveira et al. 2002). One of the traditional ways is to suppress nitrification and maintenance of N fertilizer in the reduced form so that fertilizer-N retention will be more in the soils thus improving the NUE. Sustainable primary production in agro ecosystems and fertility enhancement can be done by inhibiting nitrification. Synthetic inhibitors such as

**Table 7.1** Various strategies employed in plants to enhance their NUE

| Approach | Plant species | Underlying Mechanism | References |
|---|---|---|---|
| Synthetic nitrification inhibitors | Grasslands and general crop application | Inhibiting nitrification thus reducing nitrous oxide and methane emissions and increasing N use efficiency | Huang et al. (2014), Cahalan et al. (2015), Sun et al. (2015) |
| Biological/natural nitrification inhibitors | Wide range application | Inhibiting nitrification and increasing N use efficiency | Subbarao et al. (2007), Upadhyay et al. (2011), Subbarao et al. (2015) |
| Molecular approach | Maize | A single point mutation in male sterility 44 (*ms44*) gene increases N use and grain yield | Fox et al. (2017) |
| Chlorophyll meter technology (SPAD) based Site Specific Nutrient Management | Rice | Optimum synchronization between supply and demand of N for plant growth | Dobermann et al. (2004) |
| Genetic approach using molecular mechanism | Maize | Mapping of genes encoding proteins and enzymes involved in nitrogen assimilation and recycling for further QTL detection. | Hirel et al. (2001) |
| Transgenic approach | Maize | Expression of agrobacterium isopentenyl transferase | Robson et al. (2004) |
| Transgenic approach | Rice | Overexpression of NADH-GOGAT | Tabuchi et al. (2007) |
| Transgenic approach | Arabidopsis | Enhanced expression of asparagine synthetase | Lam et al. (2003) |
| Transgenic approach | Canola and Rice | Overexpression of alanine aminotransferase | Good et al. (2007) |

nitrapyrin, dicyandiamide (DCD) and 3,4-dimethylpyrazole phosphate (DMPP) are widely used to retard nitrification but their use is restricted due to higher cost, unavailability, inconvenience of application and the risk of environmental contamination. In particular, inhibitors which are water-soluble can cause contamination to the surface and below-ground water (Huang et al. 2014; Cahalan et al. 2015; Sun et al. 2015). Due to these drawbacks of synthetic inhibitors, it is important to develop plant-derived nitrification inhibitors, which are termed as either natural nitrification inhibitors (NNIs) or biological nitrification inhibitors (BNIs) (Subbarao et al. 2007; Upadhyay et al. 2011; Subbarao et al. 2015). Some such environmentally friendly and easily available compounds from plants have been reported which reduces the nitrogen loss and enhances NUE of crops by controlling the leaching of nitrates into the water bodies and emissions of $N_2O$ into the atmosphere. Nitrification inhibitory potential has been identified in two plants, *Brachiaria humidicola* and *Sorghum bicolor* which are characterized for their effects and modes of inhibition on the bacterium *Nitrosomonas europaea,* which oxidizes the ammonia (Gopalakrishnan et al. 2007; Subbarao et al. 2007; Zakir et al. 2008). These inhibitors were known to block either the ammonia monooxygenase (AMO) pathway or both the AMO and hydroxylamine oxidoreductase (HAO) pathways which are involved in ammonia oxidation. Cereal crops, especially rice, are not very efficient at absorbing the soil nitrogen, since the ratio of absorption to the amount of nitrogen applied as fertilizer is typically only 30–40%.

Another approach to improve NUE is to increase either uptake and assimilation or partitioning of nitrogen to grain production. Numerous studies using transgenic approaches have shown that improved NUE can be achieved by manipulating the candidate genes that are involved in the uptake, assimilation and signalling of nitrogen and also root development. Apart from overexpressing various structural genes, an improved NUE can be achieved by expressing the regulatory genes. The most advanced lead is that of alanine aminotransferase wherein the expression of this gene from barley into canola and rice under low nitrogen conditions increased not only the biomass but also the seed yield (Good et al. 2007). Improved nitrogen use for grain production was observed in Ms44 wherein a single point mutation caused male sterility. While this male sterility does not change the total content of nitrogen but it improves the utilization efficiency by reducing the nitrogen use in tassel and pollen development and allocating more nitrogen to immature ear development, as a result there is an increase in the number of maize kernel (Fox et al. 2017).

## 7.6 Conclusions and Way Forward

The environmental foot print of nitrogen fertilizer and the cost to the grower can be reduced along with increase in grain yields which is critical for food security, by improving NUE. Many efforts have been made to breed crop varieties that grow well under low nitrogen soil conditions. However, improving NUE in major crops through genetic engineering is still in its initial stages and far from ready for commercialization. Many studies have reported an improvement in NUE in Arabidopsis, rice, wheat and maize using transgenic approaches, but the NUE of the transgenic lines using elite germplasm has not been validated in fields. The varietal development of crops with either efficient N acquisition from the soil or with traits to use the acquired nitrogen more efficiently or with both traits would be beneficial to improve NUE especially in the N-limited conditions. Also, genetic mapping through QTLs of various traits linked to the physiological and biochemical pathways would help in identifying the key genes involved in overall NUE of the plants. Research and development in 'omics' including genomics, transcriptomics, proteomics and metabolomics along with physiological data can be a way forward for improving the NUE in crop plants.

**Acknowledgements** R.V.S. acknowledges an INSPIRE Faculty grant (DST/INSPIRE/04/2018/000214) from DST, New Delhi and also wish to thank the Director, CSIR- Indian Institute of Chemical Technology (CSIR-IICT) for supporting this work (IICT/Pubs./2021/064).

## References

AbdElgawad H, Farfan-Vignolo ER, de Vos D, Asard H (2015) Elevated $CO_2$ mitigates drought and temperature-induced oxidative stress differently in grasses and legumes. Plant Sci 231:1–10

Adams MA, Buckley TN, Salter WT, Buchmann N, Blessing CH, Turnbull TL (2018) Contrasting responses of crop legumes and cereals to nitrogen availability. New Phytol 217(4):1475–1483

Ainsworth EA, Rogers A (2007) The response of photosynthesis and stomatal conductance to rising [CO2]: mechanisms and environmental interactions. Plant Cell Environ 30(3):258–270

Amthor JS (1994) Plant respiratory responses to the environment and their effects on the carbon balance. In: Wilkinson RE (ed) Plant–environment interactions. Marcel Dekker, New York, pp 501–554

Amthor JS (2000) The McCree–de wit–penning de Vries–Thornley respiration paradigms: 30 years later. Ann Bot 86(1):1–20

Amthor JS, Baldocchi DD (2001) Terrestrial higher plant respiration and net primary production. In: Roy J, Saugier B, Mooney HA (eds) Terrestrial global productivity. Academic Press, San Diego, CA, pp 33–59

Atkin OK, Tjoelker MG (2003) Thermal acclimation and the dynamic response of plant respiration to temperature. Trends Plant Sci 8(7):343–351

Bi YM, Kant S, Clark J, Gidda S, Ming F, Xu J, Rochon A, Shelp BJ, Hao L, Zhao R (2009) Increased nitrogen use efficiency in transgenic rice plants over-expressing a nitrogen-responsive early nodulin gene identified from rice expression profiling. Plant Cell Environ 32:1749–1760

Bolan N, Hedley MJ, White RE (1991) Processes of soil acidification during nitrogen cycling with emphasis on legume based pastures. Plant Soil 134(1):53–63

Bouma TJ (2005) Understanding plant respiration: separating respiratory components versus a process-based approach. In: Plant respiration. Springer, Dordrecht, pp 177–194

Butterly CR, Phillips LA, Wiltshire JL, Franks AE, Armstrong RD, Chen D, Mele PM, Tang C (2016) Long-term effects of elevated $CO_2$ on carbon and nitrogen functional capacity of microbial communities in three contrasting soils. Soil Biol Biochem 97:157–167

Cahalan E, Ernfors M, Muller C, Devaney D, Laughlin RJ, Watson CJ, Hennessy D, Grant J, Khalil MI, McGeough KL, Richards KG (2015) The effect of the nitrification inhibitor dicyandiamide (DCD) on nitrous oxide and methane emissions after cattle slurry application to Irish grassland. Agric Ecosyst Environ 199:339–349

Cannell M, Thornley J (2000) Modelling the components of plant respiration: some guiding principles. Ann Bot 85(1):45–54

Cassman KG, Dobermann A, Walters DT (2002) Agroecosystems, nitrogen-use efficiency, and nitrogen management. AMBIO J Hum Environ 31(2):132–141

Cernusak LA, Winter K, Martinez C, Correa E, Aranda J, Garcia M, Jaramillo C, Turner BL (2011) Responses of legume versus nonlegume tropical tree seedlings to elevated $CO_2$ concentration. Plant Physiol 157(1):372–385

Chardon F, Barthélémy J, Daniel-Vedele F, Masclaux-Daubresse C (2010) Natural variation of nitrate uptake and nitrogen use efficiency in Arabidopsis thaliana cultivated with limiting and ample nitrogen supply. J Exp Bot 61(9):2293–2302

Coque M, Gallais A (2006) Genomic regions involved in response to grain yield selection at high and low nitrogen fertilization in maize. Theor Appl Genet 112(7):1205–1220

Dawson J, Huggins D, Jones S (2008) Characterizing nitrogen use efficiency to improve crop performance in organic and sustainable agricultural systems. Field Crop Res 107:89–101

Dechorgnat J, Nguyen CT, Armengaud P, Jossier M, Diatloff E, Filleur S, Daniel-Vedele F (2010) From the soil to the seeds: the long journey of nitrate in plants. J Exp Bot 62(4):1349–1359

Dobermann A, Blackmore S, Cook S, Adamchuk V (2004) Precision farming: challenges and future directions. In: Fischer T, Turner N, Angus J, McIntyre L, Robertson M, Borrell A, Lloyd D (eds) Proceedings of the fourth international crop science congress. Brisbane, Queensland, 26 September–1 October, CD-ROM or, http://www.cropscience.org.au/icsc2004/symposia/4/3/217_dobermanna.htm

Duvick DN (2005) The contribution of breeding to yield advances in maize (Zea mays L.). Adv Agron 86:83–145

Duvick DN, Cassman KG (1999) Post–green revolution trends in yield potential of temperate maize in the north-Central United States. Crop Sci 39(6):1622–1630

Edwards J, Barber S (1976) Nitrogen flux into corn roots as influenced by shoot requirement 1. Agron J 68(3):471–473

Fageria N, Baligar V (2005) Enhancing nitrogen use efficiency in crop plants. Adv Agron 88:97–185

Feng YL, Lei YB, Wang RF, Callaway RM, Valiente-Banuet A, Inderjit LYP, Long ZY (2009) Evolutionary tradeoffs for nitrogen allocation to photosynthesis versus cell walls in an invasive plant. Proc Natl Acad Sci 106(6):1853–1856

Forde BG (2002) The role of long-distance signalling in plant responses to nitrate and other nutrients. J Exp Bot 53(366):39–43

Foulkes MJ, Hawkesford MJ, Barraclough PB, Holdsworth MJ, Kerr S, Kightley S, Shewry PR (2009) Identifying traits to improve the nitrogen economy of wheat: recent advances and future prospects. Field Crop Res 114(3):329–342

Fox T, DeBruin J, Collet KH, Trimnell M, Clapp J, Leonard A, Li B, Scolaro E, Collinson S, Glassman K, Miller M, Schussler J, Dolan D, Liu L, Gho C, Albertsen M, Loussaert D, Shen B (2017) A single point mutation in Ms44 results in dominant male sterility and improves nitrogen use efficiency in maize. Plant Biotechnol J 15(8):942–952

Fuentes SI, Alen DJ, Ortiz-Lopez A, Hernandez G (2001) Overexpression of cytosolic glutamine synthetase increases photosynthesis and growth at low nitrogen concentrations. J Exp Bot 52:1071–1081

Gallais A, Coque M (2005) Genetic variation and selection for nitrogen use efficiency in maize: a synthesis. Maydica 50:531–537

Garnett T, Conn V, Kaiser BN (2009) Root based approaches to improving nitrogen use efficiency in plants. Plant Cell Environ 32(9):1272–1283

Gastal F, Lemaire G, Durand JL, Louarn G (2015) Quantifying crop responses to nitrogen and avenues to improve nitrogen-use efficiency. In: Crop physiology. Elsevier, Amsterdam, pp 161–206

Ghannoum O, Evans JR, Chow WS, Andrews TJ, Conroy JP, Caemmerer SV (2005) Faster Rubisco is the key to superior nitrogen-use efficiency in NADP-malic enzyme relative to NAD-malic enzyme C4 grasses. Plant Physiol 137(2):638–650

Gifford RM (2003) Plant respiration in productivity models: conceptualisation, representation and issues for global terrestrial carbon-cycle research. Funct Plant Biol 30(2):171–186

Good AG, Johnson SJ, Pauw MD, Carroll RT, Savidov N, Vidmar J, Lu Z, Taylor G, Stroehen V (2007) Engineering nitrogen use efficiency with alanine aminotransferase. Botany 85 (3):252–262

Gopalakrishnan S, Subbarao GV, Nakahara K, Yoshihashi T, Ito O, Maeda I, Ono H, Yoshida M (2007) Nitrification inhibitors from the root tissues of Brachiaria humidicola, a tropical grass. J Agric Food Chem 55(4):1385–1388

Grant CA, Peterson GA, Campbell CA (2002) Nutrient considerations for diversified cropping systems in the northern Great Plains. Agron J 94(2):186–198

Han M, Okamoto M, Beatty PH, Rothstein SJ, Good AG (2015) The genetics of nitrogen use efficiency in crop plants. Annu Rev Genet 49:269–289

Hatch M, Osmond C (1976) Compartmentation and transport in C4 photosynthesis. In: Transport in plants III. Springer, Berlin, pp 144–184

Hirel B, Bertin P, Quillere I, Bourdoncle W, Attagnant C, Dellay C, Gouy A, Cadiou S, Retailliau C, Falque M, Gallais A (2001) Towards a better understanding of the genetic and physiological basis for nitrogen use efficiency in maize. Plant Physiol 125(3):1258–1270

Hirel B, Le Gouis J, Ney B, Gallais A (2007) The challenge of improving nitrogen use efficiency in crop plants: towards a more central role for genetic variability and quantitative genetics within integrated approaches. J Exp Bot 58(9):2369–2387

Houghton R (2007) Balancing the global carbon budget. Annu Rev Earth Planet Sci 35:313–347

Huang T, Gao B, Hu X-K, Lu X, Well R, Christie P, Bakken LR, Ju X-T (2014) Ammonia-oxidation as an engine to generate nitrous oxide in an intensively managed calcareous Fluvo-aquic soil. Sci Rep 4:3950

Jansson S, Persson J (1982) Mineralization and immobilization of soil nitrogen. In: Nitrogen in agricultural soils (nitrogeninagrics). Wiley, New York, pp 229–252

Jing ZP, Gallardo F, Pascual MB, Sampalo R, Romero J, Torres de Vavarra A, Canovas FM (2004) Improved growth in a field trial of transgenic hybrid poplar overexpressing glutamine synthetase. New Phytol 164:137–145

King AW, Gunderson CA, Post WM, Weston DJ, Wullschleger SD (2006) Plant respiration in a warmer world. Science 312(5773):536–537

Krapp A, Truong HN (2005) Regulation of C/N interaction in model plant species. J Crop Improv 15:127–173

Kumar S, Sreeharsha RV, Mudalkar S, Sarashetti PM, Reddy AR (2017) Molecular insights into photosynthesis and carbohydrate metabolism in Jatropha curcas grown under elevated $CO_2$ using transcriptome sequencing and assembly. Sci Reports 7(1):11066

Lam H-M, Wong P, Chan H-K, Yam K-M, Chen L, Chow C-M, Coruzzi GM (2003) Overexpression of the ASN1 gene enhances nitrogen status in seeds of Arabidopsis. Plant Physiol 132(2):926–935

Lambers H, Robinson SA, Ribas-Carbo M (2005) Regulation of respiration in vivo. In: Plant respiration. Springer, Dordrecht, pp 1–15

Lemaire G, Millard P (1999) An ecophysiological approach to modelling resource fluxes in competing plants. J Exp Bot 50(330):15–28

Lynch JP (2013) Steep, cheap and deep: an ideotype to optimize water and N acquisition by maize root systems. Ann Bot 112(2):347–57

Mae T (1997) Physiological nitrogen efficiency in rice: nitrogen utilization, photosynthesis, and yield potential. Plant Soil 196(2):201–210

Makino A (2011) Photosynthesis, grain yield, and nitrogen utilization in rice and wheat. Plant Physiol 155(1):125–129

Man HM, Boriel R, El-Khatib R, Kirby EG (2005) Characterization of transgenic poplar with ectopic expression of pine cytosolic glutamine synthetase under conditions of varying nitrogen ability. New Phytol 167:31–39

McAllister CH, Beatty PH, Good AG (2012) Engineering nitrogen use efficient crop plants: the current status. Plant Biotech J 10(9):1011–1025

Mudalkar S, Sreeharsha RV, Reddy AR (2016) A novel aldo-keto reductase from *Jatropha curcas* L.(JcAKR) plays a crucial role in the detoxification of methylglyoxal, a potent electrophile. J Plant Physiol 195:39–49

Mudalkar S, Sreeharsha RV, Reddy AR (2017) Involvement of glyoxalases and glutathione reductase in conferring abiotic stress tolerance to *Jatropha curcas* L. Environ Exp Bot 134:141–150

Oliveira IC, Brears T, Knight TJ, Clark A, Coruzzi GM (2002) Overexpression of cytosolic glutamine synthetase. Relation to nitrogen, light, and photorespiration. Plant Physiol 129:1170–1180

Parry MAJ, Andralojc PJ, Mitchell RAC, Madgwick PJ, Keys AJ (2003) Manipulation of Rubisco: the amount, activity, function and regulation. J Exp Bot 54(386):1321–1333

Prasad PVV, Staggenborg S, Ristic Z (2008) Impacts of drought and/or heat stress on physiological, developmental, growth, and yield processes of crop plants. In: Response of crops to limited water: Understanding and modeling water stress effects on plant growth processes (response of crops). Wiley, New York, pp 301–355

Reich PB, Tjoelker MG, Machado J-L, Oleksyn J (2006) Universal scaling of respiratory metabolism, size and nitrogen in plants. Nature 439(7075):457

Reich PB, Tjoelker MG, Pregitzer KS, Wright IJ, Oleksyn J, Machado J-L (2008) Scaling of respiration to nitrogen in leaves, stems and roots of higher land plants. Ecol Lett 11(8):793–801

Reddy KS, Sekhar KM, Sreeharsha RV, Reddy AR (2019) Hydraulic dynamics and photosynthetic performance facilitate rapid screening of field grown mulberry (Morus spp.) genotypes for drought tolerance. Environ Exp Bot 157:320–330

Robson PRH, Donnison IS, Wang K, Frame B, Pegg SE, Thomas A, Thomas H (2004) Leaf senescence is delayed in maize expressing the agrobacterium IPT gene under the control of a novel maize senescence-enhanced promoter. Plant Biotechnol J 2(2):101–112

Rogers A, Ainsworth EA, Leakey ADB (2009) Will elevated carbon dioxide concentration amplify the benefits of nitrogen fixation in legumes? Plant Physiol 151(3):1009–1016

Rogers A, Gibon Y, Stitt M, Morgan PB, Bernacchi CJ, Donald RORT, Long SP (2006) Increased C availability at elevated carbon dioxide concentration improves N assimilation in a legume. Plant Cell Environ 29(8):1651–1658

Sadras VO, Slafer GA (2012) Environmental modulation of yield components in cereals: heritabilities reveal a hierarchy of phenotypic plasticities. Field Crop Res 127:215–224

Sekhar KM, Rachapudi VS, Mudalkar S, Reddy AR (2014) Persistent stimulation of photosynthesis in short rotation coppice mulberry under elevated $CO_2$ atmosphere. J Photochem Photobiol B 137:21–30

Sekhar KM, Sreeharsha RV, Reddy AR (2015) Differential responses in photosynthesis, growth and biomass yields in two mulberry genotypes grown under elevated $CO_2$ atmosphere. J Photochem Photobiol B Biol 151:172–179

Serraj R, Sinclair TR, Purcell LC (1999) Symbiotic $N_2$ fixation response to drought. J Exp Bot 50 (331):143–155

Spreitzer RJ, Salvucci ME (2002) Rubisco: structure, regulatory interactions, and possibilities for a better enzyme. Plant Biol 53:449–475

Sreeharsha RV, Mudalkar S, Sengupta D, Unnikrishnan DK, Reddy AR (2019) Mitigation of drought-induced oxidative damage by enhanced carbon assimilation and an efficient antioxidative metabolism under high $CO_2$ environment in pigeonpea (*Cajanus cajan* L.). Photosynth Res 139:425–439

Sreeharsha RV, Reddy AR (2015) Dynamics of vegetative and reproductive growth patterns in Pigeonpea (Cajanus cajan L.) grown under elevated $CO_2$. Procedia Environ Sci 29:147–148

Sreeharsha RV, Sekhar KM, Reddy AR (2015) Delayed flowering is associated with lack of photosynthetic acclimation in pigeon pea (*Cajanus cajan* L.) grown under elevated $CO_2$. Plant Sci 231:82–93

Stevenson FJ, Cole MA (1999) Cycles of soils: carbon, nitrogen, phosphorus, sulfur, micronutrients. Wiley, New York

Streeter J (2003) Effects of drought on nitrogen fixation in soybean root nodules. Plant Cell Environ 26(8):1199–1204

Subbarao G, Rondon M, Ito O, Ishikawa T, Rao IM, Nakahara K, Lascano C, Berry WL (2007) Biological nitrification inhibition (BNI)—is it a widespread phenomenon? Plant Soil 294 (1-2):5–18

Subbarao GV, Yoshihashi T, Worthington M, Nakahara K, Ando Y, Sahrawat KL, Rao IM, Lata J-C, Kishii M, Braun H-J (2015) Suppression of soil nitrification by plants. Plant Sci 233:155–164

Sun H, Zhang H, Powlson D, Min J, Shi W (2015) Rice production, nitrous oxide emission and ammonia volatilization as impacted by the nitrification inhibitor 2-chloro-6-(trichloromethyl)-pyridine. Field Crop Res 173:1–7

Tabuchi M, Abiko T, Yamaya T (2007) Assimilation of ammonium ions and reutilization of nitrogen in rice (Oryza sativa L.). J Exp Bot 58(9):2319–2327

Taub DR, Lerdau MT (2000) Relationship between leaf nitrogen and photosynthetic rate for three NAD-ME and three NADP-ME C4 grasses. Am J Bot 87(3):412–417

Taylor SH, Hulme SP, Raes M, Ripley BS, Woodward FI, Osborne CP (2010) Ecophysiological traits in C3 and C4 grasses: a phylogenetically controlled screening experiment. New Phytol 185 (3):780–791

Uhart SA, Andrade FH (1995) Nitrogen deficiency in maize: I. effects on crop growth, development, dry matter partitioning, and kernel set. Crop Sci 35(5):1376–1383

Upadhyay RK, Tewari S, Patra DD (2011) Natural nitrification inhibitors for higher nitrogen use efficiency, crop yield, and for curtailing global warming. J Trop Agric 49:19–24

Vance C, Heichel G (1991) Carbon in $N_2$ fixation: limitation or exquisite adaptation. Annu Rev Plant Biol 42(1):373–390

Xu G, Fan X, Miller AJ (2012) Plant nitrogen assimilation and use efficiency. Annu Rev Plant Biol 63:153–182

Yanagisawa S, Akiyama A, Kisaka H, Uchimiya H, Miwa T (2004) Metabolic engineering with Dof1 transcription factor in plants: improved nitrogen assimilation and growth under low-nitrogen conditions. Proc Natl Acad Sci U S A 101:7833–7838

Zakir HA, Subbarao GV, Pearse SJ, Gopalakrishnan S, Ito O, Ishikawa T, Kawano N, Nakahara K, Yoshihashi N, Ono H, Yoshida M (2008) Detection, isolation and characterization of a root-exuded compound, methyl 3-(4-hydroxyphenyl) propionate, responsible for biological nitrification inhibition by sorghum (Sorghum bicolor). New Phytol 180(2):442–451

Zanetti S, Hartwig UA, Luscher A, Hebeisen T, Frehner M, Fischer BU, Hendrey GR, Blum H, Nosberger J (1996) Stimulation of symbiotic $N_2$ fixation in Trifolium repens L. under elevated atmospheric $pCO_2$ in a grassland ecosystem. Plant Physiol 112(2):575–583

Zinta G, AbdElgawad H, Domagalska MA, Knapen VL, Ivans N, Janssens IA, GTS B, Asard H (2014) Physiological, biochemical, and genome-wide transcriptional analysis reveals that elevated $CO_2$ mitigates the impact of combined heat wave and drought stress in Arabidopsis thaliana at multiple organizational levels. Glob Chang Biol 20(12):3670–3685

# 8 Omics Approaches for Elucidating Abiotic Stress Responses in Plants

Israr Ahmed, Pawan Shukla, and Ranjana Gautam

**Abstract**

Abiotic stresses are the limiting factors that adversely affect the sustainable food production to meet the global food demand. However, plant has evolved an inbuilt mechanism to sense and respond these abiotic stresses. Unraveling these mechanisms of abiotic response is important for developing climate resilient crop. In last two decades, the advent of omics platform has made significant progress in this direction. Omics technologies like Next Generation sequencing (NGS), Transcriptomics, Proteomics, and Metabolomics have been used in crop plants under abiotic stresses. The quantum of information generated through omics platform has been utilized by plant breeders for quantitative trait loci (QTL) analysis, marker development, and deciphering abiotic stress pathways. This chapter focused on different omics technology has been adopted for understanding abiotic stress response in agriculturally important crop.

**Keywords**

Abiotic Stress · Omics · Genomics · Proteomics · Transcriptomics · Metabolomics

I. Ahmed (✉)
Department of Plant Molecular Biology, University of Delhi South Campus, Dhaula Kuan, New Delhi, India

P. Shukla
Biotechnology Division, Seribiotech Research Laboratory (SBRL), Bangalore, Karnataka, India

R. Gautam
Department of Plant Sciences, University of Hyderabad, Hyderabad, Telangana, India



A. Kumar et al. (eds.), *Omics Technologies for Sustainable Agriculture and Global Food Security Volume 1*, https://doi.org/10.1007/978-981-16-0831-5_8

## 8.1 Introduction

Plants being sessile in nature have to endure various environmental cues. These environmental cues may be biotic or abiotic in nature. Biotic stresses include infestation by different insect pests and diseases caused by pathogens. Abiotic stresses include water scarcity, salinity, temperature (high or low), nutrient limitation, or heavy metal stress. These stresses together limit crop productivity. It is estimated that in the next 30 years, global human population will increase by 25% and reach 10 billion (Hickey et al. 2019). Drought alone affects 40% of global area, whereas 7% of total area is affected by salinity (Rozema and Flowers 2008). The decrease in arable land due to one or more abiotic stresses poses a serious threat to food supply for ever increasing human population. To ensure enough food supply to feed this huge population, agricultural production needs to be increased by two folds. Climate change is further exacerbating the situation. Scientists need to develop climate resilient crops that can sustain their yield under these adverse conditions. To develop such climate resilient crop varieties, understanding of plants response to different abiotic stresses is important.

During course of evolution, plants have developed an inbuilt mechanism to sense and respond to different kind of abiotic stresses. They undergo various types of changes at morphological, physiological, biochemical, and at molecular levels (Lata 2015). One of the key questions that remain important to plant biologists is to know how plants perceive the stress, transduce the signal to reprogram the gene expression, and finally bring an adaptive response. Although the progress in plant molecular biology has helped in decoding the molecular mechanisms behind it, still we are far behind the complete understanding of stress responses.

Conventional breeding approaches such as inbreeding, back crossing, mutational breeding, distant hybridization, and marker assisted selections were successful in developing crop varieties with enhanced stress tolerance. Moreover, most of the stress responses were studied individually, but in field condition these stresses often occur in combination. For example, salinity stress is often associated with osmotic stress. Similarly, flooding stress is often associated with hypoxia. But in the post-genomic era especially advent of omics, the breeding approaches were changed substantially.

The term "omic" is derived from Latin suffix "ome" that means mass or many (Lay et al. 2006). Progress in genome sequencing technologies and high throughput data analysis has changed the approach to study stress responses. Now, it is easy to study the change in transcriptome, proteome, or metabolome at given time period. In this approach, first one has to understand at the molecular level how plants sense and response to a particular stress. This can be facilitated by dissecting the response a genomic, transcriptomic, proteomic, or metabolomics levels (Fig. 8.1). Recent research has made efficient use of transcriptomic, proteomic, and molecular approaches to identify the complex networks linked to stress perception and response in the model and crop plants.

Abiotic stress signaling can be subdivided into four major steps: signal perception, signal transduction, transcriptional regulation of stress responsive genes, and

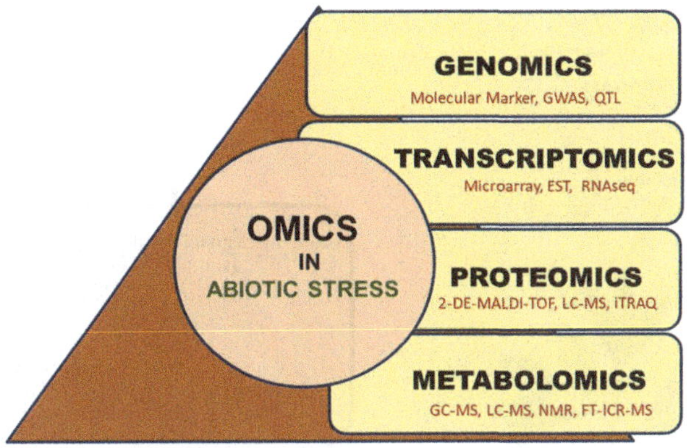


**Fig. 8.1** Omics approaches for studying abiotic stress responses in plants

finally expression of functional proteins that generate physiological and metabolic responses (Wang et al. 2016). A general signal transduction pathway in response to abiotic stresses has been shown in Fig. 8.2. Stress signals are perceived by receptors or sensors present on plasma membrane or cell wall. For example, in Arabidopsis OSCA1 (reduced hyperosmolality-induced calcium increase 1 (OSCA1), a plasma membrane bound channel protein acts as a sensor for osmotic stress (Yuan et al. 2014). Another sensor, COLD1 senses cold stress in rice (Ma et al. 2015). Upon stress perception, plant cells generate second messengers such as calcium $Ca^{2+}$, ROS Cyclic nucleotides (cAMP and cGMP), inositol tri phosphate, NO, and sugars. These second messengers initiate a signaling cascade via series of phosphorylation and dephosphorylation of proteins mediated by different types of kinases and phosphatases. At the end of phosphorylation events, transcription factor or other genes involved in stress acclimation gets activated. The transcription factors such as DREB, NAC, MYB, and WRKY bind to the promoter region of stress responsive proteins and regulate their transcription (Wang et al. 2016).

Phytohormones form an integral part of various signaling pathways activated in response to abiotic stresses (Peleg and Blumwald 2011). These molecules are synthesized at their site of action or reach to distant tissues by specific transporters (Peleg and Blumwald 2011). These compounds form an integral part of the plant signaling network. The mechanisms by which these signals are generated and translated into adaptations to counter the unfavorable environment are being intensively studied. Major class of phytohormones that are widely studied in stress responses include abscisic acid (ABA), ethylene (ET), cytokinin (CK), auxin (IAA), gibberellin (GA), jasmonate (JA), salicylic acid (SA), and strigolactone (SL) (Peleg and Blumwald 2011). The signaling pathways mediated by these phytohormones do not operate independently but crosstalk at several points (Peleg and Blumwald 2011). Omics has been utilized to study this complex networking of crosstalk among different signaling pathways.

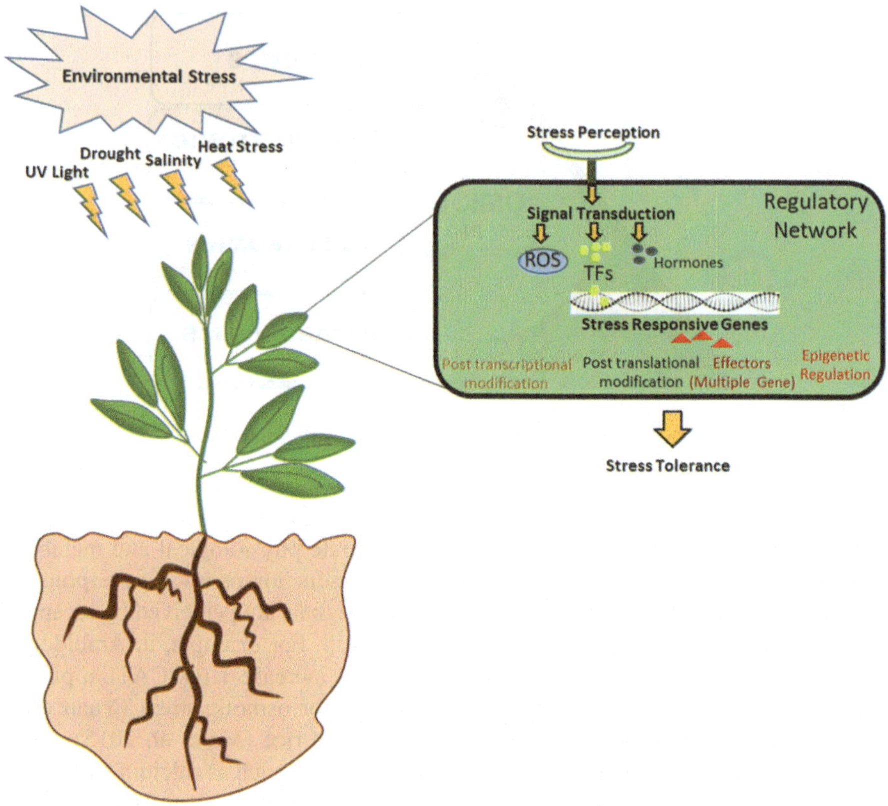


**Fig. 8.2** Schematic representation of general signal transduction pathway in response to abiotic stresses in plants

This chapter presents an overview of different omics technology like genomics, transcriptomics, proteomics, and metabolomics which were utilized for understanding abiotic stress response of the plants in the last two decades.

## 8.2 Genomics for Identification of Abiotic Stress Related Genes

The availability of whole genome sequences of Arabidopsis and other crops made it easy to study the expression profile of a large set of genes at a time in response to different abiotic stresses. Transcription factors are proteins consisting of sequence specific DNA binding domain that binds to the regulatory region of a gene and controls their expression. In Arabidopsis, 1500 genes code for transcription factors which constitute about 5% of the genome (Riechmann et al. 2000). In rice 1611 genes code for transcription factors (Xiong et al. 2005). These transcription factors play important role in plant abiotic stress responses. Major classes of transcription factors involved in abiotic stress tolerance include AP2/ERF, MYB, WRKY, NAC, and bZIP gene families (Wang et al. 2016). Genome wide identification and expression analysis of these transcription factors under different abiotic stresses have helped in narrowing down the candidate genes that can be functionally characterized for crop improvement. (Chen et al. 2014, Nakashima et al. 2012, Wang et al. 2018a, Wang et al. 2015, Xie et al. 2019, Zhang et al. 2012). Genomics has been widely used for identification of molecular marker, quantitative trait loci, and genomics selection associated with abiotic stress.

### 8.2.1 Molecular Markers Linked with Abiotic Stress Response

Molecular breeding of crops for the development of elite varieties with enhanced stress tolerance depends on the availability of molecular markers, genetic maps, and sequence information (Lata 2015). The advancement in the sequencing technologies and the availability of whole genome sequences of different crop germplasms expedite identification of single nucleotide polymorphisms (SNP) and simple sequence repeats markers (SSR). These polymorphisms within the germplasm are responsible for genetic diversity. Rice SNP-Seek database was developed using sequence information from 3000 Rice Genome Database containing 20 million SNPs (Alexandrov et al. 2015). Rice stress resistant SNP database was developed that focuses on SNPs related to biotic and abiotic stresses. It contains 9.5 million SNPs that were obtained from more 400 stress resistant rice varieties (Woldegiorgis et al. 2019).

### 8.2.2 Abiotic Stress Responsive Quantitative Trait Loci

Quantitative trait locus/loci are the locus in the genome which governs a particular trait on a quantitative scale. Complex traits such as tolerance to drought, salt, temperature, and other abiotic stresses are governed by a number of quantitative trait loci (QTLs). Identification and manipulation of these QTLs are crucial for successful implementation of genomic approaches for any marker assisted breeding programs. The identification of QTLs governing abiotic stress tolerance relies on combination of phenotypic data and genotypic data. QTLs can be identified using genetic fingerprinting, linkage maps, and QTL mapping (Deshmukh et al. 2014). These techniques need extensive genotypic information.

Smita et al. (2011) have developed a QlicRice database that can be used to retrieve information about abiotic stress responsive quantitative trait loci (QTLs) and genes present within that loci. SolQTL database was developed to upload raw genotype and phenotype data to Sol genomics Network repository. This database allows Solanaceae research community carry out QTL mapping and helps to identify to candidate genes responsible for phenotypic variation. It can also be used to get information about the markers more closely linked to QTLs (Tecle et al. 2010).

Another crucial technique that has been used to associate the genetic variation to a particular trait is genome-wide association studies (GWAS). The availability of whole genome sequence information has enabled breeders to dissect the genetic basis of agronomic traits such as drought, salt, and cold tolerance using GWAS (Li et al. 2019a). It has been used for identifying the relevant genes in several crop species like rice, wheat, maize, and barley (Guo et al. 2020, Li et al. 2019b, Mwando et al. 2020, Yano et al. 2016, Zhang et al. 2020). Recently, Hazzouri et al. (2018) have carried out a GWAS study comprising of 2671 barley lines and identified HKT1.5 ion transporter as one of the candidate genes involved in salt tolerance.

### 8.2.3 Genomic Selection Strategy for Improving Abiotic Stress Tolerance

Conventional marker assisted selection method makes use of molecular markers that are closely linked to trait of interest and accordingly, plants with desirable alleles are selected (Bhat et al. 2016). This method proved to be successful for the traits that are controlled by few numbers of QTLs. However, complex traits such as yield, quality and abiotic stress tolerance are governed by a large number of QTLs that may have large or small effects on the trait of interest (Bhat et al. 2016). Identification and quantification of the small QTLs with minor effects are not taken into consideration in MAS. To overcome this shortcoming, an advanced method of MAS known as Genomic Selection has been developed that takes into consideration all the molecular markers available across the genome affecting a trait of interest (Bhat et al. 2016; Crossa et al. 2017).

Genomic selection method uses a training population that has been genotyped and phenotyped to obtain the genomic estimated breeding values (GEBV) of individuals in a testing population (Crossa et al. 2017, Wang et al. 2018b). It allows one to identify the individuals based on its GEBV that can perform better as a parent in hybridization or for next generation advancement in a breeding program.

This technique was initially proposed by (Meuwissen 2003) for breeding plants in animals with desirable complex traits. But the major constraint associated with this technique is the non-availability of extensive genotyping platforms. In recent years, advances in next generation sequencing technologies have revolutionized the sequencing of entire genome of large number of accessions of different crops. It has significantly reduced the cost and time of sequencing different germplasms. This has helped in the development of cost effective and high throughput genotyping platforms for SNP discovery.

## 8.3 Depicting Transcriptome for Abiotic Stress Response

The study of total RNAs of a cell, tissue, or organism at a particular time is referred as transcriptome and accordingly, the field of study that deals with the quantification or abundance of the complete set of transcripts in a specific developmental, physiological, or stress condition is known as transcriptomics (Wang et al. 2009). The study of whole transcriptome began in early 1990s and is continuously transformed with parallel technological advancement. Broadly, there are two techniques involved: microarrays, which depend on the availability of predetermined sequences and RNA-Seq which involves the sequencing of all the transcripts.

Plants sense and respond to different kind of stresses by regulating the expression of the different kind of genes. Initially, most of the transcriptome studies were carried out using Sanger sequencing of expressed sequence tags or microarray-based gene expression analysis. These techniques helped in the identification of a number of candidate genes that are involved in salt, drought, temperature, or heavy metal stress response (Ding et al. 2011, Fernandez et al. 2008, Luo et al. 2010, Rabbani et al. 2003).

Genome wide expression profiling of genes under different kind of stresses such as drought, salinity, temperature, and metal stress resulted in the identification of different kind of stress responsive genes. Some of these get induced within the second or minutes while others take several hours to get induced. Accordingly, these genes are categorized as early responsive and late responsive genes, respectively (Lata 2015, Ramanjulu and Bartels 2002). Early responsive genes generally include sensors, protein kinases, and transcription factors which are involved in the stress perception and transduce the signal to the downstream pathway to initiate the response. Late responsive genes encode for ROS scavengers, transporters, heat shock proteins, and late embryogenesis proteins. Together, these proteins help in the maintenance of membrane integrity and ionic homeostasis within cell (Lata 2015, Ramanjulu and Bartels 2002). Transcriptomic changes vary with tissues in response to a particular stress. It has been observed that under drought stress, there is more reduction in the shoot growth, whereas the roots continue to grow in search of water. This differential growth pattern is due to difference in the expression of genes in these organs (Bashir et al. 2019). The tissues which perceive the stress undergo rapid or fast transcriptomic changes compared to the tissues that sense them at later stages (Bashir et al. 2019).

With the development of NGS, it is easy to study the whole transcriptomic changes during a particular stress or a combination of stresses (Ma et al. 2012). Comparative transcriptomic study of a tolerant and sensitive genotypes were carried out to identify the genes that are differently expressed in these genotypes and how they are linked to the tolerance level. This facilitates in the elucidation of molecular mechanism of stress response. Apart from transcript quantification, new genes or splice variants could be identified (Jain 2012, Ma et al. 2012).

Plants within a family differ in their response to abiotic stresses. For example, rice and barley belong to same family Poaceae, but differs in their ability to salt stress. Rice is salt susceptible while barley is tolerant. A microarray-based transcriptome study revealed that gene inducibility is better in barley compared to rice within the 1 hour of salt stress treatment. Also, $Na^+$ partitioning is better in barley compared to rice (Ueda et al. 2006). The physiological response and partitioning of $Na^+$ in the roots are correlated with differences in transcript abundances of genes involved in adaptive responses (Ueda et al. 2006). Similarly, various studies were carried out to get insight into the regulatory mechanisms that result in genotype dependent stress tolerance. Lenka et al. (2011) carried out a comparative transcriptomic study in *indica* rice genotypes, Nagina 22 (N22) and IR64 with contrasting drought tolerance. Both genotypes exhibited different global transcript profile in response to drought stress. Drought tolerance of N22 was attributed to enhanced expression of transcription factors such as *ZEF, MADS-box, LZP, WRKY, HSF, NAC, NFY,* etc. (Lenka et al. 2011). These transcription factors are involved in the activation of various signaling cascades to generate abiotic stress induced adaptive responses in plants. Examples of application of transcriptome technologies for studying abiotic stress responses have been shown in Table 8.1.

In field condition, plants often encounter multiple stresses at a time such as heat and drought together. It was observed that stress responses to a combination of stresses may be agonistic or antagonistic to the responses of individual stresses (Rasmussen et al. 2013) Transcriptome profiling under heat and drought stress combination suggested that their responses are not simply additive but may be synergistic in nature (Johnson et al. 2014, Wang et al. 2018c). Transcriptomic studies in Arabidopsis suggested that they have delineated coexpression network in response to single and stress combinations (Rasmussen et al. 2013). Transcriptomic studies helped in the identification of various stress induced genes and regulatory networks based on transcript quantification. However, many post-transcriptional changes take place before translation to functional proteins. Hence, studies at protein levels were focused to get clearer picture of abiotic stress response.

## 8.4 Proteomics Approaches Used in Abiotic Stress

Proteomics is more advanced and reliable technique compared to genomics or transcriptomics as it deals with the study of functional molecules, proteins. Many genes undergo posttranscriptional changes that could not be detected with transcriptomic approaches. Hence, although the study of alterations in the transcript levels gives ideas of the plant stress response, it could not detect the changes at post-transcriptional and post-translational levels. Proteins are the functional molecules that carry out various cellular processes. Hence, proteomics complement other omics approaches in elucidating the more complex stress responses (Kosová et al. 2018, Wu et al. 2016).

Like transcriptome, proteome is also dynamic and varies under different conditions (Ghosh and Xu 2014). Proteomic approaches have allowed the identification and quantification of novel stress related proteins (Kumar et al. 2014). Understanding the function of these proteins, their interaction with other proteins studies and post-translational modifications will help in elucidating the various stress responsive pathways in plants (Ghosh and Xu 2014).

Rapid progress in proteome profiling technologies has revolutionized the field which enabled us to compare stress induced responses in crop plants. Advanced proteomic methodologies such as Isobaric tags for relative and absolute quantitation

**Table 8.1** Example of applications of transcriptomic technologies for studying abiotic stress tolerance in plants

| Species | Sequencing platform | Gene name | Gene function | Stress | References |
|---|---|---|---|---|---|
| *Malus* × *domestica* | Illumina HiSeq 2500 platform | *DGK1* | Diacylglycerol kinase | Drought, cold and high salinity | Li et al. (2019c) |
| | | *ABCB13* | ABC transporter | | |
| | | *PP2C-37b* and *PP2C-77a* | Protein phosphatase 2C, | | |
| | | *ABI5-5b* | ABA pathways | | |
| | | *SAPK3* | Serine/threonine-protein kinase pathway | | |
| | | *HPt3a* | Histidine containing phosphotransferase | | |
| *Phalaris arundinacea* and *Dactylis glomerata* | Illumina HiSeq 2000 sequencer | Genes with NB-ARC domains | Bind and hydrolyze ATP, responses to pathogen attacks | Flooding and drought | Klaas et al. (2019) |
| | | Leucine-rich repeat domains | Protein-protein interactions | | |
| | | Protein kinase domains | Cellular regulation | | |
| *Zygophyllum xanthoxylum* | Illumina HiSeq™ 2000 sequencing platform | *SnRKs*, *PP2C* | ABA signaling | Salt and osmotic | Yin et al. (2019) |
| | | *ARF*, *3 GH3* and 8 *SAUR*, *TIR1/AFB* | Auxin signaling | | |
| Rice | Illumina sequencing platform | *NCED3*, *NCED4* | ABA biosynthesis, | Drought | Li et al. (2019b) |
| | | *Malate dehydrogenase* | TCA cycle | | |
| | | *SAL1*, *3′-phosphoadenosine 5′-phosphate (PAP)* and *ascorbate peroxidase (APX)* | Chloroplast retrograde signaling | | |
| | | *P5CS1* | Proline biosynthesis | | |
| Pearl millet | Isoform sequencing protocol | *HSP70*, *HSP90*, *HSP100* | Heat stress response | Heat and drought stress | Sun et al. (2020) |
| | | *PYL8* | ABA receptor | | |
| | | *SOD*, *APX*, *GPX* | ROS scavenging | | |
| | | *Asr* | Adaption to drought stress | | |

(continued)

**Table 8.1** (continued)

| Species | Sequencing platform | Gene name | Gene function | Stress | References |
|---|---|---|---|---|---|
| Maize | BGISEQ-500 platform | *Peroxidase 2* and *L-ascorbate peroxidase 1* | ROS accumulation | Salt tolerance | Chen et al. (2020) |
| | | *CBL-interacting protein kinase 14* | ABA signaling | | |
| | | *Mitogen-activated protein kinase kinase kinase 18* | Enzyme activation and inactivation through phosphorylation/ dephosphorylation | | |
| | | *NAC* TF | Salt stress response | | |
| *Arabidopsis* | Arabidopsis gene Chip microarray | *COR15B* | Cold response | Salt, osmotic, and cold | Kreps et al. (2002) |
| Rice | Illumina HiSeq. 2000 platform | *OsHAK21, OsCYL2* | Salt stress response | Alkaline stress | Li et al. (2018) |
| | | *ALMT* | Abiotic stress resistance | | |
| | | *CRKs* | Plant response to abiotic stress | | |
| *Solanum tuberosum* L. | Illumina HiSeq X ten platform | *RLKs* | Signal transducer or receptor | Salt | Li et al. (2020) |
| | | *RPK* | MAPK signaling pathway | | |
| | | *POD, CAT3, SOD, APX* | Antioxidant pathway | | |
| | | *CIPK11, CIPK14* | Calcium signaling | | |
| | | *STH-2-like* | Pathogenesis | | |
| Rice | Affymetrix 22 K Barley1 gene Chip microarray | *HvSUT1* | Sucrose transport | High-temperature | Mangelsen et al. (2011) |
| | | *HvSTP3* | Sugar transport | | |
| | | *HvUGE1* | Raffinose biosynthesis | | |
| | | *HvAPL2* | Large subunit of AGPase enzyme | | |
| | | *ACD1* | ROS scavenger | | |
| | | *SRO2* | ROS scavenger | | |

(iTRAQ), 2-dimensional difference gel electrophoresis method (2D-DIGE), and high-resolution tandem mass spectroscopy provide faster and more reliable results.

Initially, the stress responsive proteins were identified by a comparative study of the stressed and unstressed plant samples. In this approach, the stress responsive proteins were identified by their relative abundance in the proteome. The study of whole plant proteome at a time is complex, therefore gradually focus has been shifted towards cell or tissue specific responses (Gong et al. 2015).

Plant proteome undergoes different types of changes in response to stresses. This could change in the protein abundances or post-translation changes in the proteins that affect protein stability, their localization and activity (Kosová et al. 2011). Post-translational changes are often more important than protein abundance. Till date, nearly 300 different types of post-translational modifications were reported. Among these phosphorylation, ubiquitination, nitrosylation, glycosylation, and carbonylation are most commonly studied (Wu et al. 2016). There are several studies that were carried out to find out how the proteins undergo post-translational changes under abiotic stresses (Wu et al. 2016). ABA is one of the important hormones regulating plant development and stress responses. Qiu et al. (2017) carried out a comprehensive proteomic study to identify proteins that get phosphorylated in response to ABA treatments in rice. They reported that 1060 proteins are differentially phosphorylated in response to ABA. In another phosphoproteomic study, 482 proteins were found to be differentially phosphorylated in response to Cd stress. The differentially phosphorylated proteins belong to ABA signaling pathway, ROS scavengers, and transcription factors (Zhong et al. 2017). A list of studies on application of proteomics to study abiotic stress response in various plants has been shown in Table 8.2.

## 8.5 Role of Metabolomics in Mitigating Abiotic Stress

Abiotic stresses bring different kinds of metabolic changes in plants such as modification in enzymatic activities, substrates scarcity for essential metabolic reactions, increased ROS levels (Obata and Fernie 2012; Kumar et al. 2017). Most stress conditions perturb the metabolic profile of plants. This is due to uncontrolled production of reactive oxygen species (ROS) within cells which in turn oxidizes and damages the cellular components and finally leads to cell death (Obata and Fernie 2012). Plants produce different kinds of ROS as by-products of various metabolic processes. These molecules are confined to specific cellular components such as chloroplasts, mitochondria, and peroxisomes under optimal growth conditions (Huang et al. 2019). Abiotic stresses perturb the cellular homeostasis and negatively affect plant growth and development. In response to abiotic stresses, plants reprogram its metabolic network to attain a new steady state. The metabolic reprogramming ensures the production of various stress responsive compounds. These are attributed to the change in the enzymatic activities, substrate concentration, or production of antioxidants to scavenge the excessive ROS produced as a result of unfavorable growth conditions. ROS scavenging mechanisms could be enzymatic or non-enzymatic antioxidant defense system (Huang et al. 2019). Both

**Table 8.2** Recent examples of application of proteomic technologies for studying abiotic stress response in plants

| S. No. | Plant name | Abiotic stress | Proteomic technique | Identified proteins | Regulated function | References |
|---|---|---|---|---|---|---|
| 1 | Wheat (cv. Bahar/ Kavir) | Drought | 2D gel electrophoresis and LC-MS/MS analysis | 85/20 | Photosynthesis, carbohydrate metabolic process, and nitrogen | Michaletti et al. (2018) |
| 2 | Maize | Drought | iTRAQ | 111 | Mitochondrial electron transport, photosynthesis, carbon fixation, photorespiration, ATP synthesis coupled electron transport | Jiang et al. (2019) |
| 3 | Soybean | Drought and heat | 2D-DIGE and MALDI-TOF MS analysis | 44 | Photosynthesis, ATP synthesis, and protein biosynthesis | Das et al. (2016) |
| 4 | Rice | Drought | 2D | 42 | Drought-responsive proteins, an actin depolymerizing factor, S-like RNase homologue, actin depolymerizing factor and rubisco activase | Salekdeh et al. (2002) |
| 5 | Cassava | Cold | 2-DE | 44 | Photosynthesis metabolism, defense system, carbohydrate and energy metabolism, inorganic ion transport and metabolism | An et al. (2016) |
| 6 | Soybean | Chilling temperature | 2-DE | 40 | Metabolic pathways including, cell defense, energy, protein synthesis, cell growth/division, storage, transcription and transport | Cheng et al. (2010) |
| 7 | Coconut | Low temperature | iTRAQ | 193 | Metabolism, stress response, photosynthesis and respiration | Yang et al. (2020) |
| 8 | Brassica napus | Drought | iTRAQ LC–MS/MS | 1976 | Metabolism, protein folding and degradation, and signaling, energy (photosynthesis), protein synthesis, and stress and defense | Koh et al. (2015) |
| 9 | Radish | Heat | 2-DE and MALDI-TOF MS analysis | 11 | Heat shock proteins (HSPs), energy and metabolism, redox homeostasis, and signal transduction | Zhang et al. (2013) |

| | | | | | | |
|---|---|---|---|---|---|---|
| 10 | Rice | Ozone stress | 2-DE | 52 | Pathogenesis related (PR) class 5 protein, ascorbate peroxidase(s), superoxide dismutase, calcium-binding protein, calreticulin, a novel ATP-dependent CLP protease | Agrawal et al. (2002) |
| 11 | Wheat | Hydrogen peroxide stress | iTRAQ | 157 | Stress/defense/detoxification, signal transduction, and carbohydrate metabolism | Ge et al. (2013) |
| 12 | Foxtail millet | Salt | 2-DE and MALDI-TOF MS analysis | 175 | Signal transduction, photosynthesis, cell wall biogenesis, stress related and several metabolisms like energy, lipid, nitrogen, carbohydrate, and nucleotide metabolism | Veeranagamallaiah et al. (2008) |
| 13 | *Pinellia ternata* | Heat | 2-DE and MALDI-TOF MS analysis | 24 | Small heat shock proteins, RNA processing, photosynthesis, chlorophyll biosynthetic processes, protein degradation and defense | Zhu et al. (2013) |
| 14 | Alfalfa | Salt and drought stress | 2-DE and MALDI-TOF-MS/MS | 24 | Salt- and PEG-mediated stress included defense response, energy metabolism, protein synthesis and degradation, oxidative stress, carbohydrate metabolism-associated proteins | Ma et al. (2017) |

mechanisms work together to scavenge the ROS produced in plants. Enzymatic method involves various enzymatic reactions catalyzed by super oxide dismutase (SOD), catalase (CAT), ascorbate peroxidase (APX), and glutathione peroxidase (GPX) (Apel and Hirt 2004). Non-enzymatic mechanism involves low molecular weight antioxidants such as glutathione, ascorbic acid (AsA), amino acids (proline), sugar, carotenoids, glucosinolates, and flavonoids (Nakabayashi and Saito 2015).

Metabolic adjustment in plants varies with genotype, duration, intensity, and type of stress encountered (Obata and Fernie 2012). For example, salt tolerant genotypes of rice and barley accumulate higher levels of sugars and amino acids compared to salt sensitive genotype (Zhao et al. 2014; Shen et al. 2016). Similarly, stress type also decides the metabolite accumulation. Higher proline accumulation favors tolerance to salt, drought, cold, and heavy metal stress in many plant species. However, higher temperature induced proline accumulation did not confer stress tolerance in Arabidopsis (Lv et al. 2011).

Metabolomics is one of the omics approaches which give a snapshot of the metabolite pool of the cell at a particular stage. The metabolic pool of a plant determines its phenotype. Accumulation of various metabolites depends on the expression of various genes in a particular condition. Plant metabolomics aims to study the total metabolite present in a tissue at a particular developmental stage. The metabolite content of the plants varies with external environment. Total metabolic pool exhibits enormous diversity in the chemical nature; hence it is not possible to study all the metabolites with a single technique. It is estimated that Arabidopsis may contain around 5000 different metabolites. KNApSAcK database dedicated for species metabolite relationship contains more than 51,000 metabolite entries (Afendi et al. 2012). Most common techniques employed for metabolic profiling in plants include gas chromatography mass spectrometry (GC-MS), liquid chromatography mass spectrometry (LC-MS), capillary electrophoresis mass spectrometry (CE-MS), Fourier transform ion cyclotron resonance mass spectrometry (FT-ICR-MS), and nuclear magnetic resonance (NMR) (Jorge et al. 2016).

Abiotic stresses affect the metabolic networks in plants. There are numerous efforts made to elucidate the effect of a stress on the metabolic state of crops. In rice, potential metabolite markers associated with abiotic stress tolerance were identified (Degenkolbe et al. 2013, Maruyama et al. 2014). Flavonoids are secondary metabolites with strong in vitro antioxidant property. Integrated transcriptomic and metabolic approaches proved the in vivo antioxidant property of flavonoids. Arabidopsis transgenic lines accumulating flavonoids were tolerant to oxidative and drought stress (Nakabayashi and Saito 2015). Another report, Pi et al. (2016) used integrated phosphoproteomic and metabolic approach to study the salt stress response in Soybean. They found that salt stress induces phosphorylation of transcription factor GmMYB173 that regulates biosynthesis of dihydroxy B-ring flavonoids which in turn contribute to salt tolerance.

Xiong et al. (2019) carried out a comprehensive proteomic and metabolic analysis to dissect the drought and submergence response in rice during panicle differentiation stage. They found that several metabolic pathways such as energy metabolism, carbon fixation, ROS metabolisms are affected by drought and submergence stress.

Drought and heat stresses affect the reproductive ability of crops. Using metabolic and transcriptomic approaches, it was found that sugar starvation is the primary determinant of reproductive success of crops (Li et al. 2015, Zhang et al. 2010).

Recently, integrated omics approaches were applied to understand the complex genetic interaction that regulate plant metabolism. For example, transcriptomic and metabolic approaches are integrated together. Recently, a study has been carried out in sesame using a combination of transcriptomic and metabolomics approaches to understand drought tolerance mechanism (You et al. 2019). Integrated omic approaches suggested that ABA, amino acids (proline, arginine, lysine, and aromatic branched amino acids) GABA, organic acids, and sugars are more abundant in drought tolerant genotypes compared to drought sensitive genotype. Comparison between transcriptome and metabolomics data suggests that there is a correlation between the expression and metabolic profile of these metabolites (You et al. 2019). Similarly there are several reports where proteomic and metabolomics were integrated to understand the stress responses. Adaptive responses in response to water deficit conditions have been studied in spring wheat using proteomic and metabolomics approaches in two contrasting genotypes (Michaletti et al. 2018). It was observed that leaf proteome of drought sensitive cultivar is severely affected by water deficit conditions. Proteins involved in maintaining photosynthetic machinery, ROS detoxification, and sugar and nitrogen metabolism were drastically reduced in drought sensitive genotype. Metabolomics data revealed that metabolites such as amino acids, organic acids, and sugars showed increased abundance in response to water deficit (Michaletti et al. 2018).

A new approach called Metabolomic quantitative trait loci (mQTL) mapping is being used to associate a genetic loci with the variability in the metabolic profile (Abdelrahman et al. 2018). Linkage mapping of mQTL to the genetic variant is essential component of metabolite assisted breeding as it helps in the identification of genes that are associated with production of metabolites (Abdelrahman et al. 2018). Mqtl mapping has been used for genotype to phenotype association in several cereal crops (Gong et al. 2013, Lisec et al. 2011, Matsuda et al. 2012, Shi et al. 2020).

Integrated metabolic approaches and mQTL mapping will be of great use in crop breeding. In conventional breeding programs, selection of individuals based on genetic markers is affected by environmental factors.

## 8.6 Conclusions

Abiotic stresses severely reduce crop yields and serve as a major threat to food security in future. Elucidation of abiotic stress response in crop plants is important to develop climate resilient crop varieties to ensure enough food supply to the growing population. Different omics tools like genomics, transcriptomics, proteomics, and metabolomics have been employed to understand the molecular mechanism underlying the adaptive responses. The technological advances in various omics tools have made them cost effective and easily accessible for different studies. Numerous genes, proteins, and metabolites were identified that are involved in conferring stress tolerance in crops. Although there are several reports on effect of abiotic stresses on particular tissues of the contrasting genotypes, studies with multi-tissue, developmental stages, or at whole plant level are very limited. This is one of the major limitations in understanding the complex nature of abiotic stress responses as it varies with tissue, developmental stage, and at whole plant level. Abiotic stress tolerance traits are complex and regulated at several levels like post-transcriptional, translational, and post-translational levels. Hence, more integrated omics studies should be undertaken for the better understanding of stress responses at molecular level. In this chapter, we discussed role of different omics tools in understanding abiotic stress responses.

**Acknowledgements** IA acknowledges University Grants Commission, Government of India, for providing DS Kothari Post-Doctoral Fellowship.

**Conflict of Interest** The authors have no competing interest to declare.

## References

Abdelrahman M, Burritt DJ, Tran L-SP (2018) The use of metabolomic quantitative trait locus mapping and osmotic adjustment traits for the improvement of crop yields under environmental stresses. In: Seminars in cell & developmental biology. Elsevier, Amsterdam, pp 86–94

Afendi FM, Okada T, Yamazaki M, Hirai-Morita A, Nakamura Y, Nakamura K, Ikeda S, Takahashi H, Altaf-Ul-Amin M, Darusman LK (2012) KNApSAcK family databases: integrated metabolite–plant species databases for multifaceted plant research. Plant Cell Physiol 53:e1

Agrawal GK, Rakwal R, Yonekura M, Kubo A, Saji H (2002) Proteome analysis of differentially displayed proteins as a tool for investigating ozone stress in rice (*Oryza sativa* L.) seedlings. Proteomics 2:947–959

Alexandrov N, Tai S, Wang W, Mansueto L, Palis K, Fuentes RR, Ulat VJ, Chebotarov D, Zhang G, Li Z (2015) SNP-seek database of SNPs derived from 3000 rice genomes. Nucleic Acids Res 43:D1023–D1027

An F, Li G, Li QX, Li K, Carvalho LJ, Ou W, Chen S (2016) The comparatively proteomic analysis in response to cold stress in cassava plantlets. Plant Mol Biol Report 34:1095–1110

Apel K, Hirt H (2004) Reactive oxygen species: metabolism, oxidative stress, and signal transduction. Annu Rev Plant Biol 55:373–399

Bashir K, Matsui A, Rasheed S, Seki M (2019) Recent advances in the characterization of plant transcriptomes in response to drought, salinity, heat, and cold stress. F1000Research 8:F1000

Bhat JA, Ali S, Salgotra RK, Mir ZA, Dutta S, Jadon V, Tyagi A, Mushtaq M, Jain N, Singh PK (2016) Genomic selection in the era of next generation sequencing for complex traits in plant breeding. Front Genet 7:221

Chen F, Fang P, Zeng W, Ding Y, Zhuang Z, Peng Y (2020) Comparing transcriptome expression profiles to reveal the mechanisms of salt tolerance and exogenous glycine betaine mitigation in maize seedlings. PLoS One 15:e0233616

Chen N, Yang Q, Pan L, Chi X, Chen M, Hu D, Yang Z, Wang T, Wang M, Yu S (2014) Identification of 30 MYB transcription factor genes and analysis of their expression during abiotic stress in peanut (*Arachis hypogaea* L.). Gene 533:332–345

Cheng L, Gao X, Li S, Shi M, Javeed H, Jing X, Yang G, He G (2010) Proteomic analysis of soybean [*Glycine max* (L.) Meer.] seeds during imbibition at chilling temperature. Mol Breed 26:1–17

Crossa J, Pérez-Rodríguez P, Cuevas J, Montesinos-López O, Jarquín D, de los Campos G, Burgueño J, González-Camacho JM, Pérez-Elizalde S, Beyene Y (2017) Genomic selection in plant breeding: methods, models, and perspectives. Trends Plant Sci 22:961–975

Das A, Eldakak M, Paudel B, Kim D-W, Hemmati H, Basu C, Rohila JS (2016) Leaf proteome analysis reveals prospective drought and heat stress response mechanisms in soybean. Biomed Res Int 2016:23

Degenkolbe T, Do PT, Kopka J, Zuther E, Hincha DK, Köhl KI (2013) Identification of drought tolerance markers in a diverse population of rice cultivars by expression and metabolite profiling. PLoS One 8:e63637

Deshmukh R, Sonah H, Patil G, Chen W, Prince S, Mutava R, Vuong T, Valliyodan B, Nguyen HT (2014) Integrating omic approaches for abiotic stress tolerance in soybean. Front Plant Sci 5:244

Ding Y, Chen Z, Zhu C (2011) Microarray-based analysis of cadmium-responsive microRNAs in rice (*Oryza sativa*). J Exp Bot 62:3563–3573

Fernandez P, Di Rienzo J, Fernandez L, Hopp HE, Paniego N, Heinz RA (2008) Transcriptomic identification of candidate genes involved in sunflower responses to chilling and salt stresses based on cDNA microarray analysis. BMC Plant Biol 8:11

Ge P, Hao P, Cao M, Guo G, Lv D, Subburaj S, Li X, Yan X, Xiao J, Ma W (2013) iTRAQ-based quantitative proteomic analysis reveals new metabolic pathways of wheat seedling growth under hydrogen peroxide stress. Proteomics 13:3046–3058

Ghosh D, Xu J (2014) Abiotic stress responses in plant roots: a proteomics perspective. Front Plant Sci 5:6

Gong F, Hu X, Wang W (2015) Proteomic analysis of crop plants under abiotic stress conditions: where to focus our research? Front Plant Sci 6:418

Gong L, Chen W, Gao Y, Liu X, Zhang H, Xu C, Yu S, Zhang Q, Luo J (2013) Genetic analysis of the metabolome exemplified using a rice population. Proc Natl Acad Sci 110:20320–20325

Guo J, Li C, Zhang X, Li Y, Zhang D, Shi Y, Song Y, Li Y, Yang D, Wang T (2020) Transcriptome and GWAS analyses reveal candidate gene forseminal root length of maize seedlings under drought stress. Plant Sci 292:110380

Hazzouri KM, Khraiwesh B, Amiri K, Pauli D, Blake T, Shahid M, Mullath SK, Nelson D, Mansour AL, Salehi-Ashtiani K (2018) Mapping of *HKT1; 5* gene in barley using GWAS approach and its implication in salt tolerance mechanism. Front Plant Sci 9:156

Hickey LT, Hafeez AN, Robinson H, Jackson SA, Leal-Bertioli SC, Tester M, Gao C, Godwin ID, Hayes BJ, Wulff BB (2019) Breeding crops to feed 10 billion. Nat Biotechnol 37:744–754

Huang H, Ullah F, Zhou D-X, Yi M, Zhao Y (2019) Mechanisms of ROS regulation of plant development and stress responses. Front Plant Sci 10:800

Jain M (2012) Next-generation sequencing technologies for gene expression profiling in plants. Brief Funct Genomics 11:63–70

Jiang Z, Jin F, Shan X, Li Y (2019) iTRAQ-based proteomic analysis reveals several strategies to cope with drought stress in maize seedlings. Int J Mol Sci 20:5956

Johnson SM, Lim F-L, Finkler A, Fromm H, Slabas AR, Knight MR (2014) Transcriptomic analysis of *Sorghum bicolor* responding to combined heat and drought stress. BMC Genomics 15:456

Jorge TF, Rodrigues JA, Caldana C, Schmidt R, van Dongen JT, Thomas-Oates J, António C (2016) Mass spectrometry-based plant metabolomics: metabolite responses to abiotic stress. Mass Spectrom Rev 35:620–649

Klaas M, Haiminen N, Grant J, Cormican P, Finnan J, Arojju SK, Utro F, Vellani T, Parida L, Barth S (2019) Transcriptome characterization and differentially expressed genes under flooding and drought stress in the biomass grasses *Phalaris arundinacea* and *Dactylis glomerata*. Ann Bot 124:717–730

Koh J, Chen G, Yoo M-J, Zhu N, Dufresne D, Erickson JE, Shao H, Chen S (2015) Comparative proteomic analysis of *Brassica napus* in response to drought stress. J Proteome Res 14:3068–3081

Kosová K, Vítámvás P, Prášil IT, Renaut J (2011) Plant proteome changes under abiotic stress—contribution of proteomics studies to understanding plant stress response. J Proteome 74:1301–1322

Kosová K, Vítámvás P, Urban MO, Prášil IT, Renaut J (2018) Plant abiotic stress proteomics: the major factors determining alterations in cellular proteome. Front Plant Sci 9:122

Kreps JA, Wu Y, Chang H-S, Zhu T, Wang X, Harper JF (2002) Transcriptome changes for Arabidopsis in response to salt, osmotic, and cold stress. Plant Physiol 130:2129–2141

Kumar A, Bimolata W, Kannan M, Kirti PB, Qureshi IA, Ghazi IA (2014) Comparative proteomics reveals differential induction of both biotic and abiotic stress response associated proteins in rice during Xanthomonas oryzae pv. Oryzae infection. Funct Integr Genomics 15(4):425–437

Kumar R, Bohra A, Pandey AK, Pandey MK, Kumar A (2017) Metabolomics for plant improvement: status and prospects. Front Plant Sci 8:1302

Lata C (2015) Advances in omics for enhancing abiotic stress tolerance in millets. Proc Indian Natl Sci Acad 81(2):397–417

Lay JO Jr, Liyanage R, Borgmann S, Wilkins CL (2006) Problems with the "omics". TrAC Trends Anal Chem 25:1046–1056

Lenka SK, Katiyar A, Chinnusamy V, Bansal KC (2011) Comparative analysis of drought-responsive transcriptome in *Indica* rice genotypes with contrasting drought tolerance. Plant Biotechnol J 9:315–327

Li L, Mao X, Wang J, Chang X, Reynolds M, Jing R (2019b) Genetic dissection of drought and heat-responsive agronomic traits in wheat. Plant Cell Environ 42:2540–2553

Li N, Liu H, Sun J, Zheng H, Wang J, Yang L, Zhao H, Zou D (2018) Transcriptome analysis of two contrasting rice cultivars during alkaline stress. Sci Rep 8:1–16

Li P, Yang H, Wang L, Liu H, Huo H, Zhang C, Liu A, Zhu A, Hu J, Lin Y (2019c) Physiological and transcriptome analyses reveal short-term responses and formation of memory under drought stress in rice. Front Genet 10:55

Li Q, Qin Y, Hu X, Li G, Ding H, Xiong X, Wang W (2020) Transcriptome analysis uncovers the gene expression profile of salt-stressed potato (*Solanum tuberosum* L.). Sci Rep 10:1–19

Li X, Lawas LM, Malo R, Glaubitz U, Erban A, Mauleon R, Heuer S, Zuther E, Kopka J, Hincha DK (2015) Metabolic and transcriptomic signatures of rice floral organs reveal sugar starvation as a factor in reproductive failure under heat and drought stress. Plant Cell Environ 38:2171–2192

Li X, Li M, Zhou B, Yang Y, Wei Q, Zhang J (2019a) Transcriptome analysis provides insights into the stress response crosstalk in apple (*Malus*× *domestica*) subjected to drought, cold and high salinity. Sci Rep 9:1–10

Lisec J, Römisch-Margl L, Nikoloski Z, Piepho HP, Giavalisco P, Selbig J, Gierl A, Willmitzer L (2011) Corn hybrids display lower metabolite variability and complex metabolite inheritance patterns. Plant J 68:326–336

Luo M, Liu J, Lee RD, Scully BT, Guo B (2010) Monitoring the expression of maize genes in developing kernels under drought stress using oligo-microarray. J Integr Plant Biol 52:1059–1074

Lv W-T, Lin B, Zhang M, Hua X-J (2011) Proline accumulation is inhibitory to Arabidopsis seedlings during heat stress. Plant Physiol 156:1921–1933

Ma Q, Kang J, Long R, Zhang T, Xiong J, Zhang K, Wang T, Yang Q, Sun Y (2017) Comparative proteomic analysis of alfalfa revealed new salt and drought stress-related factors involved in seed germination. Mol Biol Rep 44:261–272

Ma Y, Dai X, Xu Y, Luo W, Zheng X, Zeng D, Pan Y, Lin X, Liu H, Zhang D (2015) COLD1 confers chilling tolerance in rice. Cell 160:1209–1221

Ma Y, Qin F, Tran L-SP (2012) Contribution of genomics to gene discovery in plant abiotic stress responses. Mol Plant 5:1176–1178

Mangelsen E, Kilian J, Harter K, Jansson C, Wanke D, Sundberg E (2011) Transcriptome analysis of high-temperature stress in developing barley caryopses: early stress responses and effects on storage compound biosynthesis. Mol Plant 4:97–115

Maruyama K, Urano K, Yoshiwara K, Morishita Y, Sakurai N, Suzuki H, Kojima M, Sakakibara H, Shibata D, Saito K (2014) Integrated analysis of the effects of cold and dehydration on rice metabolites, phytohormones, and gene transcripts. Plant Physiol 164:1759–1771

Matsuda F, Okazaki Y, Oikawa A, Kusano M, Nakabayashi R, Kikuchi J, Yonemaru JI, Ebana K, Yano M, Saito K (2012) Dissection of genotype–phenotype associations in rice grains using metabolome quantitative trait loci analysis. Plant J 70:624–636

Meuwissen T (2003) Genomic selection: the future of marker assisted selection and animal breeding. In: Marker Assisted selection: a fast track to increase genetic gain in plants and animal breeding. FAO/Turín, Italy, pp 54–59

Michaletti A, Naghavi MR, Toorchi M, Zolla L, Rinalducci S (2018) Metabolomics and proteomics reveal drought-stress responses of leaf tissues from spring-wheat. Sci Rep 8:1–18

Mwando E, Han Y, Angessa, TT, Zhou G, Hill CB, Zhang XQ, Li C (2020) Genome-wide association study of salinity tolerance during germination inbarley (Hordeum vulgare L.). Front Plant Sci 11:118

Nakabayashi R, Saito K (2015) Integrated metabolomics for abiotic stress responses in plants. Curr Opin Plant Biol 24:10–16

Nakashima K, Takasaki H, Mizoi J, Shinozaki K, Yamaguchi-Shinozaki K (2012) NAC transcription factors in plant abiotic stress responses. Biochim Biophys Acta (BBA) 1819:97–103

Obata T, Fernie AR (2012) The use of metabolomics to dissect plant responses to abiotic stresses. Cell Mol Life Sci 69:3225–3243

Peleg Z, Blumwald E (2011) Hormone balance and abiotic stress tolerance in crop plants. Curr Opin Plant Biol 14:290–295. https://doi.org/10.1016/j.pbi.2011.02.001

Pi E, Qu L, Hu J, Huang Y, Qiu L, Lu H, Jiang B, Liu C, Peng T, Zhao Y (2016) Mechanisms of soybean roots' tolerances to salinity revealed by proteomic and phosphoproteomic comparisons between two cultivars. Mol Cell Proteomics 15:266–288

Qiu J, Hou Y, Wang Y, Li Z, Zhao J, Tong X, Lin H, Wei X, Ao H, Zhang J (2017) A comprehensive proteomic survey of ABA-induced protein phosphorylation in rice (*Oryza sativa* L.). Int J Mol Sci 18:60

Rabbani MA, Maruyama K, Abe H, Khan MA, Katsura K, Ito Y, Yoshiwara K, Seki M, Shinozaki K, Yamaguchi-Shinozaki K (2003) Monitoring expression profiles of rice genes under cold, drought, and high-salinity stresses and abscisic acid application using cDNA microarray and RNA gel-blot analyses. Plant Physiol 133:1755–1767

Ramanjulu S, Bartels D (2002) Drought-and desiccation-induced modulation of gene expression in plants. Plant Cell Environ 25:141–151

Rasmussen S, Barah P, Suarez-Rodriguez MC, Bressendorff S, Friis P, Costantino P, Bones AM, Nielsen HB, Mundy J (2013) Transcriptome responses to combinations of stresses in Arabidopsis. Plant Physiol 161:1783–1794

Riechmann JL, Heard J, Martin G, Reuber L, Jiang C-Z, Keddie J, Adam L, Pineda O, Ratcliffe O, Samaha R (2000) Arabidopsis transcription factors: genome-wide comparative analysis among eukaryotes. Science 290:2105–2110

Rozema J, Flowers T (2008) Crops for a salinized world. Science 322(5907):1478–1480

Salekdeh GH, Siopongco J, Wade LJ, Ghareyazie B, Bennett J (2002) Proteomic analysis of rice leaves during drought stress and recovery. Proteomics 2:1131–1145

Shen Q, Fu L, Dai F, Jiang L, Zhang G, Wu D (2016) Multi-omics analysis reveals molecular mechanisms of shoot adaption to salt stress in Tibetan wildbarley. Bmc Genomics 17(1):1–15

Shi T, Zhu A, Jia J, Hu X, Chen J, Liu W, Ren X, Sun D, Fernie AR, Cui F (2020) Metabolomics analysis and metabolite-agronomic trait associations using kernels of wheat (*Triticum aestivum*) recombinant inbred lines. Plant J 103(1):279–292

Smita S, Lenka SK, Katiyar A, Jaiswal P, Preece J, Bansal KC (2011) QlicRice: a web interface for abiotic stress responsive QTL and loci interaction channels in rice. Database 2011:bar037

Sun M, Huang D, Zhang A, Khan I, Yan H, Wang X, Zhang X, Zhang J, Huang L (2020) Transcriptome analysis of heat stress and drought stress in pearl millet based on Pacbio full-length transcriptome sequencing. BMC Plant Biol 20:1–15

Tecle IY, Menda N, Buels RM, van der Knaap E, Mueller LA (2010) solQTL: a tool for QTL analysis, visualization and linking to genomes at SGN database. BMC Bioinform 11:1–9

Ueda A, Kathiresan A, Bennett J, Takabe T (2006) Comparative transcriptome analyses of barley and rice under salt stress. Theor Appl Genet 112:1286–1294

Veeranagamallaiah G, Jyothsnakumari G, Thippeswamy M, Reddy PCO, Surabhi G-K, Sriranganayakulu G, Mahesh Y, Rajasekhar B, Madhurarekha C, Sudhakar C (2008) Proteomic analysis of salt stress responses in foxtail millet (*Setaria italica* L. cv. Prasad) seedlings. Plant Sci 175:631–641

Wang H, Wang H, Shao H, Tang X (2016) Recent advances in utilizing transcription factors to improve plant abiotic stress tolerance by transgenic technology. Front Plant Sci 7:67

Wang L, Liu L, Ma Y, Li S, Dong S, Zu W (2018a) Transcriptome profilling analysis characterized the gene expression patterns responded to combined drought and heat stresses in soybean. Comput Biol Chem 77:413–429

Wang X, Xu Y, Hu Z, Xu C (2018b) Genomic selection methods for crop improvement: current status and prospects. Crop J 6:330–340

Wang Y, Zhang Y, Zhou R, Dossa K, Yu J, Li D, Liu A, Mmadi MA, Zhang X, You J (2018c) Identification and characterization of the bZIP transcription factor family and its expression in response to abiotic stresses in sesame. PLoS One 13:e0200850

Wang Z, Cheng K, Wan L, Yan L, Jiang H, Liu S, Lei Y, Liao B (2015) Genome-wide analysis of the basic leucine zipper (bZIP) transcription factor gene family in six legume genomes. BMC Genom 16:1053

Wang Z, Gerstein M, Snyder M (2009) RNA-Seq: a revolutionary tool for transcriptomics. Nat Rev Genet 10:57–63

Woldegiorgis ST, Wang S, He Y, Xu Z, Chen L, Tao H, Zhang Y, Zou Y, Harrison A, Zhang L (2019) Rice stress-resistant SNP database. Rice 12:97

Wu X, Gong F, Cao D, Hu X, Wang W (2016) Advances in crop proteomics: PTMs of proteins under abiotic stress. Proteomics 16:847–865

Xie Z, Nolan TM, Jiang H, Yin Y (2019) AP2/ERF transcription factor regulatory networks in hormone and abiotic stress responses in Arabidopsis. Front Plant Sci 10:228

Xiong Q, Cao C, Shen T, Zhong L, He H, Chen X (2019) Comprehensive metabolomic and proteomic analysis in biochemical metabolic pathways of rice spikes under drought and submergence stress. Biochimica et Biophysica Acta (BBA)-proteins and. Proteomics 1867:237–247

Xiong Y, Liu T, Tian C, Sun S, Li J, Chen M (2005) Transcription factors in rice: a genome-wide comparative analysis between monocots and eudicots. Plant Mol Biol 59:191–203

Yang Y, Saand MA, Abdelaal WB, Zhang J, Wu Y, Li J, Fan H, Wang F (2020) iTRAQ-based comparative proteomic analysis of two coconut varieties reveals aromatic coconut cold-sensitive in response to low temperature. J Proteomics 220:103766

Yano K, Yamamoto E, Aya K, Takeuchi H, Lo PC, Hu L, Yamasaki M, Yoshida S, Kitano H, Hirano K, Matsuoka M (2016) Genome-wideassociation study using whole-genome sequencing rapidly identifies new genes influencing agronomic traits in rice. Nat Genet 48(8):927

Yin H, Li M, Li D, Khan S-A, Hepworth SR, Wang S-M (2019) Transcriptome analysis reveals regulatory framework for salt and osmotic tolerance in a succulent xerophyte. BMC Plant Biol 19:1–15

You J, Zhang Y, Liu A, Li D, Wang X, Dossa K, Zhou R, Yu J, Zhang Y, Wang L (2019) Transcriptomic and metabolomic profiling of drought-tolerant and susceptible sesame genotypes in response to drought stress. BMC Plant Biol 19:1–16

Yuan F, Yang H, Xue Y, Kong D, Ye R, Li C, Zhang J, Theprungsirikul L, Shrift T, Krichilsky B (2014) OSCA1 mediates osmotic-stress-evoked $Ca^{2+}$ increases vital for osmosensing in Arabidopsis. Nature 514:367–371

Zhang X, Guan Z, Wang L, Fu J, Zhang Y, Li Z, Ma L, Liu P, Zhang Y, Liu M, Li P (2020) Combined GWAS and QTL analysis for dissecting thegenetic architecture of kernel test weight in maize. Mol Genet Gen 295(2):409–420

Zhang H, Liang W, Yang X, Luo X, Jiang N, Ma H, Zhang D (2010) Carbon starved anther encodes a MYB domain protein that regulates sugar partitioning required for rice pollen development. Plant Cell 22:672–689
Zhang L, Zhao G, Jia J, Liu X, Kong X (2012) Molecular characterization of 60 isolated wheat MYB genes and analysis of their expression during abiotic stress. J Exp Bot 63:203–214
Zhang Y, Xu L, Zhu X, Gong Y, Xiang F, Sun X, Liu L (2013) Proteomic analysis of heat stress response in leaves of radish (*Raphanus sativus* L.). Plant Mol Biol Report 31:195–203
Zhao X, Wang W, Zhang F, Deng J, Li Z, Fu B (2014). Comparative metabolite profiling of two rice genotypes with contrasting salt stress tolerance atthe seedling stage. PloS one 9(9):e108020
Zhong M, Li S, Huang F, Qiu J, Zhang J, Sheng Z, Tang S, Wei X, Hu P (2017) The phosphoproteomic response of rice seedlings to cadmium stress. Int J Mol Sci 18:2055
Zhu Y, Zhu G, Guo Q, Zhu Z, Wang C, Liu Z (2013) A comparative proteomic analysis of *Pinellia ternata* leaves exposed to heat stress. Int J Mol Sci 14:20614–20634

# Integrating Omics Technologies to Understand Microbial Systems

9

Debashish Dey, Lakshmi Prasuna Mekala, and Mujahid Mohammed

**Abstract**

Microorganisms, an integral part of all ecological systems, with their unprecedented genetic and metabolic diversity influence all ecosystems. However, our understanding of microbial cellular functions from a system perspective is limited. Thus, studying the different biological layers of a cell as a single integrated system is crucial to unveil the microbial functions. To this end, omics approaches have become an indispensable tool to study the microbial processes by profiling the entire pool of DNA, RNA, proteins, and metabolites. Thanks to omics tools for capturing the true snapshot of cellular events and multi-omics approaches are far more powerful than any single omic study in deciphering the microbial functions. The main advantage of the multi-omics approach is its inherent nature of connecting the various biological layers and thereby uncovering the dynamic interplay of the various cellular processes which governs the cellular phenotype. Significance of multi-omics approaches is quite evident from the increasing number of studies employing integrated omics in various areas of microbiology. Integrated omics can resolve cellular events not only at the single cell or population level but also at the community level(meta-omics). The multi-omics approaches brought a paradigm shift from a tunnel view to a holistic view of

D. Dey
School of Biotechnology, Institute of Science, Banaras Hindu University, Varanasi, Uttar Pradesh, India

L. P. Mekala
Department of Plant Sciences, Avvaiyar Government College for Women, Affiliated to Pondicherry University, Karaikal, Puducherry, India

M. Mohammed (✉)
Department of Botany, Bharathidasan Government College for Women, Affiliated to Pondicherry University, Puducherry, India


A. Kumar et al. (eds.), *Omics Technologies for Sustainable Agriculture and Global Food Security Volume 1*, https://doi.org/10.1007/978-981-16-0831-5_9

microbial functions. The current chapter briefly describes different omics-platforms and recent trends of omics in studying the microbes. The chapter highlighted recent studies wherein integrated omics is employed to understand the microbial systems. The chapter also briefly discusses the applications of omics in various fields of microbiology and potential challenges and future prospects.

**Keywords**

Microorganisms · Omics · Omics-platforms · Omics integration · Metabolism · Meta-omics

## 9.1 Introduction

Microorganisms are the most diverse and abundant life forms on planet earth. They are widely distributed and thrive in some of the most inhospitable extreme ecological niches. Rich genetic diversity and metabolic plasticity enable them to occupy diverse niches and use various growth modes to survive. Microorganisms play a pivotal role in shaping global ecosystems to human health and disease (Segata et al. 2013). They cycle nutrients through biogeochemical cycles thereby sustaining the ecosystems which are vital for the survival of life forms (Abram 2015). Recent studies have highlighted the role of the microbiome in human health that was underappreciated until now (Gilbert et al. 2018; Segata et al. 2013). Microorganisms are considered as another organ of the human body and there are more number of microbes associated with the human body than the total number of cells in a body (Abram 2015; Baquero and Nombela 2012). Though microorganisms influence global ecosystems to human health, yet our understanding of microbial systems from a systems perspective is limited.

Traditionally microbial cellular processes and biochemical events were uncovered by a reductionist approach wherein each cellular process is studied as an individual entity (De Keersmaecker et al. 2006; Zhang et al. 2010). Although in the early days the reductionist approach was successful in dissecting these microbial systems to some extent, this approach underestimates the complexity of microbial processes from a holistic point of view (De Keersmaecker et al. 2006; Zhang et al. 2010). Microorganisms are tiny and simple yet their cellular processes are highly complex and dynamic. For example, an individual cell has many levels of regulatory circuits at genome, transcriptome, proteome, and metabolic levels and a coordinated symphony of these levels enables the optimum management of cellular resources (Zhang et al. 2010). Thus deciphering such intricate systems requires a holistic approach wherein all cellular events are studied as a single integrated system (Zhang et al. 2010).

In the past decades, it was a daunting task to decipher the complex cellular processes due to the lack of high-throughput platforms. However, the dawn of new next-generation sequencing techniques and advancements in mass spectrometry led to the rapid development of high-throughput platforms such as genomics, transcriptomics, proteomics, and metabolomics. These omic platforms captured the complex cellular events at different layers of central dogma and these tools made it possible to generate large datasets from biological systems (De Keersmaecker et al. 2006; Zhang et al. 2010). Advanced computational tools allowed the integration of such large omics datasets thereby uncovering the underlying regulatory networks and metabolic events associated with a phenotype (Segata et al. 2013). The recent omics studies revealed that integration of multi-omics approaches has high potential to dissect intricate biological systems than using a single omics tool (De Keersmaecker et al. 2006; Zhang et al. 2010).

Integrated omics approach has made tremendous progress and emerged as a powerful strategy enabling the scientific community to unravel the complex regulatory mechanisms and metabolic networks in microbial systems (Segata et al. 2013; Teitzel 2014). Considering the pivotal role of the microorganisms in everyday life, holistic understanding of these tiny life forms would have a greater impact in the field of medicine, agriculture, biotechnology, and environmental sciences (Misra et al. 2018; Teitzel 2014). In the present chapter, we reviewed the recent advancements in the field and highlighted the application of the integrated omics in understanding the intricate microbial systems at an organism and community (meta-omics) level. The chapter also discussed briefly on various current omics platforms.

## 9.2 Omic Tools: A Roadway to Uncover the Different Functionalities of Cell

### 9.2.1 Genomic Approaches

Genome is the haploid set of DNA including all the genes present in a cell of an organism. Genomics is the study of structure, function, mapping, editing, determination, and analysis of the sequence of the whole genome of a single organism. Comparative study of whole genomes of different organisms using computational tools constitutes comparative genomics. Functional genomics deals with the assignment of function to genes mined from the whole-genome sequencing data through gene expression analysis or mutational studies. The first DNA sequencing technique used was Sanger's chain termination method which was superseded by recent next-generation sequencing (NGS) techniques which offer high-throughput large sequencing data (Buermans and den Dunnen 2014). Different high-throughput NGS platforms have been used catering to differing sequencing requirements such as Roche-454, SOLiD, PacBio, Nanopore MinION, Ion Torrent, and Illumina (Buermans and den Dunnen 2014). Out of these, the Illumina MiSeq and the Ion Torrent are most suitable for amplicon sequencing, Illumina NextSeq 500 and the Illumina HiSeq 2500 are useful for whole metagenome sequencing and metatranscriptomics while the PacBio and MinION are suitable for de novo genome sequencing as in case of microbial genomes. The availability of complete draft genomes of several microorganisms and the use of advanced bioinformatic applications paved the way for comparative genomic approaches. Comparative genomics studies of *Pseudomonas syringae* revealed the convergent gain of virulence-associated effectors such as *hopAB and hopC1* through horizontal gene transfer and it was confirmed by *in planta* avirulence activity of these effector proteins (Hulin et al. 2018). Genome mining approach followed by expression of these genes in a surrogate host (*Streptomyces lividans)* and screening for potential antimicrobial activity (Xu et al. 2016) lead to the discovery of two novel linear lipopeptides 8D1–1 and 8D1–2 from *Streptomyces rochei.*

### 9.2.2 Transcriptomic Approaches

One of the main objectives of the genomic approaches is to infer the functional potential of the microbial genomes; however, expression of genome or sub-set of genome is true measure of function. Synthesis of all RNA molecules is the first phase of the gene expression and collection of a complete set of RNA molecules of a cell constitutes transcriptome (Zhang et al. 2010). Transcriptomics refers to the study of the transcriptome of a cell under a specific set of conditions. Using this approach, the gene expression could be measured at different time points or in tissue types. Transcriptomics study mainly captures the coding sequences (mRNA) and of late it is possible to capture the non-coding sequences (snRNA) as well. Transcriptomics has emerged as a powerful tool in the postgenomic era to validate the annotated genes and assigning the gene functions of unannotated genes (Creecy and Conway 2015). Capturing both coding and non-coding RNAs would allow identification of unknown gene functions and possible snRNA mediated regulatory mechanisms (Wang et al. 2009). Various transcriptomic approaches could be broadly categorized into two main categories: chip-based microarrays which require knowledge of pre-determined gene sequences and most popular RNA sequencing (RNA-Seq) which does not require the prior knowledge of gene sequences.

Microarray-based transcriptomics employs a DNA chip which is an ordered array of probes designed from the gene sequences and gene expression is measured by hybridizing the cDNA samples with the probes on the chip (Wang et al. 2009). The raw data is normalized and analyzed for checking the relative gene expression. Although microarray was popular method in early days, due to inherent limitations such as low sensitivity, inability to detect the splice variants/novel transcripts and regulatory RNAs nowadays it is not a preferred choice of the researchers (Wang et al. 2009).

RNA-Sequencing based transcriptomics has emerged as a powerful tool to study gene expression. It involves a combination of NGS approaches and computational tools to identify and quantify the transcripts present in an RNA sample. This technique has become a major success due to the advent of NGS approaches which can sequence billions of bases as well as massively parallel sequencing in a single day (Wang et al. 2009). Typical RNA-sequencing involves isolation of total RNA from the samples, depletion of the rRNA fractions from the total RNA pool followed by library preparation, and RNA sequencing using any of the NGS platforms. Postsequencing, these fragments are assembled into original RNA transcript either though computational tools by aligning to a reference genome or through the de novo assembly. Finally, the RNA-Seq data analysis is carried out utilizing bioinformatic tools in four sequential steps, viz. quality control, alignment, quantification, and differential expression.

Using RNA-Seq it is possible to quantify even low abundance RNAs, requires a lower quantity of input RNA, and generates lower background noise as compared to the microarray (Creecy and Conway 2015; Wang et al. 2009). RNA-Seq helps in analysis of different RNA isoforms which arise due to alternative splicing of the same gene (Wang et al. 2009). It is possible to study gene expression patterns in organisms whose genome sequence is known using the RNA-Seq technique (Wang et al. 2009). Recent technologies employ direct RNA sequencing (DRS) such as nanopore sequencing which offers direct sequencing of RNA without an intervening cDNA synthesis step and helps in detection of modified bases in RNA which was otherwise impossible through traditional NGS technologies (Zhang et al. 2010). The RNA-Seq studies were beneficial in identifying SNPs associated with diseases, splicing mechanisms, identification of transcription start sites, gene fusions, understanding of promoter and other regulatory mechanisms, resistance to drought and biotic stress, biofilm formation (Creecy and Conway 2015; Wang et al. 2009).

### 9.2.3 Proteomics Capturing the Protein Dynamics of Microbial Systems

The second phase of gene expression is translation wherein information of mRNA is used to synthesize proteins. Proteins are one of the vital macromolecules of living organisms which are involved in defining the cell structures, catalyzing biochemical reactions and gene regulation and carry out different cellular functions (Zhang et al. 2010; Kumar et al. 2014). Therefore, the cumulative microbial proteins from a microorganism should be the true potential of activity of the microorganisms in a given environment (Chao and Hansmeier 2012). Proteome refers to the total complement of proteins that are expressed by cell or tissue or organism. Proteomics involves the use of systematic, high-throughput technologies for the identification and quantification of total proteins present in a cell or tissue of an organism. Recent advances in high-throughput technologies have enabled large scale identification and measurement of protein molecules. Two major approaches employed in various proteomic studies are (1) gel-based proteomics: 2-D PAGE for separation of proteins based on their isoelectric point and mass followed by mass spectrometric analysis and (2) gel-free proteomics: 2-D liquid chromatography (LC) for separation of proteins or peptides followed by tandem mass spectrometric analysis (Zhang et al. 2010). The gel-based proteomics are employed for the identification of proteins from low complexity proteomes. The gel-based approaches have been unsuccessful for the detection of low abundance proteins and fail to resolve complex proteome samples (Otto et al. 2014).

The gel-free approaches are successful in detection of low abundance proteins and membrane-spanning proteins besides detecting several thousands of proteins from a sample in a single run. Both the above approaches are useful for the large scale high-throughput detection of protein molecules but they fail to provide accurate quantitative information of proteins. Recently advanced quantitative proteomic approaches were developed wherein proteins are isotopically labeled in vivo or in vitro for obtaining accurate quantitative measurements of proteins such as stable isotope labellled amino acid incorporation (SILAC), Isotope-coded affinity tags (ICAT), Mass-Coded Abundance Tagging (MCAT), Isobaric tags for relative and absolute quantitation (*iTRAQ*), Global internal standard technology (GIST), etc. (Otto et al. 2014). Now it is possible to measure the proteins in the complex sample without the use of the isotope labeling as well as absolute quantitation by using GIST method (Otto et al. 2014). Quantitative proteomics is widely employed to capture the protein dynamics of microbial systems and these approaches revolutionized the area of proteomics thereby our understanding of microbial cellular functions in proteomics context (Chao and Hansmeier 2012; Otto et al. 2014). Microbial proteomics is extensively used to understand the host–pathogen interactions, bioremediation process, new metabolic routes, molecular responses to changing environment, and molecular physiology of microbes (Chao and Hansmeier 2012; Otto et al. 2014).

### 9.2.4 Metabolomics to Understand the Metabolisms of Microbial Systems

Many of the proteins expressed by cell consist of catalytic domains indicating their role in biochemical reactions and the ultimate products of these reactions are metabolites, pivotal for cellular functions (Prosser et al. 2014). Complete set of small molecules (<1Kd) of a biological sample constitutes metabolome and metabolomics is a qualitative and quantitative analysis of a complete set of metabolites. Metabolites constitute substrates and reaction products of biochemical reactions of a cell and are end products of gene expression (Johnson et al. 2016; Sharma et al. 2018). Thus metabolic profiling provides a snapshot of the biochemical phenotype of the cell under a defined set of conditions (Prosser et al. 2014; Kumar et al. 2017). Metabolomics has emerged as a powerful functional genomics tool and an integral part of the systems biology. Typical metabolomic profiling is primarily performed by employing mass spectrometry or NMR (nuclear magnetic resonance) tools. Most commonly LC or GC-MS tools are widely used for the metabolomics studies (Reaves and Rabinowitz 2010). Two metabolomic approaches which are largely used are (1) untargeted profiling: measuring the abundance of metabolites without any prior specification and (2) targeted approach: measuring the absolute concentrations of a pre-defined set of metabolites (Johnson et al. 2016).

A typical metabolomics study involves the extraction of the metabolites, analysis of sample on NMR or MS-based analytical tools and data curation/filtration, data analysis (a multivariate/univariate statistical tool to identify the significant metabolic features defining a group)(Reaves and Rabinowitz 2010). Finally, the metabolic

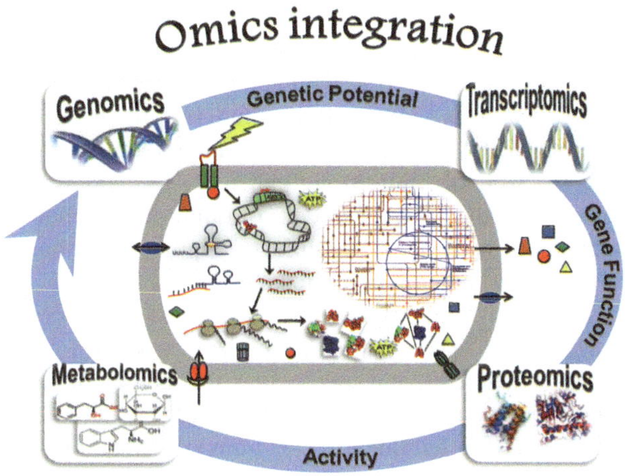


**Fig. 9.1** Schematic diagram of multi-omics integration (four principle biological layers) revealing the complex cellular processes and inherent interplay of biological layers to an external cue

features are identified by searching against spectral libraries followed by identifying the metabolic pathway perturbations. Although metabolomics could be an excellent tool in understanding the microbial systems still there are some inherent challenges such as experimental design, extraction methods, analytical tools, data processing, etc. which may bring a lot of heterogeneity in the data. However recent studies demonstrated that metabolomics could be a potential tool to unravel the intricate metabolisms of microbial systems, new metabolic routes, and novel enzyme systems (Prosser et al. 2014; Sevin et al. 2017). Further, stable isotope assisted metabolomics revealed metabolic flux (Reaves and Rabinowitz 2010) and differential metabolism (Mekala et al. 2018) which otherwise is not possible to capture by other omics tools. Hence, metabolomics has become an indispensable tool in understanding microbial systems and their functions from a systems perspective. Microbial systems are highly dynamic and their functions are regulated by different inherently connected biological layers such as genome, transcriptome, proteome, and metabolome. Thus, integrating all the omics has potential to reveal the true picture of dynamic microbial cellular processes (Fig. 9.1) and would expand our knowledge on how microbial systems function.

## 9.3 Meta-Omics: Linking Microbial Communities and Ecological Functions

In natural environments, microorganisms exist in poly-microbial communities and they thrive in diverse ecosystems ranging from marine sediments, hot springs to human bodies. Their cooperative biochemical activities are essential for sustenance of all the ecosystems (Abram 2015; Segata et al. 2013). These microbial communities recycle the nutrients and sustain the ecosystems. In addition, they interact with host systems including the human body and influence the host health and disease (Abram 2015; Singer et al. 2017). In spite of their crucial role in the sustenance of any ecosystem, capturing their biochemical activities has been a challenging task in-situ (Abram 2015). Our understanding of microbial communities structure and function was limited mainly because (1) much of our understanding of microorganisms comes from studying them as monocultures under artificial environments; (2) many of the members are unculturable under laboratory setup, and (3) lack of high-throughput tools to capture the biochemical events in-situ. However, the advancement of high-throughput sequencing and proteomic technologies made it possible to capture microbial activities at the community level under natural conditions using meta-omics studies (Abram 2015; Segata et al. 2013). Meta-omics refers to the study of microbial activity at different levels such as DNA (metagenomics), RNA (metatranscriptomics), protein (metaproteomics), and metabolites (metametabolomics) in heterogenous microbial communities (Abram 2015). A typical meta-omics study involves isolation of DNA/RNA/Protein/metabolites from an environmental sample, analysis of the sample by sequencing or mass spectrometry tools, computational analysis of data (data filtration, data interpretation, and data integration). With this advancement in the recent past, a paradigm shift has come from studying monocultures to microbial communities to better understand the concerted biochemical activities of microbes vital for shaping the ecosystems. Here we briefly summarize different meta-omics tools used in understanding microbial communities.

### 9.3.1 Metagenomic Approaches in Microbiology

The determination of DNA sequence information directly from the environmental samples without culturing is known as metagenomics. This method removes culture-based biases and provides an understanding of the microbial system in terms of its diversity, population dynamics, and genetic potential (Abram 2015; Segata et al. 2013). Shotgun sequencing approaches are widely being utilized for metagenomics studies wherein the DNA is sequenced directly from the samples after extraction without cloning into libraries using various sequencing platforms. Mainly two metagenomics approaches are practiced catering to different objectives: (1) 16s/18s rRNA profiling to know the community structure and (2) shotgun sequencing to profile entire DNA molecules of the sample which would reveal the genetic potential of the community. The 16s and 18s rRNA profiling is carried out to determine the different bacterial and eukaryotic diversity (Segata et al. 2013). A typical microbial community profiling study involves isolation of DNA from the environment, PCR-amplification of phylogenetic marker genes (16S/18S, ITS), and sequencing of amplicons, the sequencing data is subjected to quality control to remove chimeras. Finally, the clustering of sequences into operational taxonomic units (OTUs), taxonomic classification, and statistical analysis are carried out (Abram 2015; Segata et al. 2013). Microbial community profiling has emerged as a powerful tool in deciphering the microbial community of natural habitats and these studies revealed a hidden wealth of microbial diversity (Abram 2015; Cui et al. 2018; Grob et al.

2015; Moitinho-Silva et al. 2017) which was not revealed by traditional culture-based studies.

In shotgun sequencing approach, total DNA of the sample is sequenced and this generates millions of reads depending upon the platform used for sequencing and complexity of the sample. These reads are assembled using computational tools and further analyzed for functional annotations. Computational tools specifically designed for metagenomic applications are MG-RAST and RAMMCAP whereas specific databases such as FOAM were built for functional analysis of metagenomic data (Abram 2015). COG (cluster of orthologous groups) categories, eggNOG orthologous groups, SEED subsystems, InterPro entries, etc. and BlastX tool of NCBI are used to find its functional role in the microorganisms (Segata et al. 2013). Shotgun sequencing would reveal both microbial diversity and potential functions of a microbial community (Gilbert et al. 2018; Segata et al. 2013).

### 9.3.2 Metatranscriptomics

Metagenomics data provides insights into the genetic potential of a microbial community; however, the expressed RNA and proteins are the true measure of the community function and activity (Abram 2015). Metatranscriptome refers to the complete set of expressed RNA molecules in the natural community and the identification and quantification of expressed RNA are known as metatranscriptomics. Though different approaches such as microarray-based, Sanger-cDNA libraries sequencing are available, RNA sequencing is superior and preferred tool for studying the metatranscriptomics (Bashiardes et al. 2016). The typical metatranscriptomic study is similar to that of transcriptomics; however, the RNA is isolated from environmental samples and requires deep sequencing. All the sequencing reads are then assembled and annotated using various computational tools similar to that of metagenomics (Bashiardes et al. 2016). Using metatranscriptomics studies one can enumerate the active microbial communities and their possible functions in-situ. However, extraction of RNA from environmental samples, data filtering and computational analysis of large datasets are some of the challenges in metatranscriptomic studies (Bashiardes et al. 2016).

### 9.3.3 Metaproteomics and Metabolomics

The analysis of end products of gene expression, i.e., proteins and metabolites from a microbial community would provide information about true microbial function and activity in a given habitat (Abram 2015). Metaproteomics refers to the identification and quantification of the entire set of proteins expressed by the microbial community (Siggins et al. 2012) and it has become the main choice of microbial ecologists to understand the community functions (Abram 2015; Siggins et al. 2012). Proteomic tools and methodology that is used to study the pure cultures are applied for metaproteomics studies. However computational tools and quantitation methods

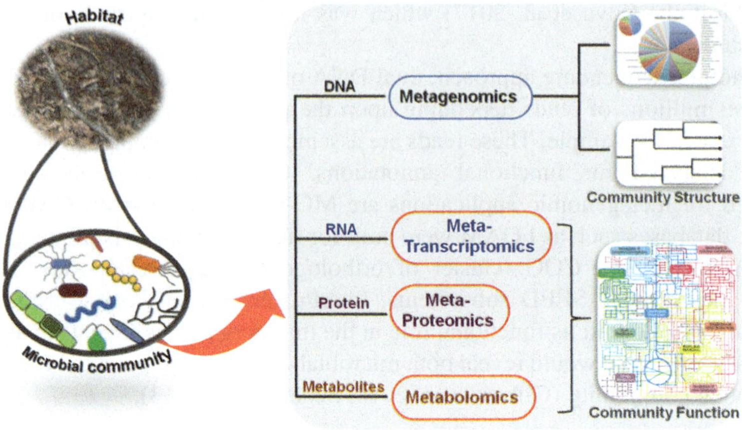


**Fig. 9.2** Integrated meta-omics workflows; uncovering the microbial diversity and underlying cellular processes and functions of microbial community

are less developed as compared to similar tools for metagenomics and metatranscriptomics studies. Nonetheless, a number of studies demonstrated how metaproteomics can link the community composition to function under diverse ecosystems (Abram 2015; Siggins et al. 2012).

On the other hand, metabolomics is also used to study the microbial community functions and identification of metabolites of a community would reveal active metabolic pathways (Abram 2015). GC-MS or LC-MS based analytical tools are used to analyze the community metabolomes. Although metabolomics is informative, it is impossible to assign metabolite productions to a species in a community and this limitation can be partially resolved when metabolomics is combined with metaproteomics/metatranscriptomics studies (Zhang et al. 2010). Metabolomics integrated with the other meta-omics studies has high potential to capture the active metabolic processes and thereby community functions in an ecosystem (Abram 2015; Bargiela et al. 2015; Cui et al. 2018). Integration of more than one meta-omic tools will provide a more comprehensive understanding of microbial diversity as well as potential functions and underpinning microbial processes of an ecosystem. A typical integrated meta-omics workflow is illustrated (Fig. 9.2) using leaf litter sample as an example.

### 9.3.4 Stable Isotope Probing (SIP) Approach: Capturing Active Microbial Players

Although meta-omics has tremendously enhanced our understanding of microbial diversity and potential genetic and metabolic diversity, the real potential of omic tools is to understand the microbial functions in natural conditions. Combining the stable isotope probing and meta-omic tools resulted in the development of SIP-based omics and it has emerged as an excellent tool to understand metabolically active communities (Singer et al. 2017). SIP omics allows the capturing of the active microbial members of the community and their functions responding to a particular stimulus in-situ (Abram 2015). In a typical SIP approach, stable isotope labeled substrates are added to the natural community and active microbes consume the labeled substrates and thereby incorporate the labeled atoms into macromolecules (DNA, RNA, proteins, and metabolites) (Berry and Loy 2018; Singer et al. 2017). Finally labeled macromolecules are separated and subjected to conventional omics tools (sequencing/mass spectrometry) to generate data followed by data analysis (Singer et al. 2017). Depending upon the labeled macromolecule used for the analysis, these are called as DNA-SIP, RNA-SIP, and Protein-SIP and DNA-SIP is the most widely used SIP tool (Berry and Loy 2018). SIP tools brought a paradigm shift from cataloging the microbial diversity and potential functions to capturing the active microbial members and their ecological roles (Singer et al. 2017). Although the concerted effort of microbial communities has long been appreciated, it is now possible to capture their interplay in-situ. SIP approaches are extensively used to unravel the active communities of contaminated sites, the human microbiome (Berry and Loy 2018), animal microbiome (Berry and Loy 2018), the soil microbiome (Pepe-Ranney et al. 2016), and aquatic sediment microbes(Fortunato and Huber 2016), etc.

## 9.4 Integrated Multi-Omics: Painting the Holistic Picture of Microbial Systems

Recent advancement of high-throughput sequencing tools and hyphenated mass spectrometry tools enabled the capturing of comprehensive genetic information and expressed cellular entities (transcripts, proteins, and metabolites) (Misra et al. 2018). These revolutionary tools which are able to capture the different levels of cellular functions have marked the beginning of the omics era. There has been an upsurge in the number of studies employing omic tools to understand the functions of microbial systems. Genomics is the first omics tool which rapidly expanded and provided a wealth of information on microbial diversity, complex genetic make-up, and potential functions (Zhang et al. 2010). With the advent of huge genomic information, other functional omics tools such as transcriptomics, proteomics, and metabolomics have gained importance in microbial studies for functional annotation of genomic information (Prosser et al. 2014).

Although the single omic studies enhanced our understanding of microbial systems, they could not provide a holistic snapshot of cellular functions (De Keersmaecker et al. 2006; Zhang et al. 2010). Though it is convenient to study RNAs, proteins, and metabolites as individual entities, in the real-time scenario; these are intricately connected in biological systems. To paint a truly holistic picture, omics tools should be integrated to the extent possible (Reaves and Rabinowitz 2010; Zhang et al. 2010) and it has already emerged as an indispensable part of microbial systems biology (Abram 2015). This would be like, putting together each piece of jigsaw puzzle to get a complete picture. Similarly, here each omics study acts as a piece of jigsaw puzzles and provides crucial information on cellular functions. In the recent past, realizing the potential of integrated omics, a number of studies have emerged discovering the cellular functions of microbial systems at an organism as well as community level.

To this date, there are many studies on integrated omics to understand the microbial systems and the list is in-exhaustive, hence we highlighted a few of these studies in this chapter (Table 9.1). These studies range from a single organism to natural microbial communities and these have provided unprecedented knowledge on microbial functions and hitherto unknown interplay at a community level, refer Table 9.1 for more information. Understanding the microbial systems using integrated omics will have a direct impact on health and disease (Khan et al. 2019), agriculture, ecology and environment, and biotechnology fields (Marco and Abram 2019) and a snapshot of these application depicted in Fig. 9.3. These applications include biomarker discovery, novel drug targets (Aderem et al. 2011), increasing plant productivity (Meena et al. 2017) and discovery of novel antibiotics, etc. (Genilloud 2018; Marco and Abram 2019).

The integration of omics is possible due to the advancements of computational tools and currently there are number of integrating tools available both on private and public domains. A recent review has extensively discussed the available tools for the integration of different omics approaches, we recommend readers to read this review for more information (Misra et al. 2018). Although there are many successful omics integration studies, designing of omics studies, data acquisition, mining, and interpretation of large datasets are still a big challenge (Misra et al. 2018). These tools are constantly being evolved and improved upon to address these bottlenecks. A typical omics integration workflow is illustrated here (Fig. 9.4) and integrated omics reveals a holistic picture of microbial systems from a functional point of view. The new working model or mechanism revealed by omics study must be validated and tested whenever possible by other methods. The validation of omics data must be an intrinsic part of the omics workflow and this would not only strengthen the omics driven hypothesis but also highlight the importance of interconnecting the disciplines to understand cellular functions.

## 9.5 Conclusions

Rapid advancements in sequencing technologies, mass spectrometry, and computational power have propelled the research in bio-sciences from reductionist to a holistic systemic approach. These technological advancements brought a paradigm shift in our understanding how microbial systems work. It is evident from increasing

**Table 9.1** Application of various integrated omics studies in microbial systems

| Omics tools employed | Species/microbial Community | Goals and key findings | References |
|---|---|---|---|
| Metagenomic, metatranscriptomic | Marine benthic group D (MBG-D) archaea | Deciphering the ecological roles of the uncultured MBG-D archaea through integrated meta-omics tools. The study revealed the mixotrophic metabolic potential of the MBG-D and possibility of transforming organic matter in the sediment and thereby contributing to the global sedimentary carbon cycle | Zhou et al. (2018) |
| | Sponge-microbiome | Deciphering the metabolic interactions between sponge host and symbionts. The meta omics tools unveiled the unprecedented metabolic potential and metabolic interactions in sponge *Cymbastela concentrica* microbiome | Moitinho-Silva et al. (2017) |
| Metagenomic, metaproteomic | Natural cellulolytic microbial consortium | Exploring the mechanism of degradation of recalcitrant biomass by natural soil microbial community. The study revealed the presence of cooperative microbial consortium expressing complementary enzymes for the deconstruction of complex plant biomass | Zhou et al. (2014) |
| | Rat gut microbiota | To study the effect of polyphenol-hesperidin on the gut microbiota of obese rats and implication in disease and health. Multi-omics data revealed changes in gut microbial diversity and metabolic functions and these complex changes may have implications in host health | Guirro et al. (2018) |
| Metagenomic, metabolomic | Chronic heart failure (CHF) patient gut microbiota | Uncovering the metabolite pattern and gut microbiota dysbiosis of CHF patients. Metagenomics revealed a significant change in the gut microbial community and indicated an altered metabolic profile in CHF | Cui et al. (2018) |

(continued)

**Table 9.1** (continued)

| Omics tools employed | Species/microbial Community | Goals and key findings | References |
|---|---|---|---|
| | | patients. The study also highlighted the correlation of altered metabolic pattern with changes in the gut microbial community | |
| | *Microglena, Chloromonas* | To explore the red and green snow microbial physiology and ecology. Integrated metagenomics and metabolite profiling revealed a change in the microbial community and metabolic profile of red and green snow communities. Green snow is a nutrient-rich environment and metabolites present in supports proliferation of *Microglena* sp. conversely, the red snow nutrient-poor environment dominated by *Chloromonas* sp. and it survives severe conditions by accumulating storage and reserve metabolites | Lutz et al. (2015) |
| Transcriptomic, proteomic | *Aeromonas hydrophila* | To gain insights into the role of iron on the physiology of *Aeromonas hydrophila.* mRNA and protein profiles obtained from iron limting cultures using RNA-seq and iTRAQ. The study showed that iron limitation leads to the expression of genes related to iron absorption and transport and pathogenicity factors. The study also demonstrated that iron limitation enhanced the virulence by activating the enterobactin synthesis and other virulence factors | Teng et al. (2018) |
| Transcriptomic, metabolomic | *Saccharomyces* strains | To explore the association between gene expression and aroma related compounds production during fermentation of wine. The integrated study revealed the expression of genes such as ADH6 and ADH7 involved | Mendes et al. (2017) |

(continued)

**Table 9.1** (continued)

| Omics tools employed | Species/microbial Community | Goals and key findings | References |
|---|---|---|---|
| | | in the synthesis of higher alcohols and other genes related to aroma compound synthesis. Expression of these genes correlated to increased levels of higher alcohols, acetate esters and ethyl esters required for the better quality wines | |
| Metaproteomic, metabolomic | Petroleum-polluted sediment microbial community | To decipher the metabolic processes of oxygen-depleted petroleum-polluted sediments microbiomes across cost line of Italy. Integrated omics study revealed high metabolic heterogeneity and the prevalence of anaerobic metabolism in oil-polluted sites. The study also suggested that suppression of the hydrocarbon degradation possible due to low oxygen levels | Bargiela et al. (2015) |
| | *Rubrivivax benzoatilyticus* JA2 | To decipher the systemic responses of the photosynthetic bacterium to major environmental pollutant, aniline. Integrated proteomic and metabolomics study revealed that strain employs multiple stress response mechanisms, energy conservation and metabolic remodeling to combat aniline stress. Multi-omics for the first time revealed survival mechanisms of an aniline non-degrading bacterium | Mujahid et al. (2014) |
| Metagenomic, metatranscriptomic, metaproteomic | Bisphenol A (BPA) degrading microbial community | To understand the microbial interactions in a BPA degrading model community. Integrated multi meta-omics revealed metabolic potential at the community as well as individual level. Interestingly study also unraveled the metabolic synergy in BPA | Yu et al. (2019) |

(continued)

**Table 9.1** (continued)

| Omics tools employed | Species/microbial Community | Goals and key findings | References |
|---|---|---|---|
| | | degrading microbial community and metabolic cooperation in BPA degradation | |
| Genomic, transcriptomic, metabolomic | *Propionibacterium acidipropionici* | To dissecting the acid stress response mechanism in acid-tolerant *Propionibacterium acidipropionici.* The bacterial cells showed modulation of cellular pH by controlling protons, ion transport, and by activating the amino acid metabolism. The study indicated that down regulation of lactate and acetate synthesis leading to increased production of propionate | Guan et al. (2018) |
| Genomic, proteomic, metabolomic | *Aspergillus nidulans* | Study of adaptive mechanisms of *A. nidulans* in space station conditions revealed positive selection by genome alterations through missense and intergenic mutations. There was loss of stop codon in the *lecA* gene, a key regulator of secondary metabolites leading to increased production of asperthecin which shows a protective role from ISS radiations | Romsdahl et al. (2019) |
| Transcriptomic, proteomic, metabolomic | *Phaeodactylum tricornutum (diatom)* | To decipher the molecular mechanisms underlying lipid accumulation under nitrogen limiting conditions in *Phaeodactylum tricornutum.* The integrated multi-omic study revealed nitrogen limitation downregulated photosynthesis, ribosomal and protein synthesis thus directly effecting the growth. Further, the study indicated that membrane lipid rearrangement and degradation of lipids possibly lead to accumulation of triacylglycerol | Remmers et al. (2018) |

(continued)

**Table 9.1** (continued)

| Omics tools employed | Species/microbial Community | Goals and key findings | References |
|---|---|---|---|
| RNA-stable isotope probing and metatranscriptomics. DNA-SIP and metaproteomics | Hydrothermal vent microbial community | To uncover the microbial diversity and metabolic potential of the hydrothermal vent community. RNA-SIP analysis revealed the change in microbial diversity across the temperature layers. Metatranscriptomics indicated activity of unique chemolithoautotrophic communities across the thermal layers of the vent and key autotrophic metabolism operating in deep-sea geothermal conditions | Fortunato and Huber (2016) |
| | *Uncultivated marine Methylotrophs* | To characterize uncultivated methylotrophic metabolism. DNA-SIP based metagenomic analysis retrieved almost complete genome of methanol metabolizing uncultivated Methylophaga species. Metaproteome revealed metabolic pathways involved in methanol metabolism of this naturally occurring uncultivated bacterium | Grob et al. (2015) |

numbers of omics studies highlighting the significance of microbial dark matter which is otherwise hidden. In the era of omics, microbiology has moved from single omics to multi omics, from a single organism to community study and the most significant of all is capturing the microbial functions from laboratory setup to natural environments. Multi-omics based studies are rapidly expanding our understanding of microbial function in different fields such as medicine, environment, ecology, and biotechnology. Omics-based system approaches are highly promising in addressing the pressing problems currently humanity is facing such as health and disease, fuel and food, ecological and environmental sustenance.

Although omics approaches are highly promising, integrating and interpretation of the large data sets to extract meaningful biological information is the biggest challenge. Development of user-friendly integrating tools and building of mathematical models are crucial to understand the complex biochemical processes. On the

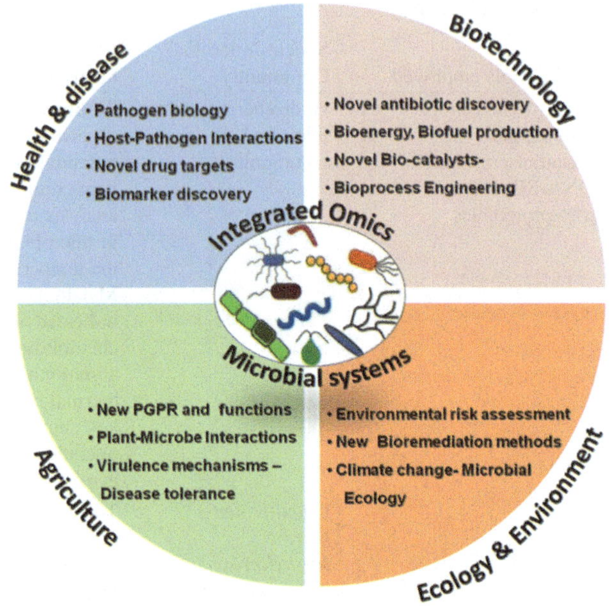


**Fig. 9.3** Potential applications of integrated microbial omics in various fields

other hand, experimental design, sampling methods, and quality control of data are some of the factors which have the potential to introduce lots of heterogeneity in the data. Although omics studies are exiting, some of the omics are highly challenging compared to others such as meta-omics studies are highly challenging compared to omics studies due to the complexity of microbial communities and highly diverse habitats which poses major challenges.

However, integration of omics data, extraction and interpretation of data in a biological context require a multidisciplinary approach. To fully extract meaningful information from complex data sets, the data needs to be subjected to different tools related to different layers of disciplines such as biology, mathematics/statistics, and computational biology. Integrated omics generates huge data if interpreted correctly, would provide novel mechanisms and biological functions which would help in postulating new hypothesis. However, these newly generated hypotheses must be validated by other classical tools such as biochemical and gene disruption studies. On the contrary, mere cataloging the expressed entities (transcripts, proteins, and metabolites) and their relative levels would have less significance and will not enhance our understanding of microbial systems. The full potential and impact of integrated omics tools in the understanding of the microbial systems can only be

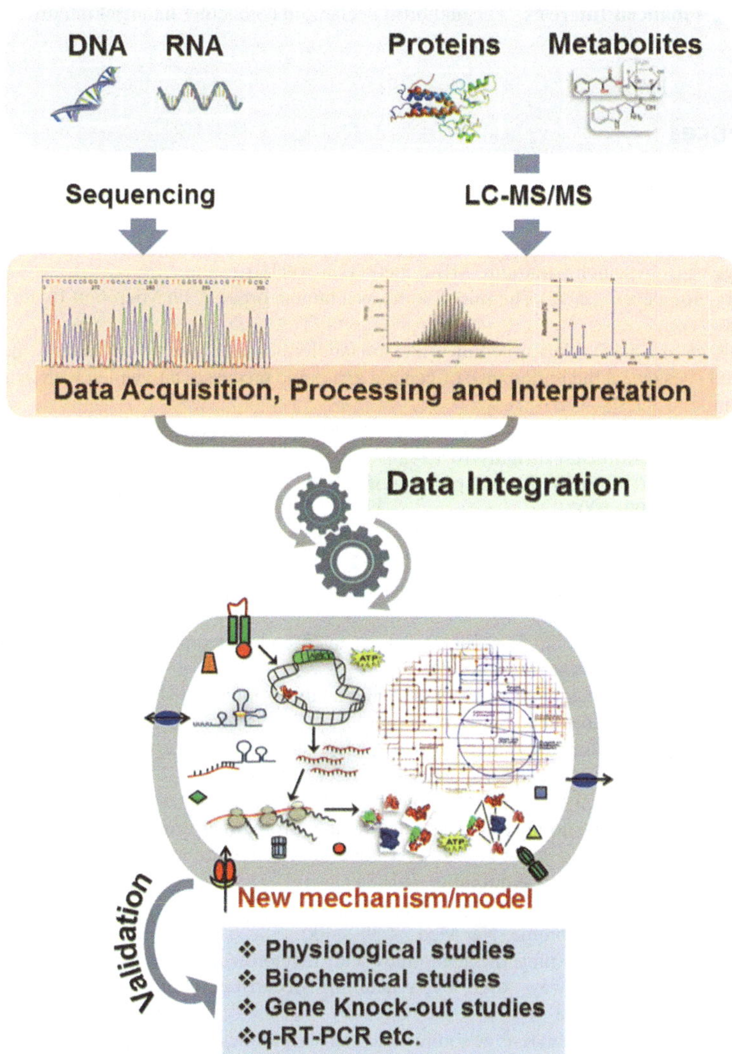


**Fig. 9.4** A complete integrated multi-omics workflow: A journey from biomolecules to complex cellular events and discovery of new mechanisms

appreciated when we move from descriptive omics to discovery and functional omics and this trend has great potential in uncovering new microbial functionalities.

**Acknowledgements** Dr. Debashish Dey would like to acknowledge the financial support in the form of 'Seed Grant' under Institute of Eminence (IoE) scheme from Banaras Hindu University (BHU), Varanasi. The financial support from the Department of Biotechnology (DBT) to the School of Biotechnology, BHU is also acknowledged.

**Competing Financial Interests** The author(s) declare no competing financial interests.

## References

Abram F (2015) Systems-based approaches to unravel multi-species microbial community functioning. Comput Struct Biotechnol J 13:24–32

Aderem A et al (2011) A systems biology approach to infectious disease research: innovating the pathogen-host research paradigm. MBio 2:e00325–e00310

Baquero F, Nombela C (2012) The microbiome as a human organ. Clin Microbiol Infect 18(Suppl 4):2–4

Bargiela R et al (2015) Metaproteomics and metabolomics analyses of chronically petroleum-polluted sites reveal the importance of general anaerobic processes uncoupled with degradation. Proteomics 15:3508–3520

Bashiardes S, Zilberman-Schapira G, Elinav E (2016) Use of metatranscriptomics in microbiome research. Bioinform Biol Insights 10:19–25

Berry D, Loy A (2018) Stable-isotope probing of human and animal microbiome function. Trends Microbiol 26:999–1007

Buermans HP, den Dunnen JT (2014) Next generation sequencing technology: advances and applications. Biochim Biophys Acta 1842:1932–1941

Creecy JP, Conway T (2015) Quantitative bacterial transcriptomics with RNA-seq. Curr Opin Microbiol 0:133–140

Chao TC, Hansmeier N (2012) The current state of microbial proteomics: where we are and where we want to go. Proteomics 12:638–650

Cui X et al (2018) Metagenomic and metabolomic analyses unveil dysbiosis of gut microbiota in chronic heart failure patients. Sci Rep 8:635

De Keersmaecker SC, Thijs IM, Vanderleyden J, Marchal K (2006) Integration of omics data: how well does it work for bacteria? Mol Microbiol 62:1239–1250

Fortunato CS, Huber JA (2016) Coupled RNA-SIP and metatranscriptomics of active chemolithoautotrophic communities at a deep-sea hydrothermal vent. ISME J 10:1925–1938

Genilloud O (2018) Mining actinomycetes for novel antibiotics in the omics era: are we ready to exploit this new paradigm? Antibiotics 7(4):85

Gilbert JA, Blaser MJ, Caporaso JG, Jansson JK, Lynch SV, Knight R (2018) Current understanding of the human microbiome. Nat Med 24:392–400

Grob C et al (2015) Combining metagenomics with metaproteomics and stable isotope probing reveals metabolic pathways used by a naturally occurring marine methylotroph. Environ Microbiol 17:4007–4018

Guan N et al (2018) Comparative genomics and transcriptomics analysis-guided metabolic engineering of *Propionibacterium acidipropionici* for improved propionic acid production. Biotechnol Bioeng 115:483–494

Guirro M et al (2018) Multi-omics approach to elucidate the gut microbiota activity: Metaproteomics and metagenomics connection. Electrophoresis 39:1692–1701

Hulin MT et al (2018) Comparative genomics of *Pseudomonas syringae* reveals convergent gene gain and loss associated with specialization onto cherry (*Prunus avium*). New Phytol 219:672–696

Johnson CH, Ivanisevic J, Siuzdak G (2016) Metabolomics: beyond biomarkers and towards mechanisms. Nat Rev Mol Cell Biol 17:451–459

Khan MM, Ernst O, Manes NP, Oyler BL, Fraser IDC, Goodlett DR, Nita-Lazar A (2019) Multi-omics strategies uncover host-pathogen interactions. ACS Infect Dis 5:493–505

Kumar A, Bimolata W, Kannan M, Kirti PB, Qureshi IA, Ghazi IA (2014) Comparative proteomics reveals differential induction of both biotic and abiotic stress response associated proteins in rice during Xanthomonas oryzae pv. Oryzae infection. Funct Integr Genomics 15(4):425–437

Kumar R, Bohra A, Pandey AK, Pandey MK, Kumar A (2017) Metabolomics for plant improvement: status and prospects. Front Plant Sci 8:1302

Lutz S, Anesio AM, Field K, Benning LG (2015) Integrated 'Omics', targeted metabolite and single-cell analyses of arctic snow algae functionality and adaptability. Front Microbiol 6:1323

Marco DE, Abram F (2019) Editorial: using genomics, metagenomics and other "Omics" to assess valuable microbial ecosystem services and novel biotechnological applications. Front Microbiol 10:151

Meena KK et al (2017) Abiotic stress responses and microbe-mediated mitigation in plants: the omics strategies. Front Plant Sci 8:172

Mekala LP, Mohammed M, Chintalapati S, Chintalapati VR (2018) Stable isotope-assisted metabolic profiling reveals growth mode dependent differential metabolism and multiple catabolic pathways of l-Phenylalanine in *Rubrivivax benzoatilyticus* JA2. J Proteome Res 17:189–202

Mendes I, Sanchez I, Franco-Duarte R, Camarasa C, Schuller D, Dequin S, Sousa MJ (2017) Integrating transcriptomics and metabolomics for the analysis of the aroma profiles of *Saccharomyces cerevisiae* strains from diverse origins. BMC Genomics 18:455

Misra BB, Langefeld CD, Olivier M, Cox LA (2018) Integrated omics: tools, advances, and future approaches. J Mol Endocrinol 2018:JME-18-0055

Moitinho-Silva L, Diez-Vives C, Batani G, Esteves AI, Jahn MT, Thomas T (2017) Integrated metabolism in sponge-microbe symbiosis revealed by genome-centered metatranscriptomics. ISME J 11:1651–1666

Mujahid M, Prasuna ML, Sasikala C, Ramana Ch V (2014) Integrated metabolomic and proteomic analysis reveals systemic responses of *Rubrivivax benzoatilyticus* JA2 to aniline stress. J Proteome Res 14:711–727

Otto A, Becher D, Schmidt F (2014) Quantitative proteomics in the field of microbiology. Proteomics 14:547–565

Pepe-Ranney C, Campbell AN, Koechli CN, Berthrong S, Buckley DH (2016) Unearthing the ecology of soil microorganisms using a high resolution DNA-SIP approach to explore cellulose and xylose metabolism in soil. Front Microbiol 7:703

Prosser GA, Larrouy-Maumus G, de Carvalho LP (2014) Metabolomic strategies for the identification of new enzyme functions and metabolic pathways. EMBO Rep 15:657–669

Reaves ML, Rabinowitz JD (2010) Metabolomics in systems microbiology. Curr Opin Biotechnol 22:17–25

Remmers M et al (2018) Orchestration of transcriptome, proteome and metabolome in the diatom *Phaeodactylum tricornutum* during nitrogen limitation. Algal Res 35:33–49

Romsdahl J et al (2019) International Space Station conditions alter genomics, proteomics, and metabolomics in *Aspergillus nidulans*. Appl Microbiol Biotechnol 103:1363–1377

Segata N, Boernigen D, Tickle TL, Morgan XC, Garrett WS, Huttenhower C (2013) Computational meta'omics for microbial community studies. Mol Syst Biol 9:666

Sevin DC, Fuhrer T, Zamboni N, Sauer U (2017) Nontargeted in vitro metabolomics for high-throughput identification of novel enzymes in Escherichia coli. Nat Methods 14:187–194

Sharma K, Sarma S, Bohra A, Mitra A, Sharma NK, Kumar A (2018) Plant metabolomics: an emerging technology for crop improvement. In: Çelik O (ed) New visions in plant science, 1st edn. IntechOpen, London, pp 65–79

Siggins A, Gunnigle E, Abram F (2012) Exploring mixed microbial community functioning: recent advances in metaproteomics. FEMS Microbiol Ecol 80:265–280

Singer E, Wagner M, Woyke T (2017) Capturing the genetic makeup of the active microbiome in situ. ISME J 11:1949–1963

Teitzel G (2014) Harnessing the power of omics in microbiology. Trends Microbiol 22:227–228

Teng T, Xi B, Chen K, Pan L, Xie J, Xu P (2018) Comparative transcriptomic and proteomic analyses reveal upregulated expression of virulence and iron transport factors of *Aeromonas hydrophila* under iron limitation. BMC Microbiol 18:52

Wang Z, Gerstein M, Snyder M (2009) RNA-Seq: a revolutionary tool for transcriptomics. Nat Rev Genet 10:57–63

Xu M et al (2016) Functional genome mining for metabolites encoded by large gene clusters through heterologous expression of a whole-genome bacterial artificial chromosome library in *Streptomyces* spp. Appl Environ Microbiol 82:5795–5805

Yu K et al (2019) An integrated meta-omics approach reveals substrates involved in synergistic interactions in a bisphenol a (BPA)-degrading microbial community. Microbiome 7:16

Zhang W, Li F, Nie L (2010) Integrating multiple 'omics' analysis for microbial biology: application and methodologies. Microbiology 156:287–301

Zhou Y et al (2014) Omics-based interpretation of synergism in a soil-derived cellulose-degrading microbial community. Sci Rep 4:5288

Zhou Z, Liu Y, Lloyd KG, Pan J, Yang Y, Gu JD, Li M (2018) Genomic and transcriptomic insights into the ecology and metabolism of benthic archaeal cosmopolitan, Thermoprofundales (MBG-D archaea). ISME J 13:885–901

# Genome Editing Technologies for Plant Improvement: Advances, Applications and Challenges

# 10

Rakesh Kumar, Narasimha Rao Nizampatnam, Minhaj Alam, Tarun K. Thakur, and Anirudh Kumar

**Abstract**

The current rate of genetic gains in crop improvement should rise to match growing need for sustainable food production and environmental safety. Recent years have seen genome editing being emerged as a promising tool to tailor a variety of traits that improve plant performance. In the context, sequence-specific nucleases like zinc finger nuclease (ZFN), transcription activator-like effector nucleases (TALENs) and more recently, clustered regularly interspaced short palindromic repeats (CRISPR/Cas) have enabled rapid and precise modification of the genomes. The CRISPR/Cas system has revolutionized targeted gene modification approaches owing of its capacity to produce allelic series with high precision in both domesticated and crop wild species. Recent examples demonstrating simultaneous mutagenesis of multiple genes lends credence to targeted genome editing for tailoring complex quantitative traits. In parallel,

R. Kumar
Present Address: Department of Life Science, Central University of Karnataka, Kalaburagi, Karnataka, India

N. R. Nizampatnam
Chemveda Life Sciences India Pvt. Ltd. Uppal, Hyderabad, Telangana, India

M. Alam
Department of Botany and Biotechnology, MG College, Gaya, Bihar, India

T. K. Thakur
Department of Environmental Science, Indira Gandhi National Tribal University (IGNTU), Amarkantak, Madhya Pradesh, India

A. Kumar (✉)
Present Address: Department of Botany, Indira Gandhi National Tribal University (IGNTU), Amarkantak, Madhya Pradesh, India
e-mail: anirudh.kumar@igntu.ac.in





A. Kumar et al. (eds.), *Omics Technologies for Sustainable Agriculture and Global Food Security Volume 1*, https://doi.org/10.1007/978-981-16-0831-5_10

oligogenic traits like disease resistance can be improved by precise base editing by accurate protein remodelling. Notwithstanding encouraging results on plant genome editing, adoption of gene-edited plants remains a moot point. To realize immense potential of genome editing, emphasis should be given on resolving the technical and regulatory apprehensions associated with the adoption of gene-edited plant products. This article presents latest advances in techniques grouped under "genome editing", with a brief discussion on the current status of genome edited plants. We also highlight current challenges that limit widespread applications of targeted genome modification in crop improvement for sustainable food security.

**Keywords**

Genome editing · CRISPR/Cas · TALENs · ZFNs · Plant breeding · Intellectual property rights

## Abbreviations

| | |
|---|---|
| CRISPR | Clustered regularly interspaced short palindromic repeats |
| EU | European union |
| FAO | Food and agriculture organization |
| FTO | Freedom to operate |
| GMO | Genetically modified organism |
| IPR | Intellectual property rights |
| NCA | National competent authority |
| NPBTs | New plant breeding techniques |
| NTWG | New technique working group |
| SDN | Site-directed nuclease |
| SG | Synthetic genomics |
| TALEN | Transcription activator-like effector nucleases |
| ZFN | Zinc finger nuclease |

## 10.1 Introduction

Practise of plant breeding started nearly 10,000 years ago that brought first grain crops under domestication and selective breeding (Hickey et al. 2019). Subsequent discovery of Mendel's law, hybrid vigour and experimental designs not only improved the understanding of genetic elements underlying various plant traits, but also rendered plant breeding more systematic and efficient. Conventional plant breeding remains a key technology to facilitate crop improvement; however, it has limitations such as polyploidy, zygosity and longer generation time. Also, trait introgression from wild to cultivated varieties through hybridization and selection is extremely difficult (Zamir 2001; Warschefsky et al. 2014). Similarly, utilization of mutants generated through chemicals and/or irradiation is restricted either due to the mutational load or low mutation frequency in the targeted genomic region controlling trait(s) (Jung et al. 2018). Now molecular breeding approaches that integrate genomics and high-throughput phenomics and multipotent genetic material offer faster delivery of improved varieties (Varshney et al. 2009, 2018; Appels et al. 2013, 2015). Though biotechnological tools that could precisely engineer plant traits are available such as genetic transformation, these face challenges from regulators and policy makers. Furthermore, the cost of regulating GMOs is much higher than non-GMO crops and the entire process consumes considerable time even after developing improved products (Sprink et al. 2016). To address these challenges, precise modification of crop genes and/or regulatory elements has now become possible through genome editing. Recent years have seen genome editing gaining attention of researchers because it offers predictable allelic series to optimize both quantitative and qualitative traits (Kumar et al 2020; Scheben and Edwards 2018; Biswal et al. 2019).

Domestication and modern breeding practices favouring certain genomic regions have eroded genetic variation in current cultivated pools of different crop species. For instance, transition of domesticated rice from prostrate (wild rice) to erect growth (modern rice cultivars) resulted from the selection of an important single mutation *prostrate growth 1* (*PROG1*) gene (Jin et al. 2008; Tan et al. 2008). Therefore, endeavouring precise modification of crop gene(s) by generating beneficial alleles with site-specific nucleases for desire phenotype will make huge impact on trait discovery and accelerate domestication of crop species (Scheben and Edwards 2018; Nogué et al. 2016). And with genome editing tools in place, it is possible to achieve this in much shorter duration (Scheben and Edwards 2018); however, their acceptance is still in obscurity. The onus is thus on the scientific community to provide ample evidences and generate awareness regarding technically different nature of genome editing products that lack foreign DNA, thus rendering this similar to the plants improved using conventional breeding tools. Researchers argue that the edited plants developed through genome editing should not be treated as GMOs (Araki and Ishii 2015). These technologies should be kept free from the hurdle of GMO legislation to allow their speedy adoption in routine genetic improvement programmes not only in developed countries but also in developing countries.

The present review aims to underscore the potential of modern genome editing tools for developing improved crop varieties for sustainable food production. This article evaluates genome editing with respect to environment and consumer risk. Also, the constraints that limit adoption of the crops improved with genome editing are briefly discussed.

## 10.2 Introducing Mutations Through Advanced Genome Editing Tools

### 10.2.1 Zinc Finger Nuclease (ZFN)

The term ZFN was initially used by Lusser et al. (2011) and successively by new technique working group (NTWG) in 2012. In this technique, a synthetic restriction endonuclease is customized to cut double-stranded deoxyribonucleic acid (DNA) at specific sequences (Wyman and Kanaar 2006). It comprises a zinc finger domain that allows recognition of a specific DNA sequence, enabling both site-specific mutation and integration of gene(s) into the plant genome (Bibikova et al. 2002; Wyman and Kanaar 2006). ZFN acts as a heterodimer, and therefore, ZFN transcribing genes are transported in a designed expression vector to plant cells (Söllü et al. 2010). The transfer of gene through ZFN technology involves electroporation (Wright et al. 2005), transfection (Szczepek et al. 2007), whiskers (Shukla et al. 2009), microparticle bombardment (Ainley et al. 2013), and *Agrobacterium* (De Pater et al. 2009). The viral vectors are also used for gene(s) transfer into the plant genome. ZFN causes double strand breaks at unambiguous site in the genome, which activate the repair mechanism of the host plant (Petolino 2015). Afterwards, both homologous recombination (HR) and DNA inclusion take place (Fig. 10.1a, e). This technique involves three artificial restriction enzymes, namely ZFN-1, ZFN-2 and ZFN-3 (Bibikova et al. 2001). (1) ZFN-1: Here ZFN is transported to the plant genome without taking repair template. Once it reaches the plant genome, it creates double-stranded breaks (DSB) to the host DNA that leads to non-homologous end-joining (NHEJ) of DNA (Puchta 2005), which either generates site-specific random mutations or small insertion or deletion. (2) ZFN-2: In contrast to ZFN1, a homology-directed repair (HDR) along with short repair template is delivered to plant genome along with ZFN enzyme (Lusser et al. 2011). The template DNA is homologous to target DNA, which binds to specific sequence causing a double-stranded break. The template starts repairing competing with endogenous repair machinery which led to site-specific point mutations through homologous recombination (HR). (3) ZFN-3: When ZFN transcribing gene is transported to the plant genome along with large repair template (for gene addition or replacement), it is called ZFN3 (Lusser et al. 2011; Araki et al. 2014). It binds to double-stranded DNA and causes site-specific double-stranded cleavage followed by HR. The end sequence flanking the double-stranded cleavage is the homologous results insertion of DNA stretch in a site-specific manner. ZFN-3 also helps in addition or replacement of the gene of interest, and for trait stacking in crops, such as herbicide resistance in plants (Townsend et al. 2009).

### 10.2.2 Transcription Activator-like Effector Nucleases (TALEN)

Transcription activator-like effector (TALE) proteins were discovered in the bacterial *Xanthomonas* sp. (Bonas et al. 1989). Bacterial system utilizes this to infect

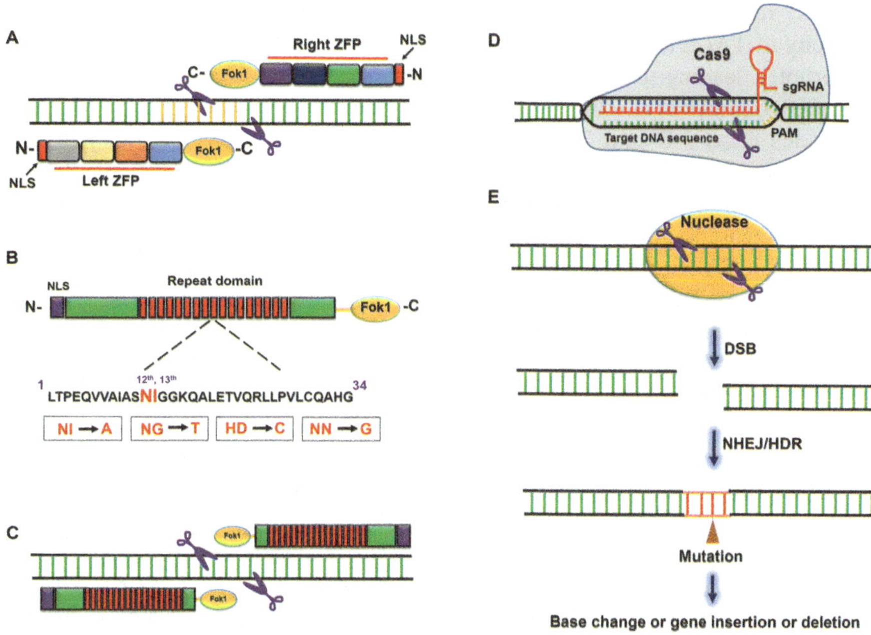


**Fig. 10.1** Structural representation of nucleases. (**a**) Structure of ZFN. ZFP represents zinc figure protein. The ZFN recognizes target site by the left and right ZFPs, and each engineered ZFP can recognize a target nucleotide. The ZFN monomer is contained a NLS (red) domain at N-terminal. The C-terminal comprises the Fok I endonuclease. The target sequence recognized by the left and right ZFPs which undergo for the dimerization of the Fok I endonuclease for activity. (**b**) TALEN contains an N-terminal domain comprising a nuclear localization signal (NLS); an essential domain typically formed of tandem TALE repeats to recognition a specific target DNA sequence; and a C-terminal domain with functional endonuclease Fok I. Each TALE repeat consists of 34-amino-acid with a variation at 12th and 13th amino acid position: NI (recognizes nucleotide A), NG (recognizes T), HD (recognizes C) or NN (recognizes G) (marked in black box). (**c**) Mode of action for TALEN. (**d**) Schematic representation of the CRISPR/Cas9 system structure and the principle for mutation induced through CRISPR. The synthetic guide RNA (sgRNA) complementary to the target DNA binding site and stem loops facilitates the binding of the Cas9 protein. The protospacer adjacent motif (PAM, NGG) is required for DSB which facilitate genome editing through error prone non-homologous end-joining (NHEJ) and homology-directed repair (HDR) repair pathway. (**e**) Mode of action for nucleases. The DNA double-stranded break DSB is repaired through HDR/NHEJ which causes base change or gene insertion or deletion in the target region (Adapted from Kumar et al. 2021)

plants through injecting TALE protein in plant cell via the Type III secretion system, triggering effector specific genes in host (Römer et al. 2007). TALEs consist of effector proteins, which facilitate localization, activation and specific DNA binding (Miller et al. 2011). The DNA binding domain consists of TALE effector proteins that are highly conserved, and possesses tandem repeats of 5–30 (average of 17.5) amino acids which specifically recognizes target DNA sequences (Boch and Bonas 2010). The highly conserved domain shows variation at 12th and 13th position called as "repeat variable di-residues" (RVDs); it primarily determines the DNA specificity of TALE (Bogdanove and Voytas 2011). However, these tandem repeats end abruptly, leading to truncated repeated are termed as "half repeats" (Boch and Bonas 2010; Miller et al. 2011). The DNA binding efficacy of RVDs of TALEN to nucleotides (A, C, G and T) depends upon the amino acids Asn-Ile, His-Asp, Asn-Asn, Asn-Lys and Asn-Gly (Moscou and Bogdanove 2009). The deeper understanding of RVDs has allowed molecular biologist to modify naturally occurring TALEs for genome editing (Römer et al. 2007). The fusion of nikase Fok I to the C-terminus of TALEs results in development of specific TALEN for genome editing (Fig. 10.1b, c, e). The Fok I enzymes work in a dimeric state, hence pair of TALENs is required to facilitate DNA binding by Fok I heterodimer (Zu et al. 2013; Shin et al. 2014). Then the Fok I dimer cuts specific DNA region at the spacer site to create DSB. These DSB are repaired through NHEJ, which often yield indels within the target site of the genome. Further, the TALE protein can be fused with activator, repressor, nuclease or methylase to improve TALE based proteins for genome editing (Chen and Gao 2013). The application of TALENs was extended for—(a) introduction of exogenous sequences, e.g. fluorescent tags, etc.; (b) conditional gene expression and specific gene knockout; (c) controllable rearrangements of genomic DNA through deletions, inversions/reversions (Quétier 2016). Though widely used in animals, limited attempts have been reported so far in case of plant system due to the complex nature of TALEN construct (Araki and Ishii 2015).

### 10.2.3 CRISPR: A Modern Editing Tool to Assist Plant Breeding

Bacterial and archaea genomes encode nucleases that trim invaders (bacteriophages) DNA. These small segments of foreign DNA are incorporated into the host genome as a long term permanent records of infectious genome (Barrangou and Doudna 2016). This yields direct repeats in bacterial genome intervened by short unique sequences (proto-spacers, 32 nucleotides), indeed representing a short sequence of foreign genome (Quétier 2016). The term CRISPR is an abbreviation for "clustered regularly interspaced short palindromic repeats", whereas Cas represents the nuclease associated with CRISPRs assembly. Recent genome sequencing experiments have revealed occurrence of CRISPRs in almost 40% of bacteria and 90% archaea species (Horvath and Barrangou 2010). The bacterial genome encodes a range of Cas proteins, of which Cas9 represents Type II CRISPR/Cas system (Song et al. 2016). The CRISPR/Cas system was initially discovered in 1987 by Ishino and colleagues (Ishino et al. 1987). The principle that underlies CRISPR/Cas9 system was elucidated later in 2011 (Fig. 10.1d). The Cas9 associates with trans-activator crRNA (tracrRNA) and CRISPR RNA (crRNA, transcript of a protospacer) to create a double strand break in foreign DNA that matches the crRNA (Fig. 10.1d) (Deltcheva et al. 2011). Interestingly, these spacers are transcribed after each invasion and aligned with complementary nucleotides bases present in the foreign DNA, causing CRISPR/Cas mediated degradation of invaded DNA. The Cas9 protein consists of RuvC and HNH domains that create a blunt end DSB at the three base pairs upstream of protospacer at the 3′ end (Garneau et al. 2010). The DSB is repaired by NHEJ or HDR (Fig. 10.1e) mechanism which often results mutation such as indels (Xiong et al. 2015). Furthermore, the specificity of Cas9 also depends on its three-dimensional conformation. The nuclear DNA regulates the differential binding and residence time. For instance, extended binding time with target DNA sequences, whereas a shorter period for off-targets (Knight et al. 2015). To make this technique more robust, researcher fused the tracrRNA and crRNA to a single guide RNA molecule (sgRNA) (Jinek et al. 2012). The Cas9 nuclease specifically cleaves the RNA/DNA complex followed by DNA repair. With this modification CRIPSR/Cas9 genome engineering is achieved with much higher efficiency. A recent modification involves development of Cas9 variant using Fok I (from Streptococcus pyogenes), Cpf1 (Cas12; from *Francisella novicida* U112) and C2c2 (Cas13; from *Alicyclobacillus acidoterrestris*) nucleases (Tsai et al. 2015; Shmakov et al. 2015; Zetsche et al. 2015). Genome editing CRISPR technology has been extended beyond site-specific mutagenesis (Barrangou and Doudna 2016). Recent research has shown transcriptional regulation by deactivating the Cas9, and fusing the guide RNA with activator or repressor (Fig. 10.2) (Qi et al. 2013; Gilbert et al. 2014). Likewise, fusion of fluorophores enables Cas9 sequence-specific DNA visualization or chromatin imaging (Chen et al. 2012; Mao et al. 2016). Additionally, RNA manipulation has been reported using CRISPR/Cas13 in eukaryotes, including plants. RNA editing is a post-transcriptional mechanism, which converts adenosine to inosine (A to I) (Matsoukas 2018). Cox et al. (2017) reported that CRISPR/Cas13 in a programmable manner to alter the coding potential in mammalian cells. Further, Abudayyeh et al. (2017) and Aman et al. (2018) used Cas13 system to target mammalian and plant cells to knockdown of either endogenous or reporter transcripts and RNA virus, clearly indicating the potential applications in agricultural biotechnology (Ali et al. 2018). In recent years, CRISPR application has been extended to epigenetic modifications in genome to activate gene through promoters and enhancers by fusing to acetyltransferases to Cas9 (Hilton et al. 2015; Kearns et al. 2014). Unlike ZFNs and TALENs, CRISPR/Cas9 offers RNA guided genome editing in an cost-efficient and user-friendly manner (Nagamangala Kanchiswamy et al. 2015). These advancements have inspired increasing use of CRISPR/Cas9 technology in crop and animal breeding (Quétier 2016; Song et al. 2016).

## 10.3 Showcasing of the Candidate Genes Through Genome Editing of Crop Plants

Site-specific nucleases have allowed the introduction of targeted sequence-specific changes in both plant and animal system. Initially adopted in animal systems for targeted genome modification, ZFN and TALEN protein-guided recognition tools were later extended to create mutations or indels in the target gene of various plant species (Table 10.1) (Gaj et al. 2013; Lor et al. 2014; Sawai et al. 2014). In model plant Arabidopsis ZFN technique was employed to generate several mutants (Qi et al. 2013). For instance, *Arabidopsis* loss of function mutant for endogenous

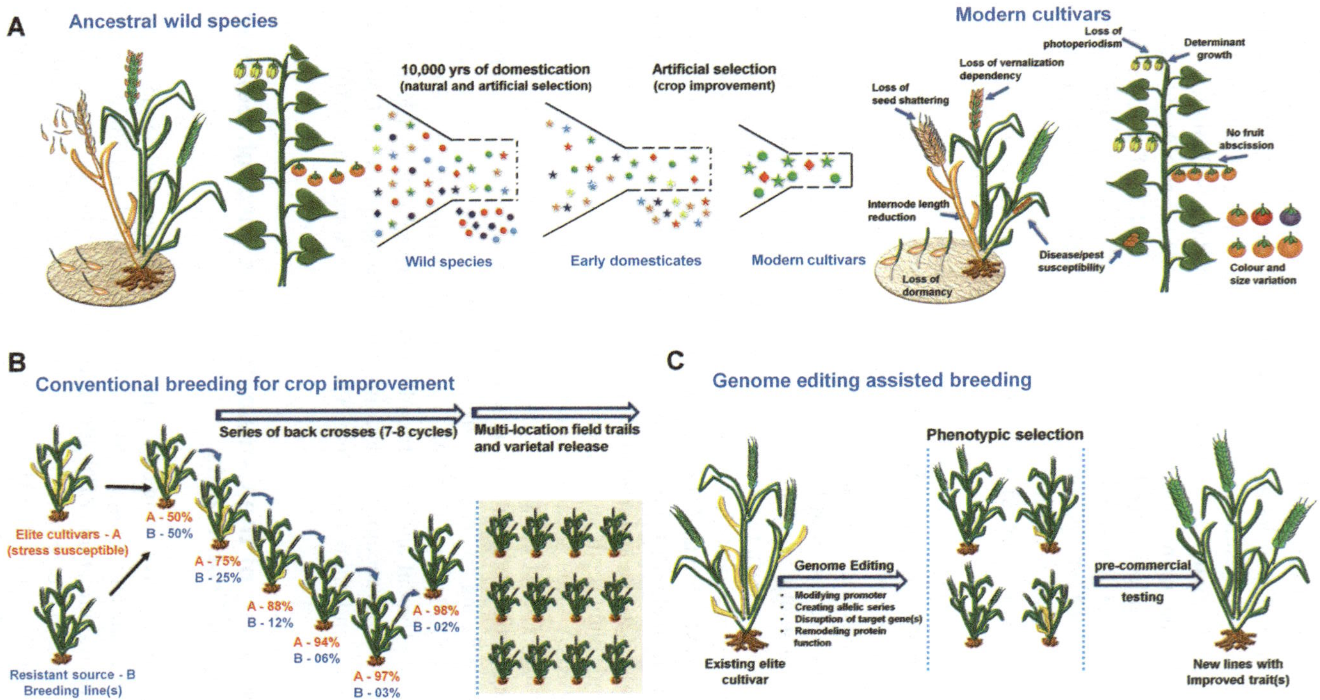


**Fig. 10.2** Schematic illustration of crop improvement. (**a**) Loss of genetic diversity prior to domestication resulting in domesticated crop with desired trait(s), (**b**) conventional breeding approach for crop improvement which involves up to 6–15 yrs., (**c**) genome editing assisted breeding for faster improvement of existing elite cultivars (Adapted from Kumar et al. 2021)

**Table 10.1** List of genomes edited crops developed through genome editing—ZFNs, TALENs and CRISPR/Cas9

| Technique | Gene modification/ addition | Effect | Plant/crop | References |
|---|---|---|---|---|
| ZFN | *SuRA, SuRB* | Herbicide resistance | Tobacco | Townsend et al. (2009) |
| | *IPK1, PAT* | PA biosynthetic pathway<br>Herbicide tolerance | Maize | Zu et al. (2013) |
| | *ADH1, TT4* | Stress responsive | Arabidopsis | Zhang et al. (2010) |
| | *ABI-4* | Stress responsive | Arabidopsis | Osakabe et al. (2010) |
| | *CHN50 PAT* | Endochitinase<br>Herbicide resistance | Tobacco | Cai et al. (2009) |
| | *GFP DCL1a, DCL1b, DCL2a, DCL2b, DCL4a, DCL4b, RDR6a, RDR6b, HEN1a* | Reporter gene<br>Mutations creation in duplicate genes | Soybean | Curtin et al. (2011) |
| | *L1L4* | Plant development | Tomato | Hilioti et al. (2016) |
| | *AG2, LFY* | Plant development | Populus | Lu et al. (2016) |
| | *RFP DSM-2 e* | Reporter gene<br>Resistance | Tobacco | Schneider et al. (2016) |
| | *UIDA* | GUS repair mutation | Apple, FIG | Peer et al. (2015) |
| TALENs | *DEP, BADH2, CKX, SD1, ABA1, CKX2, SMC6, SPL, SBP, COI1, RHT, HTA1* | Plant development and stress responsive | Rice<br>Brachy podium | Shan et al. (2013) |
| | *SWEET14* | Biotic stress resistance | Rice | Li et al. (2012) |
| | *PAPhy_a* | Phytase reduction and seed development | Barley | Wendt et al. (2013) |
| | *MLO* | Biotic stress resistance | Wheat | Wang et al. (2014) |
| | *PRO* | GA hormone regulation | Tomato | Lor et al. (2014) |
| | *VInv* | Sugar metabolism | Potato | Clasen et al. (2016) |

(continued)

**Table 10.1** (continued)

| Technique | Gene modification/ addition | Effect | Plant/crop | References |
|---|---|---|---|---|
| CRISPR/ CAS9 | *SWEET14, SWEET11* | Biotic stress resistance | Arabidopsis, tobacco, rice, sorghum | Jiang et al. (2013) |
| | *INOX, PDS* | Stress responsive<br>carotenoid biosynthesis | Wheat<br>tobacco | Upadhyay et al. (2013) |
| | *MLO* | Biotic stress resistance | Rice, wheat | Shan et al. (2013), Wang et al. (2014), Wang et al. (2017) |
| | *IPK1A, IPK, MRP4 PDS* | PA biosynthetic pathwaystress responsive | Maize | Liang et al. (2014) |
| | *CAO1 LAZY1* | Stress responsive | Rice | Miao et al. (2013) |
| | *AGO7 SLR1* | Auxin metabolism<br>Gibberellic acid metabolism | Tomato<br>Rice | Brooks et al. (2014), Lu et al. (2016) |
| | *ARGOS8* | Ethylene response | Maize | Shi et al. (2017) |
| | *PDS*<br>*PIF4*<br>*DMR6*<br>*RIN*<br>*DELLA* | Carotenoid biosynthesis<br>Phytochrome mediated response<br>Biotic stress resistance<br>Ripening<br>Hormone metabolism | Tomato | Pan et al. (2016), de Thomazella et al. (2016), Ito et al. (2015), Tomlinson et al. (2019) |
| | *FAD2* | Oil content | False flax | Jiang et al. (2016) |
| | *Avr4/6* | Biotic stress resistance | Soybean | Fang and Tyler (2016) |
| | Loss of gene cluster | Multiplex mutations | Arabidopsis | Peterson et al. (2016) |
| | *OST2* | Abiotic stress resistance | Arabidopsis | Osakabe et al. (2016) |
| | *ALS*<br>*BEIIb* | Herbicide resistance<br>Starch modification | Rice<br>Wheat | Sun et al. (2016), Baysal et al. (2016), Jiang et al. (2016) |
| | Promotor region of gene *HvPAPhy a* | Phytase reduction | Barley | Holme et al. (2017) |

gene *aba-insensitive-4* (*ABI4*) was generated for ABA and glucose insensitivity (Osakabe et al. 2010), deletion mutants for *alcohol dehydrogenase-1(ADH1)* and *transparent testa-4 (TT4) which have shown heritable behaviour* (Zhang et al. 2010). In maize, ZFN technique conferred herbicide tolerance through disruption of target gene *IPK1*, which alters inositol phosphate profile (Shukla et al. 2009). A similar approach in tobacco demonstrated disruption of an endogenous *endochitinase gene CHN50* through a ZFN construct that consisted of a herbicide resistance *PAT* gene flanked by short stretches of endochitinase (Cai et al. 2009). Similarly, mutations in *acetolactate synthase* genes—*SuRA* and *SuRB* of tobacco improved tolerance against herbicides (Townsend et al. 2009). The heritable nature of the genetic modifications caused by gene editing was confirmed in soybean for target 10 genes: a transgene "*GFP transgene*" and nine endogenous genes (*DCL1a, DCL1b, DCL2a, DCL2b, DCL4a, DCL4b, RDR6a, RDR6b* and *HEN1a*) (Curtin et al. 2011). Recent examples for ZFN mediated modifications in plants include apple and fig (Peer et al. 2015), populus (Lu et al. 2016), tomato (Hilioti et al. 2016) and tobacco (Schneider et al. 2016).

Like ZFNs, TALENs have also been implemented for the improvement of crop species (Gaj et al. 2013). In monocot species, nearly 12 genes were targeted to generate desirable knockout mutants through TALENs technique (Zhang et al. 2013). In rice, *Os11N3* (*OsSWEET14*, member of SWEET sucrose-efflux transporter family) gene is responsible for bacterial blight susceptibility (Antony et al. 2010; Chen et al. 2012). This gene in rice was mutated through TALEN and thus transgenic plants gained desired resistance to bacterial blight disease (Li et al. 2012). In barley, the promoter of *HvPAPhy_a* (from *phytase gene* family) was targeted as it accounts for the maximum of the phytase activity during seed development (Wendt et al. 2013). The *mildew-resistance locus* (*MLO*) gene was targeted which encodes for a protein that suppresses defence against powdery mildew disease (Wang et al. 2014). With TALEN technology three homoeoalleles of *MLO* were disrupted in bread wheat to confer heritable resistance against powdery mildew (Wang et al. 2014). In tomato DELLA protein is encoded by *procera* (*PRO*) gene (Carrera et al. 2012), and it negatively regulates the GA signalling pathway (Zentella et al. 2007). Tomato *pro* mutant possesses enhanced levels of GA, but it partially retained some GA response, suggesting a leaky phenotype of the mutant protein (Van Tuinen et al. 1999). In order to completely block the DELLA protein function, *PRO* gene mutants of tomato were raised through TALEN, which displayed a similar phenotype as *pro* mutant (Lor et al. 2014). TALENs have been implicated for improving postharvest quality of potato. The cold storage of potato induces formation of reducing sugars, which react with free amino acids at high temperature to form acrylamide (Kim et al. 2015). Recently, Clasen et al. (2016) obtained TALEN based knockout of the *vascular invertase* gene, whose tuber produces negligible level of reducing sugars and its processed chips consisted undetectable amount of acrylamide.

A growing body of literature indicate successful application of the CRISPR/Cas9 method in model and crop plants (Shan et al. 2014; Belhaj et al. 2015; Liu et al. 2017; Collonnier et al. 2017). This technique was effectively used to generate mutants in both monocots (rice and sorghum), and dicots (Arabidopsis and tobacco)

(Jiang et al. 2013). In wheat, CRISPR/Cas9 system is successfully applied for mutating *inositol oxygenase* and *phytoene desaturase* (Upadhyay et al. 2013), and *MLO* gene (Shan et al. 2013; Wang et al. 2014). Recently, *IPK* gene function was neutralized in maize by using two sgRNA in the CRISPR/Cas9 (Liang et al. 2014). Mutated *chlorophyll a oxygenase 1* (*CAO1*) and *LAZY1* gene in rice caused loss of Chlorophyll b in the mutant leaf and noticeable tiller-spreading during the tillering stage, respectively (Miao et al. 2013). Similar alterations in the promoter regions of *OsSWEET14* and *OsSWEET11* genes in rice yielded resistance against bacterial blight (Jiang et al. 2013).

The *ARGONAUTE7* (*AGO7*) *gene in tomato regulates biogenesis of a group of* sgRNAs which control the expression of *auxin response factor* gene (Husbands et al. 2009). Induction of *mutations in tomato AGO7* through CRISPR/Cas9 system *resulted* in leaf deformities and affected pollen viability (Brooks et al. 2014). Recent work exploring CRISPR/Cas9 system in tomato involved mutagenesis genes such as *PDS* and phytochrome interacting factor *PIF4* (Pan et al. 2016), *downy mildew resistance 6* (de Thomazella et al. 2016) and *ripening inhibitor* (Ito et al. 2015). This technique generates desired mutations at the specific site of interest that are inheritable. Mutagenesis of multiple genes by CRISPR/Cas9 through expressing more than one sgRNAs suggests its immense implications for improving quantitative traits. For example, 30% yield advantage was achieved in rice following CRISPR/Cas9-driven manipulation of 13 genes associated with abscisic acid biosynthesis (Miao et al. 2018). CRISPR/Cas9 approach has been applied in the model plant Arabidopsis (Upadhyay et al. 2013) and tomato (Brooks et al. 2014), and monocot plants like rice (Zhang et al. 2014). Interestingly, a deletion of 10–1000 nucleotides can be created through multiplexing the sgRNA (Belhaj et al. 2013), thus can also lead to deletion of gene clusters due to chromosomal deletion (Zhou et al. 2014). Other examples of CRISPR/CAS9 based modification in plants include targeting multiple loci in Arabidopsis to enhance yield and resistance (Mao et al. 2016; Osakabe et al. 2016; Peterson et al. 2016), gemini virus resistance in tobacco (Zaidi et al. 2016), disease resistance in tomato (de Thomazella et al. 2016), starch modification and herbicide resistance in rice (Baysal et al. 2016; Sun et al. 2016, Wang et al. 2017), improvised fatty acid accumulation in *Camelina* (Jiang et al. 2016), resistance against *Phytophthora sojae* in soybean (Fang and Tyler 2016), canker resistance in citrus (Peng et al. 2017), starch modification in wheat (Liang et al. 2017), gibberellins metabolism in rice (Lu et al. 2016), etc.

Change in the expression level and/or organization of the genes resulting from mutations in *cis*-regulatory regions is known to create quantitative and qualitative variation of the traits (Wittkopp and Kalay 2012). Gene expression is fine-tuned by cis-regulatory elements (CREs) present in the promoter region. Recently, Rodriguez-Leal et al. (2017) used the CRISPR/Cas9 to modify the CREs in the promoters of tomato *WUS* (*SlWUS*) and *CLV3* (*SlCLV3*) genes, that control fruit size, and inflorescence architecture. The induced novel *cis*-regulatory mutant alleles increased the tomato fruit size and locule number, similar to the natural QTL variants. Base-editors (BEs) are another CRISPR/CAS9 technique, which enables direct, irreversible conversion of one base to another at a target locus. Given the majority of the agronomic traits are controlled by point mutations (Huang et al. 2010), recent findings indicated that by fusing a nuclease-deactivated Cas9 (dCAS9) to a cytidine deaminase or adenosine deaminase induces C.G and A.T base pairs (bps) to T.A and G.C (Brooks and Gaj 2018). Though BEs approach was initially applied in mammalian systems, and the same was successfully employed in rice, wheat, maize, and tomato (Zong et al. 2017; Shimatani et al. 2017; Lu et al. 2016).

## 10.4 Challenges for Genome Editing

One of the major challenges in successfully achieving genome editing is plant genetic transformation and regeneration, which are the bottlenecks in agriculture biotechnology. Various technologies are available to improve plant transformation. For example, floral dip transformation is an attractive solution, it eliminates the plant tissue culture step, but the limitation is that not compatible to several crop plants, except in Arabidopsis and *Camelina sativa* (Lu and Kang 2008) and its transformation efficiency is very low. Further, to improve plant transformation and regeneration methods, high throughput, efficiency and novel transformation technologies are required. Development of novel methodologies could dramatically enhance plant genomics knowledge to feed the world (Altpeter et al. 2016).

Designing of ZFNs and TALENs is one of the most pain stacking jobs. The commercially available ZFNs are more efficient, but very expensive, than the publicly designed (Ramirez et al. 2008). In contrast, TALENs designing has become easy and efficient by using Golden Gate cloning—a DNA assembly technique (Engler et al. 2008). On the basis of available literature, some of the most important problems associated with genome editing are their low efficacy, regulatory vagueness and social acceptance (Shukla et al. 2009). Yet, the accurate estimation of efficiency is very difficult because efficiency depends on various factors such as the crop plant selection, methodology used, target gene and marker genes. For example, the efficiency of ZFN induced mutation in Arabidopsis is reported to be around 2% (de Pater et al. 2009) whereas, in the case of tobacco, it is 40% (Townsend et al. 2009). Over and above, one of the problems associated with ZFN and TALENs technologies is its non-specific binding which leads to create non-specific mutations (Pattanayak et al. 2011). These off target effects are also associated with CRISPR/Cas9 technology (Song et al. 2016). An improper concentration between Cas9 and sgRNA, or promiscuous PAM sites, or poor codon optimization of Cas9 during translation results off target/undesired cleavage of DNA sequence. It has been reported that high off-targets were found in humans, but low in mice, zebrafish and plants (Fu et al. 2013; Pan et al. 2016). Depending upon the species/cultivar the efficiency of the technology shows discrepancy. At the same time, T-DNA (foreign DNA) will be removed before proceeding for commercialization (Schaart et al. 2010). However, off target effects are expected with any genome editing tools as these are driven by several factors including sequence similarities. Nevertheless, researcher always selected the best phenotypic variants from genetically engineered lines, ruling out off target effect.

The regulatory cost of new plant breeding techniques is very high and the regulatory process alone takes 5–7 years, hence the acceptance of these techniques is low. When products become GMOs, it costs even higher and more time consuming compared to non-regulated classical breeding techniques (Kalaitzandonakes et al. 2007). Therefore, usage of these new techniques is limited. In particular, small companies are using these techniques only for limited traits of high value crops (Miller et al. 2011). Hence, it will be hard for plant breeders to invest in ventures where regulatory cost has a direct impact on the economic potential of the crops such as orphan crop and GM approaches-based product.

Once the plant is classified as GMOs, it has to be a method for identification and actual quantification on the newly introduced gene/s and has to be mentioned before going to the market (Kuzma and Kokotovich 2011). All contemporary available standard methods for GMO detection basically depends on the quality of DNA and efficiency of the techniques (PCR, qPCR, ELISA, etc.,). In order to evaluate the changes brought about by these genomes editing, those are mostly monitored by an expert committee, which is considered to be an important element of risk management (Glandorf et al. 2011). The prior knowledge of DNA sequence has an imperative role in the detection and identification of GMOs. The plant produced through ZFN1, ZFN2 and ZFN3 techniques can be detected by DNA based approaches only when there is prior information of flanking sequences of introduced modifications.

Genome editing has been mostly implemented in plant breeding to generate disease resistance and yield advantage for crops. Its application has to widen such as abiotic stress tolerance, nutritional quality enhancement and allergenicity elimination from various crops. There has been some report on RNAi based reduction of allergens from apple (Gilissen et al. 2005). Similarly, peanut allergens and gluten gene from various crops such as wheat, rye and barley can be reduced or abolished using these techniques (Gilissen et al. 2014; Smulders et al. 2015). Now breeders are opting for developing superior varieties through grafting. In case of grafting, plants are produced by joining of scions and rootstocks. When a non-GM scion is attached to the GM rootstocks, detection of scion derived products becomes impossible, but rootstock can be identified using usual genomics tools used for GM crops.

## 10.5 Genome Edited Crop: Social Acceptance and Regulatory Framework

Since the domestication of first agricultural plants nearly 10,000 years ago, plant breeding techniques have tremendously improved crop yields that can feed more than 70 million peoples (Palmgren et al. 2015). However, new breeding techniques and agronomic practices are required for a sustainable food future of 10 billion people by 2050. The unfavourable conditions like biotic and abiotic stress are conspicuous factors which have increased the losses of crop productivity over the years. Thus, the pressing demand for resilient crop species has invigorated researchers to discover the possibility through reverse genetic or genome editing. Although recent genetic engineering approaches in crop species have achieved considerable progress in crop improvement, its social acceptance is negligible due to lack of strong global policies (Araki and Ishii 2015). Argument for the accepting GMOs or genome edited plants occur not with the public; but surprisingly, it had also created a debate between researchers (Tanaka 2012; Freedman 2013; Lucht 2015). It is projected that acceptance of these genome edited plants will adversely affect the native crop germplasm resources and also human health. In order to broaden the public acceptance, constant discussion with the society is a prerequisite (Palmgren et al. 2015), excellent reviews have been published on regulatory vagueness and social acceptance (Jones 2015; Araki and Ishii 2015). In 2007, the European Union commission and member states decided to set up an expert committee on NPBTs to evaluate these new techniques with respect to GMO legislation (Schaart et al. 2016), and the commission highlighted an array of legal and social issues associated with GMOs (Lusser et al. 2012). According to committee view, these techniques may or may not involve genome alteration of the target plant species given their heterogeneous nature. The EU declaration defines GMO as "*any organism having altered genetic material which does not occur by natural mating or by natural recombination*" (Directive 2001/18/EC 2001).

Anthropogenic activity has dramatically changed agricultural strategies such as large-scale cultivation of new varieties in combination with affecting natural habitats of ancestral wild species of crop plants (von Wettberg et al. 2018). Notably, the important alleles and genetic variations present in the plant wild species allow sustainable growth in extreme environmental conditions and distant geographical regions (Hajjar and Hodgkin 2007; Lu 2013; Brozynska et al. 2016). Recent success in the field of genetic engineering has also overlooked the huge potential of wild species and their use in prebreeding for certain extent, as these modern techniques have potential to improve elite crops or domestication of crop wild relatives in shorter duration (Palmgren et al. 2015; Li et al. 2018). Numerous researchers believe that worldwide acceptance of these genome edited plants can severely affect the diversity of plant wild species, and even it could lead to extinction of some rare species (Stewart et al. 2003; Castañeda-Álvarez et al. 2016). Questions remain on the legal acceptance for the applied strategies, and also on the social, economic and ethical acceptance of them (Tanaka 2012). Hence, the major concern with the release of living modified organisms (LMOs) is their impact on the environment, biodiversity conservation and a human health risk due to consistent consumption of GMOs (Lucht 2015).

According to the Cartagena Protocol, plants raised using genome editing can be out boxed from GMO regulation as they do not possess any transgene (http://bch.cbd.int/protocol/text/). It is quite interesting that, the stringency of the regulation of GMO or LMOs considerably varies within countries. For example, in New Zealand and Europe, food obtained from the plants derived from the precision mutagenesis techniques should be compared similar to the food derived from the traditional mutagenic techniques (Lusser et al. 2011; Palmgren et al. 2015). However, the foremost challenge for GMOs is acceptance of them in public domain, which greatly relies on the mindset of citizen, farmers and decision makers (Araki and Ishii 2015). The controversies related to transgenic have led to their widespread public rejections, and limited commercialization. For instance, the expert committee of new plant biotechnology declares that ZFN-1 and 2 both create GMO and therefore, both fall under the directive of 2001/18/EC or Directive 2009 41/EC 2009 (Schiemann et al. 2009; Sprink et al. 2016). Also, the plant produced using ZFN-3 technology is transgenic and therefore, comes under the directive of 2001/18/EC (Schiemann et al. 2009; Araki et al. 2014). Similarly, several CRISPR/Cas9 mediated products including rice, maize, soybean, etc., are already developed and waiting for the approval of government regulatory bodies. Very recently, the Court of Justice of the European Union (ECJ) subjected CRISPR edited plants under tough GM laws by subjecting these plants to a 2001 directive, previously developed to control GM crops for food (Callaway 2018). However, researchers and plant breeders argue that CRISPR/Cas9 edited plants should be treated same as irradiation mutagenesis because it causes changes in DNA and does not involve the insertion of foreign genes, thus they can be exempt from the directive. Currently, the adoption of CRISPR system in agronomy has been remarkably increasing (Ricroch et al. 2017). As a result, several countries like USA, Canada, China, etc., have showed positive response towards CRISPR/Cas edited crop products (Lassoued et al. 2019); however, the developed edible food products from edited crops are of major concern. In fact, globally, the impact of genetically modified crops has been realized, especially due to the recent economic analysis obtained for the modified global crops such as maize and cotton (Brookes and Gaj 2018). For instance, in Spain and Portugal, over the 21-year period (1998 and 2018) the insect-resistant (IR) maize (aka corn) has increased farmers income by €285.4 million (US$322.9 million) by saving money on insecticides and producing more crop yields (Brookes and Gaj 2018). Additionally, use of IR maize maintained the required production by using lesser arable land because for the same production with conventional breeding material the farmers would have required an additional 15,240 hectares in the two countries. In 2014, the genetically modified soybean, cotton and canola saved 19, 9 and 1.5 million hectares of land globally (https://www.pgeconomics.co.uk/pdf/2017globalimpactstudy.pdf).

Recent reports have suggested that the genome editing tools have faded the boundaries between edited crops and regulatory bodies for social acceptance (Ishii and Araki 2016). Many products delivered through TALEN approach has been accepted (see Table 10.2). For instance, a TALEN mediated SU (sulfonylurea) Canola launched by Cibus (https://www.cibus.com/products.php) was commercially approved by the Canadian and United States governments in 2015 (Table 10.3). In the next 5 years other products from TALEN mediated genome edited products from Cibus such as glyphosate tolerant flax, soybean and maize breeds are under evaluation in the United States (Li et al. 2016) (Table 10.3). No wonder, more crop TALEN mediated genome edited products would be pushed to the market, as TALEN has proved its potential and critical role in genome editing breeding. It is obvious that the growing demand of food supply coupled with agronomic losses due to increased prevalence of diseases and abiotic stress needs supports of genome edited crops, which can provide elite varieties in very short duration. We speculate, in next two decades CRISPR/Cas9 mediated crop will have more products directly developed from the domestication of crop wild relatives (Li et al. 2018), which known to have several important features including higher nutritional quality and disease resistance, these products should be globally accepted to fulfil hunger need.

**Table 10.2** Examples of genome edited crops approved through regulatory agency or are in pipeline

| Technology | Crop | Trait | Developer | Current status | Reference |
|---|---|---|---|---|---|
| ZFN | Maize | Reduced phytate production | Dow AgroSciences | USDA approved | Wolt et al. (2016) |
| TALENs | Alfalfa | Improved quality alfalfa | Calyxt | USDA approved | http://www.calyxt.com |
| | Soybean | High oleic | Calyxt | USDA approved | http://www.calyxt.com |
| | Soybean | High oleic/low linolenic | Calyxt | USDA approved | http://www.calyxt.com |
| | Wheat | Powdery mildew resistant | Calyxt | USDA approved | http://www.calyxt.com |
| | Potato | Cold storable | Calyxt | USDA approved | http://www.calyxt.com |
| | Potato | Reduced browning | Calyxt | USDA approved | http://www.calyxt.com |
| | Rice | Bacterial blight resistance | Iowa state university | No information | Li et al. (2012) |
| | Potato | Consumer safety and processing attributes | Cellectis | USDA approved | Wolt et al. (2016) |
| | Potato | Reduced browning | Simplot plant sciences | Health Canada approved | http://www.simplot.com |
| | Potato | Late blight resistance | Simplot plant sciences | USDA approved | Halterman et al. (2016) |
| | Wheat | High fibre | Calyxt | USDA approved | http://www.calyxt.com |
| | Rice | Disease resistance | Iowa state university | USDA approved | Wolt et al. (2016) |
| | Canola | Herbicide tolerant | Cibus | Health Canada approved | Li et al. (2016) |
| | Flax | Herbicide tolerant | Cibus | Under pipeline | Li et al. (2016) |
| | Maize | Herbicide tolerant | Cibus | Under pipeline | Li et al. (2018) |
| | Soybean | Herbicide tolerant | Cibus | Under pipeline | Li et al. (2018) |
| CRISPR/CAS9 | Camelina | No information | Yield 10 bioscience | USDA approved | www.yield10bio.com |
| | Button mushroom | Non-browning | Penn State University | USDA approved | Parrott (2018), Waltz (2018) |

(continued)

**Table 10.2** (continued)

| Technology | Crop | Trait | Developer | Current status | Reference |
|---|---|---|---|---|---|
| | Maize | Improved waxy | DuPont Pioneer | USDA approved | www.pioneer.com; Waltz (2018) |
| | Maize | Increase yield | Benson Hill biosystems | No information | www.bensonhillbio.com |
| | Green foxtail | Flowering time | Danforth | No information | Parrott (2018), Waltz (2018) |
| | Maize | Leaf blight resistance | DuPont Pioneer | No information | Parrott (2018) |
| | Soybean | Drought tolerance | USDA-ARS | No information | Waltz (2018) |

**Table 10.3** List of crop and traits targeted by commercial biotech company Cibus

| Crop | Trait | Year |
|---|---|---|
| Canola | MOA-2<br>Pod shatter reduction<br>Oil quality<br>Disease resistance | 2020–2023<br>2020–2023<br>2020–2023<br>2020–2023<br>2020–2023 |
| Rice | MOA-1<br>MOA-2<br>Disease resistance | 2020–2023<br>2020–2023<br>2023+ |
| Flax | MOA-1 | 2020–2023 |
| Potato | MOA-1<br>Disease resistance | 2023+<br>2023+ |
| Corn | MOA-1<br>MOA-2<br>Disease resistance | 2023+<br>2023+<br>2023+ |
| Wheat | MOA-1<br>MOA-2<br>Disease resistance | 2023+<br>2023+<br>2023+ |
| Peanut | Aflatoxin | 2023+ |

## 10.6 Intellectual Property Rights (IPRs) Protection and Freedom to Operate (FTO)

Biotechnology research has increased the availability of crop plants that are high-yielding, nutritious, stress tolerant, etc. (Díaz de la Garza et al. 2004; Vinocur and Altman 2005; Storozhenko et al. 2007; Nunes et al. 2009; Tamás et al. 2009; Varshney et al. 2011). However, the success of these genetically engineered plants is greatly dependent on the inventor incentives because of IPRs. IPRs include a set of laws to provide a legal protection to inventor or innovators for a fixed period of time against direct exploitation of their product or method (Malik and Zafar 2005). IPRs protect the biotechnology material through two major systems: patents and rights in plant varieties, but for a limited period of time. Patents provide a wide range of legal rights to retain, use, transfer it by sale or as a gift, and restrict others from similar rights for a duration of 17–20 yrs. (Gold et al. 2002; Graff et al. 2003). According to International Union for the Protection of New Varieties of plants (UPOV), a plant variety can be protected only if it is unique, stable, uniform and fulfil the novelty requirements (Jördens 2005). This grant provides an exclusive right to the owner to sell the plant materials such as reproductive organ or whole plant, which can be up to a period of 20–30 years. In the field of agro-biotechnology there is an exponential increase of the counts for filed application patents in USA, Germany and Japan (Graff et al. 2003). This surge was motivated by the royalties which can be obtained from invention (Barrows et al. 2014). The seed companies protect their genetically engineered seeds by IPRs (Oczek 2000; Frison et al. 2010), and sell their seed to farmers at monopolistic prices. It has also affected the conventional plant breeding, which is slow due to their need on the availability of the desired traits from the ancestral or closely related species. However, the rapid increase of the advancement in agricultural plant biotechnologies has increased the monitoring responsibility for the crop biosecurity which protects from the bioterrorism, biopiracy, genetic erosion, etc., (Evenson 1999; Chen and Puttitanun 2005). Furthermore, the surge of proprietary protection in crop species has intricate the exchange of germplasm which are required to develop new cultivars against destructive disease and environmental stress (Graff et al. 2003; Luby and Goldman 2016). This has led to an inability of researchers and breeders to obtain seed without acquiring permission through entering into an agreement such as material transfer certificate, license, etc. (Chi-Ham et al. 2012; Luby et al. 2015), hence, deprive researchers and breeders from genetic gain. These circumstances restrict the reach of quantity breed from a breeder and the researcher, and limit FTO for the purpose of crop breeding (Binenbaum et al. 2003; Le Buanec 2005). FTO allows researcher/people to determine whether a commercialization or testing can be done without violating any valid IPRs of others (Luby et al. 2015; Bjørnstad 2016; Zanga et al. 2016). To increase the accessibility of germplasm to breeders and researcher several open resource centre available such as an open source seed initiative (OSSI), Chinese crop germplasm information system (CCGIS), national small grains collection (NSGC), USDA soybean germplasm collection, tomato genetics resource centre (tgrc.ucdavis.edu/), etc., (Sachs 2009). Recently, efforts have been made to enhance the FTO in corn and carrot breeding through development of open source populations (Luby and Goldman 2016; Zanga et al. 2016). Similar attempts are required for improving the crop breeding because today crop genetics resources are intensely secure with IPRs.

## 10.7 Conclusion

The expectations of genome editing have risen that the technology would expedite progress towards sustainable crop productivity. Some products are at the final stage of development. The commercial use of the technology is relatively new in crop improvement. Genome editing allows generation of superior plants rapidly with higher efficiency and in an environmentally safe manner. Majority of the commercial crop varieties developed over the last 20 years by transgenesis, conventional and molecular breeding are now being explored through genome editing. Considering wider applications, CRISPR/Cas9 has gained more attention from researchers and breeders as compared to ZFNs, TALENs, grafting, reverse breeding, etc. As a result, breeders are now encouraging CRISPR edited plants crops to combat climate change and associated yield loss. The genome editing is accepted by the commercial sector because of its impending financial gain over alternative traditional techniques. However, a wider adoption of products derived from these techniques depends on several factors, including regulatory jurisdiction, the efficiency of the techniques and political expediency. Genome edited crop products are now available in a few countries such as the U.S. and Canadian government has approved genome edited canola and mushroom. We anticipate that CRISPR/Cas9 technology is likely to bridge the gap between GMO and society. The genome editing will be instrumental in meeting the challenge of feeding 12 billion people by the end of the twenty-first century.

**Acknowledgement** RK acknowledges Science and Engineering Research Board (SERB), Department of Science and Technology (DST), Government of India for the financial support. A.K sincerely thank University Grants Commission (UGC) start-up grant (No.F.30-392/2017 (BSR) and Madhya Pradesh Council of Science and Technology (Endt. No. 3879/CST/R&D/BioSci/2018) for the funding to the laboratory.

**Authors Contribution** RK, NRN and AK conceived the manuscript. AM and TKT gave the technical support during the MS preparation. All authors read and approve the final MS.

## References

Abudayyeh OO, Gootenberg JS, Essletzbichler P, Han S, Joung J et al (2017) RNA targeting with CRISPR-Cas13. Nature 550:280–284

Ainley WM, Sastry-Dent L, Welter ME, Murray MG, Zeitler B, Amora R, Corbin DR, Miles RR et al (2013) Trait stacking via targeted genome editing. Plant Biotechnol J 11:1126–1134

Ali Z, Mahas A, Mahfouz M (2018) CRISPR/Cas13 as a tool for RNA interference. Trends Plant Sci 23:374–378

Altpeter F, Springer NM, Bartley LE, Blechl AE, Brutnell TP, Citovsky V, Conrad LJ, Gelvin SB, Jackson DP, Kausch AP et al (2016) Advancing crop transformation in the era of genome editing. Plant Cell 28:1510–1520

Aman R, Ali Z, Butt H, Mahas M, Aljedaani F, Khan MZ, Ding S, Mahfouz M (2018) RNA virus interference via CRISPR/ Cas13a system in plants. Genome Biol 19:1

Antony G, Zhou J, Huang S, Li T, Liu B, White F, Yang B (2010) Rice xa13 recessive resistance to bacterial blight is defeated by induction of the disease susceptibility gene Os-11N3. Plant Cell 22:3864–3876

Appels R, Barrero R, Bellgard M (2013) Advances in biotechnology and informatics to link variation in the genome to phenotypes in plants and animals. Funct Integr Genomics 13(1):1–9

Appels R, Nystrom J, Webster H, Keeble-gagnere G (2015) Discoveries and advances in plant and animal genomics. Funct Integr Genomics 15:121–129

Araki M, Ishii T (2015) Towards social acceptance of plant breeding by genome editing. Trends Plant Sci 20:145–149

Araki M, Nojima K, Ishii T (2014) Caution required for handling genome editing technology. Trends Biotechnol 32:234–237

Barrangou R, Doudna JA (2016) Applications of CRISPR technologies in research and beyond. Nat Biotechnol 34:933–941

Barrows G, Sexton S, Zilberman D (2014) Agricultural biotechnology: the promise and prospects of genetically modified crops. J Econ Perspect 28:99–120

Baysal C, Bortesi L, Zhu C et al (2016) CRISPR/Cas9 activity in the rice OsBEIIb gene does not induce off-target effects in the closely related paralog OsBEIIa. Mol Breed 36:1–11

Belhaj K, Chaparro-Garcia A, Kamoun S et al (2013) Plant genome editing made easy: targeted mutagenesis in model and crop plants using the CRISPR/Cas system. Plant Methods 9:39

Belhaj K, Chaparro-garcia A, Kamoun S et al (2015) ScienceDirect editing plant genomes with CRISPR/Cas9. Curr Opin Biotechnol 32:76–84

Bibikova M, Carroll D, Segal DJ, Trautman JK, Smith J, Kim YG, Chandrasegaran S (2001) Stimulation of homologous recombination through targeted cleavage by chimeric nucleases. Mol Cell Biol 21(1):289–297

Bibikova M, Golic M, Golic KG, Carroll D (2002) Targeted chromosomal cleavage and mutagenesis in Drosophila using zinc-finger nucleases. Genetics 161(3):1169–1175

Binenbaum E, Nottenburg C, Pardey PG et al (2003) South-north trade, intellectual property jurisdictions, and freedom to operate in agricultural research on staple crops. Econ Dev Cult Change 51:309–335

Biswal AK, Mangrauthia SK, Reddy MR et al (2019) CRISPR mediated genome engineering to develop climate smart rice: challenges and opportunities. Semin Cell Dev Biol 96:100–106

Bjørnstad Å (2016) "Do not privatize the giant" shoulders': rethinking patents in plant breeding. Trends Biotechnol 34:609–617

Boch J, Bonas U (2010) Xanthomonas AvrBs3 family-type III effectors: discovery and function. Annu Rev Phytopathol 48:419–436

Bogdanove AJ, Voytas DF (2011) TAL effectors: customizable proteins for DNA targeting. Science 333(6051):1843–1846. https://doi.org/10.1126/science

Bonas U, Stall RE, Staskawicz B (1989) Genetic and structural characterization of the avirulence gene avrBs3 from Xanthomonas campestris pv. Vesicatoria. Mol Gen Genet 218:127–136

Brooks AK, Gaj T (2018) Innovations in CRISPR technology. Curr Opin Biotechnol 52:95–101

Brooks C, Nekrasov V, Lippman ZB, Van Eck J (2014) Efficient gene editing in tomato in the first generation using the clustered regularly interspaced short palindromic repeats/CRISPR-Associated9 System1. Plant Physiol 166:1292–1297

Brozynska M, Furtado A, Henry RJ (2016) Genomics of crop wild relatives: expanding the gene pool for crop improvement. Plant Biotechnol J 14:1070–1085

Cai CQ, Doyon Y, Ainley WM et al (2009) Targeted transgene integration in plant cells using designed zinc finger nucleases. Plant Mol Biol 69:699–709

Callaway E (2018) CRISPR plants now subject to tough GM laws in European Union. Nature 560:16

Carrera E, Ruiz-Rivero O, Peres LEP et al (2012) Characterization of the procera tomato mutant shows novel functions of the SlDELLA protein in the control of flower morphology, cell division and expansion, and the auxin-signaling pathway during fruit-set and development. Plant Physiol 160:1581–1596

Castañeda-Álvarez NP, Khoury CK, Achicanoy HA, Bernau V, Dempewolf H, Eastwood RJ et al (2016) Global conservation priorities for crop wild relatives. Nat Plants 2(4):16022

Chen K, Gao C (2013) TALENs: customizable molecular DNA scissors for genome engineering of plants. J Genet Genomics 40:271–279

Chen Y, Puttitanun T (2005) Intellectual property rights and innovation in developing countries. J Dev Econ 78:474–493

Chen L-Q, Qu X-Q, Hou B-H et al (2012) Sucrose efflux mediated by SWEET proteins as a key step for phloem transport. Science 335:207–211

Chi-Ham CL, Boettiger S, Figueroa-Balderas R et al (2012) An intellectual property sharing initiative in agricultural biotechnology: development of broadly accessible technologies for plant transformation. Plant Biotechnol J 10:501–510

Clasen BM, Stoddard TJ, Luo S et al (2016) Improving cold storage and processing traits in potato through targeted gene knockout. Plant Biotechnol J 14:169–176

Collonnier C, Epert A, Mara K, Maclot F, Guyon-Debast A, Charlot F, White C, Schaefer D, Nogué F (2017) CRISPR-Cas9-mediated efficient directed mutagenesis and RAD51-dependent and RAD51-independent gene targeting in the moss Physcomitrella patens. Plant Biotechnol J 15:122–131

Cox DB, Gootenberg JS, Abudayyeh OO, Franklin B, Kellner MJ, Joung J, Zhang F (2017) RNA editing with CRISPR-Cas13. Science 358:1019–1027

Curtin SJ, Zhang F, Sander JD et al (2011) Targeted mutagenesis of duplicated genes in soybean with zinc-finger nucleases. Plant Physiol 156:466–473

De Pater S, Neuteboom LW, Pinas JE et al (2009) ZFN-induced mutagenesis and gene- targeting in Arabidopsis through agrobacterium-mediated floral dip transformation. Plant Biotechnol J 7:821–835

de Thomazella DPT, Brail Q, Dahlbeck, D, Staskawicz, BJ (2016) CRISPR-Cas9 mediated mutagenesis of a DMR6 ortholog in tomato confers broad-spectrum disease resistance. bioRxiv 064824.

Deltcheva E, Chylinski K, Sharma CM, Gonzales K, Chao Y, Pirzada ZA, Eckert MR, Vogel J, Charpentier E (2011) CRISPR RNA maturation by trans-encoded small RNA and host factor RNase III. Nature 471:602–607

Díaz de la Garza R, Quinlivan EP, Klaus SMJ et al (2004) Folate biofortification in tomatoes by engineering the pteridine branch of folate synthesis. Proc Natl Acad Sci U S A 101:13720–137205

Directive 2001/18/EC (2001) Directive 2001/18/EC of the European Parliament and of the Council on the deliberate release into the environment of genetically modified organisms and repealing Council Directive 90/220/EEC. Official Journal of the European Union, L 106, 17.4.2001

Directive 2009/41/EC (2009) Directive 2009/41/EC of the European Parliament and of the Council of 6 May 2009 on the contained use of genetically modified micro-organisms (Recast) (OJ L 125:75–97

Engler C, Kandzia R, Marillonnet S (2008) A one pot, one step, precision cloning method with high throughput capability. PLoS One 3:e3647

Evenson RE (1999) Intellectual property rights, access to plant germplasm, and crop production scenarios in 2020. Crop Sci 30(6):1630–1635

Fang Y, Tyler BM (2016) Efficient disruption and replacement of an effector gene in the oomycete Phytophthora sojae using CRISPR/Cas9. Mol Plant Pathol 17:127–139

Freedman DH (2013) The truth about genetically modified food. Sci Am 309:107–112

Frison C, Dedeurwaerdere T, Halewood M (2010) Intellectual property and facilitated access to genetic resources under the international treaty on plant genetic resources for food and agriculture. Eur Intellect Prop Rev 32:1–8

Fu Y, Foden JA, Khayter C, Maeder ML, Reyon D, Joung JK, Sander JD (2013) High-frequency off-target mutagenesis induced by CRISPR–Cas nucleases in human cells. Nat Biotechnol 31:822–826

Gaj T, Gersbach CA, Barbas CF (2013) ZFN, TALEN, and CRISPR/Cas-based methods for genome engineering. Trends Biotechnol 31:397–405

Garneau JE, Dupuis ME, Villion M, Romero DA, Barrangou R, Boyaval P, Fremaux C, Horvath P, Magadán AH, Moineau S (2010) The CRISPR/Cas bacterial immune system cleaves bacteriophage and plasmid DNA. Nature 468:67–71

Gilbert LA, Horlbeck MA, Adamson B et al (2014) Genome-scale CRISPR-mediated control of gene repression and activation. Cell 159:647–661

Gilissen LJ, Bolhaar ST, Matos CI, Rouwendal GJ, Boone MJ, Krens FA et al (2005) Silencing the major apple allergen mal d 1 by using the RNA interference approach. J Allergy Clin Immunol 115:364–369

Gilissen LJWJ, van der Meer IM, Smulders MJM (2014) Reducing the incidence of allergy and intolerance to cereals. J Cereal Sci 59:337–353

Glandorf B, de Loose M, Davies H (2011) Evaluation of changes in the genome of plants through application of new plant breeding techniques. Annex 15. In New plant breeding techniques. State-of-the-art and prospects for commercial development. JRC technical report EUR 24760 EN. (eds. Lusser M, Parisi C, Plan D, Rodríguez-Cerezo E) pp 141–155

Gold ER, Castle D, Cloutier LM et al (2002) Needed: models of biotechnology intellectual property. Trends Biotechnol 20:327–329

Graff GD, Cullen SE, Bradford KJ et al (2003) The public – private structure of intellectual property ownership in agricultural biotechnology. Nat Biotechnol 21:989–995

Hajjar R, Hodgkin T (2007) The use of wild relatives in crop improvement: a survey of developments over the last 20 years. Euphytica 156:1–13

Halterman D, Guenthner J, Collinge S, Butler N, Douches D (2016) Biotech potatoes in the 21st century: 20 years since the first biotech potato. Am J Potato Res 93:1–20

Hickey LT, Hafeez AN, Robinson H, Jackson SA et al (2019) Breeding crops to feed 10 billion. Nat Biotechnol 37(7):744–754. https://doi.org/10.1038/s41587-019-0152-9

Hilioti Z, Ganopoulos I, Ajith S et al (2016) A novel arrangement of zinc finger nuclease system for in vivo targeted genome engineering: the tomato LEC1-LIKE4 gene case. Plant Cell Rep 35:1–15

Hilton IB, D'Ippolito AM, Vockley CM et al (2015) Epigenome editing by a CRISPR/Cas9- based acetyltransferase activates genes from promoters and enhancers. Nat Biotechnol 33:510–517

Holme IB, Wendt T, Gil-Humanes J, Deleuran LC, Starker CG, Voytas DF, Brinch-Pedersen H (2017) Evaluation of the mature grain phytase candidate HvPAPhy_a gene in barley (Hordeum vulgare L.) using CRISPR/Cas9 and TALENs. Plant Mol Biol 95:111–121

Horvath P, Barrangou R (2010) CRISPR/Cas, the immune system of bacteria and archaea. Science 327:167–170

Huang X, Sang T, Zhao Q, Feng Q, Zhao Y et al (2010) Genome-wide association studies of 14 agronomic traits in rice landraces. Nat Genet 42:961–967

Husbands AY, Chitwood DH, Plavskin Y, Timmermans MC (2009) Signals and prepatterns: new insights into organ polarity in plants. Genes Dev 23:1986–1997

Ishii T, Araki M (2016) Consumer acceptance of food crops developed by genome editing. Plant Cell Rep 35:1507–1518

Ishino Y, Shinagawa H, Makino K, Amemura M, Nakata A (1987) Nucleotide sequence of the iap gene, responsible for alkaline phosphatase isozyme conversion in Escherichia coli, and identification of the gene product. J Bacteriol 169:5429–5433

Ito Y, Nishizawa-Yokoi A, Endo M et al (2015) CRISPR/Cas9-mediated mutagenesis of the RIN locus that regulates tomato fruit ripening. Biochem Biophys Res Commun 467:76–82

Jiang W, Zhou H, Bi H et al (2013) Demonstration of CRISPR/Cas9/sgRNA-mediated targeted gene modification in Arabidopsis, tobacco, sorghum and rice. Nucl Acids Res 41(20):e188

Jiang WZ, Henry IM, Lynagh PG, Comai L, Cahoon EB, Weeks DP (2016) Significant enhancement of fatty acid composition in seeds of the allohexaploid, Camelina sativa, using CRISPR/Cas9 gene editing. Plant Biotechnol J 15:648–657

Jin J, Huang W, Gao J, Yang J, Shi M, Zhu M, Luo D, Lin H (2008) Genetic control of rice plant architecture under domestication. Nat Genet 40:1365–1369

Jinek M, Chylinski K, Fonfara I, Hauer M, Doudna JA, Charpentier E (2012) A programmable dual-RNA-guided DNA endonuclease in adaptive bacterial immunity. Science 337:816–821

Jones HD (2015) Regulatory uncertainty over genome editing. Nat Plants 1:1–3

Jördens R (2005) Progress of plant variety protection based on the international convention for the protection of new varieties of plants (UPOV convention). World Pat Inf 27:232–243

Jung C, Capistrano-Gossmann G, Braatz J, Sashidhar N, Melzer S (2018) Recent developments in genome editing and applications in plant breeding. Plant Breed 137:1–9

Kalaitzandonakes N, Alston JM, Bradford KJ (2007) Compliance costs for regulatory approval of new biotech crops. Nat Biotechnol 25:509–551

Kearns NA, Genga RM, Enuameh MS, Garber M, Wolfe SA, Maehr R (2014) Cas9 effector-mediated regulation of transcription and differentiation in human pluripotent stem cells. Development 141:219–223

Kim H, Kim ST, Kim SG, Kim JS (2015) Targeted genome editing for crop improvement. Plant Breed Biotechnol 3:283–290

Knight SC, Xie L, Deng W, Guglielmi B, Witkowsky LB, Bosanac L, Zhang ET et al (2015) Dynamics of CRISPR-Cas9 genome interrogation in living cells. Science 350:823–826

Kumar A, Kumar R, Singh N, Mansoori A (2020). Regulatory framework and policy decisions for genome-edited crops. In: Concepts and strategies in plant sciences. Springer, Berlin, pp 193–201

Kumar R, Sharma V, Suryavanshi SS, Ramrao DP, Veershetty V et al (2021) Understanding omics driven plant improvement and de-novo crop domestication: some examples. Front Genet https://doi.org/10.3389/fgene.2021.637141

Kuzma J, Kokotovich A (2011) Renegotiating GM crop regulation. EMBO Rep 12:883–888

Lassoued R, Macall DM, Hesseln H, Phillips PW, Smyth SJ (2019) Benefits of genome-edited crops: expert opinion. Transgenic Res 28:247–256

Le Buanec B (2005) Plant genetic resources and freedom to operate. Euphytica 146:1–8

Li T, Liu B, Spalding MH et al (2012) High-efficiency TALEN-based gene editing produces disease-resistant rice. Nat Biotechnol 30:390–392

Li X, Yue W, Zhenqi L, Yan S (2016) Genome editing and breeding technology and domestic and international development trend analysis. Biotechnol Business 4:37–42

Li T, Yang X, Yu Y, Si X, Zhai X et al (2018) Domestication of wild tomato is accelerated by genome editing. Nat Biotechnol 36:1160–1163. https://doi.org/10.1038/nbt.4273

Liang Z, Zhang K, Chen K, Gao C (2014) Targeted mutagenesis in Zea mays using TALENs and the CRISPR/Cas system. J Genet Genomics 41:63–68

Liang Z, Chen K, Li T, Zhang Y, Wang Y, Zhao Q, Liu J, Zhang H, Liu C, Ran Y, Gao C et al (2017) Efficient DNA-free genome editing of bread wheat using CRISPR/Cas9 ribonucleoprotein complexes. Nat Commun 8:14261

Liu H, Ding Y, Zhou Y, Jin W, Xie K, Chen LL (2017) CRISPR-P 2.0: an improved CRISPR-Cas9 tool for genome editing in plants. Mol Plant 10:530–532

Lor VS, Starker CG, Voytas DF et al (2014) Targeted mutagenesis of the tomato PROCERA gene using transcription activator-like effector nucleases. Plant Physiol 166:1288–1291

Lu BR (2013) Introgression of transgenic crop alleles: its evolutionary impacts on conserving genetic diversity of crop wild relatives. J Syst Evol 51:245–262

Lu C, Kang J (2008) Generation of transgenic plants of a potential oilseed crop Camelina sativa by agrobacterium-mediated transformation. Plant Cell Rep 27:273–278

Lu H, Klocko AL, Dow M et al (2016) Low frequency of zinc-finger nuclease-induced mutagenesis in Populus. Mol Breed 36:121

Luby CH, Goldman IL (2016) Improving freedom to operate in carrot breeding through the development of eight open source composite populations of carrot (Daucus carota L. var. sativus). Sustainability 8:479

Luby CH, Kloppenburg J, Michaels TE, Goldman IL (2015) Enhancing freedom to operate for plant breeders and farmers through open source plant breeding. Crop Sci 55:2481–2488

Lucht JM (2015) Public acceptance of plant biotechnology and GM crops. Viruses 7:4254–4281

Lusser M, Parisi C, Plan D, Rodriguez-Cerezo E (2011) New plant breeding techniques. In: state-of-the-art and prospects for commercial development. Luxembourg: Joint Research Centre-Institute for prospective technological studies, publications Office of the European Union; 2011. EUR 24760 EN, 1–220

Lusser M, Parisi C, Plan D, Rodriguez-Cerezo E (2012) Deployment of new biotechnologies in plant breeding. Nat Biotechnol 30:231–239

Malik KA, Zafar Y (2005) The international union for the protection of new varieties of plants (UPOV) recommendations on variety denominations. Asian Biotechnol Dev Rev 8:7–43

Mao Y, Zhang Z, Feng Z et al (2016) Development of germ-line-specific CRISPR-Cas9 systems to improve the production of heritable gene modifications in Arabidopsis. Plant Biotechnol J 14:519–532

Matsoukas IG (2018) Commentary: RNA editing with CRISPR-Cas13. Front Genet 9:134

Miao J, Guo D, Zhang J et al (2013) Targeted mutagenesis in rice using CRISPR-Cas system. Cell Res 23:1233–1236

Miao C, Xiao L, Hua K, Zou C, Zhao Y, Bressan RA, Zhu JK (2018) Mutations in a subfamily of abscisic acid receptor genes promote rice growth and productivity. Proc Natl Acad Sci U S A 115:6058–6063

Miller JC, Tan S, Qiao G et al (2011) A TALE nuclease architecture for efficient genome editing. Nat Biotechnol 29:143–148

Moscou MJ, Bogdanove AJ (2009) A simple cipher governs DNA recognition by TAL effectors. Science 326:1501

Nagamangala Kanchiswamy C, Sargent DJ, Velasco R et al (2015) Looking forward to genetically edited fruit crops. Trends Biotechnol 33:62–64

Nogué F, Mara K, Collonnier C, Casacuberta JM (2016) Genome engineering and plant breeding: impact on trait discovery and development. Plant Cell Rep 35:1475–1486

Nunes ACS, Kalkmann DC, Aragão FJ (2009) Folate biofortification of lettuce by expression of a codon optimized chicken GTP cyclohydrolase I gene. Transgenic Res 18:661–667

Oczek JP (2000) In the aftermath of the "terminator" technology controversy: intellectual property protections for genetically engineered seeds and the right to save and replant seed. Bost Coll Law Rev 41:627

Osakabe K, Osakabe Y, Toki S (2010) Site-directed mutagenesis in Arabidopsis using custom-designed zinc finger nucleases. Proc Natl Acad Sci U S A 107:12034–12039

Osakabe Y, Watanabe T, Sugano SS et al (2016) Optimization of CRISPR/Cas9 genome editing to modify abiotic stress responses in plants. Sci Rep 6:26685

Palmgren MG, Edenbrandt AK, Vedel SE et al (2015) Are we ready for back-to-nature crop breeding? Trends Plant Sci 20:155–164

Pan C, Ye L, Qin L et al (2016) CRISPR/Cas9-mediated efficient and heritable targeted mutagenesis in tomato plants in the first and later generations. Sci Rep 6:24765

Parrott W (2018) Outlaws, old laws and no laws: the prospects of gene editing for agriculture in United States. Physiol Plant 164:406–411

Pattanayak V, Ramirez CL, Joung JK, Liu DR (2011) Revealing off-target cleavage specificities of zinc-finger nucleases by in vitro selection. Nat Methods 8:765–770

Peer R, Rivlin G, Golobovitch S et al (2015) Targeted mutagenesis using zinc-finger nucleases in perennial fruit trees. Planta 241:941–951

Peng A, Chen S, Lei T, Xu L, He Y, Wu L, Yao L, Zou X (2017) Engineering canker- resistant plants through CRISPR/Cas9-targeted editing of the susceptibility gene CsLOB1 promoter in citrus. Plant Biotechnol J 10:1011–1013

Peterson BA, Haak DC, Nishimura MT et al (2016) Genome-wide assessment of efficiency and specificity in CRISPR/Cas9 mediated multiple site targeting in Arabidopsis. PLoS One 11: e0162169

Petolino JF (2015) Genome editing in plants via designed zinc finger nucleases. Vitr Cell Dev Biol - Plant 51:1–8
Puchta H (2005) The repair of double-strand breaks in plants: mechanisms and consequences for genome evolution. J Exp Bot 56:1–14
Qi Y, Li X, Zhang Y et al (2013) Targeted deletion and inversion of tandemly arrayed genes in Arabidopsis thaliana using zinc finger nucleases. G3: Genes Genomes Genetics 3:1707–1715
Quétier F (2016) The CRISPR-Cas9 technology: closer to the ultimate toolkit for targeted genome editing. Plant Sci 242:65–76
Ramirez CL, Foley JE, Wright DA, Müller-Lerch F, Rahman SH, Cornu TI et al (2008) Unexpected failure rates for modular assembly of engineered zinc fingers. Nat Methods 5:374–375
Ricroch A, Clairand P, Harwood W (2017) Use of CRISPR systems in plant genome editing: toward new opportunities in agriculture. Emerg Top Life Sci 1(2):169–182
Rodriguez-Leal D, Lemmon ZH, Man J, Bartlett ME, Lippman ZB (2017) Engineering quantitative trait variation for crop improvement by genome editing. Cell 171:470–480
Römer P, Hahn S, Jordan T et al (2007) Plant pathogen recognition mediated by promoter activation of the pepper Bs3 resistance gene. Science 318:645–648
Sachs MM (2009) Cereal germplasm resources. Plant Physiol 149:148–151
Sawai S, Ohyama K, Yasumoto S et al (2014) Sterol side chain reductase 2 is a key enzyme in the biosynthesis of cholesterol, the common precursor of toxic steroidal glycoalkaloids in potato. Plant Cell 26:3763–3774
Schaart JG, Krens FA, Wolters AMA, Visser RGF (2010) Transformation methods for obtaining marker-free genetically modified plants. In: Stewart CN, Touraev A, Citovsky V, Tzfira T (eds) Plant transformation technologies. Wiley, Oxford, UK
Schaart JG, van de Wiel CCM, Lotz LAP, Smulders MJM (2016) Opportunities for products of new plant breeding techniques. Trends Plant Sci 21:438–449
Scheben A, Edwards D (2018) Towards a more predictable plant breeding pipeline with CRISPR/Cas-induced allelic series to optimize quantitative and qualitative traits. Curr Opin Plant Biol 45:218–225
Schiemann J, Hartung F, Kühn-institut J (2009) EU perspectives on new plant-breeding techniques. In: New DNA-editing approaches methods, applications and policy for agriculture. National Agricultural Biotechnology Council, New Delhi, India, pp 201–210
Schneider K, Schiermeyer A, Dolls A et al (2016) Targeted gene exchange in plant cells mediated by a zinc finger nuclease double cut. Plant Biotechnol J 14:1151–1160
Shan Q, Wang Y, Li J et al (2013) Targeted geome modification of crop plants using a CRISPR-Cas system. Nat Biotechnol 31:684–686
Shan Q, Wang Y, Li J, Gao C (2014) Genome editing in rice and wheat using the CRISPR/Cassystem. Nat Protoc 9:2395–2410
Shi J, Gao H, Wang H, Lafitte HR, Archibald RL, Yang M, Hakimi SM, Mo H, Habben JE (2017) ARGOS 8 variants generated by CRISPR-Cas9 improve maize grain yield under field drought stress conditions. Plant Biotechnol J 15:207–216
Shimatani Z, Kashojiya S, Takayama M, Terada R, Arazoe T, Ishii H, Teramura H, Yamamoto T, Komatsu H, Miura K et al (2017) Targeted base editing in rice and tomato using a CRISPRCas9 cytidine deaminase fusion. Nat Biotechnol 35:441–443
Shin J, Chen J, Solnica-Krezel L (2014) Efficient homologous recombination-mediated genome engineering in zebrafish using TALE nucleases. Development 141:3807–3818
Shmakov S, Abudayyeh OO, Makarova KS et al (2015) Discovery and functional characterization of diverse class 2 CRISPR-Cas systems. Mol Cell 60:385–397
Shukla VK, Doyon Y, Miller JC et al (2009) Precise genome modification in the crop species Zea mays using zinc-finger nucleases. Nature 459:437–441
Smulders MJM, Jouanin AA, Schaart JG, Visser RGF, Cockram J, Leigh F et al (2015) Development of wheat varieties with reduced contents of celiac-immunogenic epitopes through conventional and GM strategies. In: Koehler P (ed) Proceedings of the 28th meeting of the working

group on prolamin analysis and toxicity. Verlag Deutsche Forschungsanstalt für Lebensmittelchemie (DFA), Düsseldorf, pp 47–56

Song G, Jia M, Chen K et al (2016) CRISPR/Cas9: a powerful tool for crop genome editing. Crop J 4:75–82

Söllü C, Pars K, Cornu TI, Thibodeau-Beganny S et al (2010) Autonomous zinc-finger nuclease pairs for targeted chromosomal deletion. Nucleic Acids Res 38(22):8269–8276. https://doi.org/10.1093/nar/gkq720

Sprink T, Eriksson D, Schiemann J, Hartung F (2016) Regulatory hurdles for genome editing: process- vs. product-based approaches in different regulatory contexts. Plant Cell Rep 35:1493–1506

Stewart CN, Halfhill MD, Warwick SI (2003) Transgene introgression from genetically modified crops to their wild relatives. Nature 4:806–817

Storozhenko S, De Brouwer V, Volckaert M et al (2007) Folate fortification of rice by metabolic engineering. Nat Biotechnol 25:1277–1279

Sun Y, Zhang X, Wu C et al (2016) Engineering herbicide resistant rice plants through CRISPR/Cas9-mediated homologous recombination of the acetolactate synthase. Plant Physiol 170:1689–1699

Szczepek M, Brondani V, Büchel J et al (2007) Structure-based redesign of the dimerization interface reduces the toxicity of zinc-finger nucleases. Nat Biotechnol 25:786–793

Tamás C, Kisgyörgy BN, Rakszegi M et al (2009) Transgenic approach to improve wheat (Triticum aestivum L.) nutritional quality. Plant Cell Rep 28:1085–1094

Tan L, Li X, Liu F, Sun X, Li C et al (2008) Control of a key transition from prostate to erect growth in rice domestication. Nat Genet 40:1360–1364

Tanaka Y (2012) Attitude gaps between conventional plant breeding crops and genetically modified crops, and psychological models determining the acceptance of the two crops. J Risk Res 16:69–80

Tomlinson L, Yang Y, Emenecker R, Smoker M, Taylor J, Perkins S, Smith J, MacLean D, Olszewski NE, Jones JDG (2019) Using CRISPR/Cas9 genome editing in tomato to create a gibberellin-responsive dominant dwarf DELLA allele. Plant Biotechnol J 17:132–140

Townsend J, Wright D, Winfrey RJ et al (2009) High-frequency modification of plant genes using engineered zinc-finger nucleases. Nature 459:442–445

Tsai SQ, Zheng Z, Nguyen NT et al (2015) GUIDE-Seq enables genome-wide profiling of off-target cleavage by CRISPR-Cas nucleases. Nat Biotechnol 33:187–197

Upadhyay SK, Kumar J, Alok A, Tuli R (2013) RNA guided genome editing for target gene mutations in wheat. G3 3:2233–2238

Van Tuinen A, Peters AHLJ, Kendrick RE et al (1999) Characterisation of the procera mutant of tomato and the interaction of gibberellins with end-of-day far-red light treatments. Physiol Plant 106:121–128

Varshney RK, Nayak SN, May GD, Jackson SA (2009) Next-generation sequencing technologies and their implications for crop genetics and breeding. Trends Biotechnol 27:522–530

Varshney RK, Bansal KC, Aggarwal PK et al (2011) Agricultural biotechnology for crop improvement in a variable climate: Hope or hype? Trends Plant Sci 16:363–371

Varshney RK, Singh VK, Kumar A, Powell W, Sorrells ME (2018) Can genomics deliver climate-change ready crops? Curr Opin Plant Biol 45:205–211

Vinocur B, Altman A (2005) Recent advances in engineering plant tolerance to abiotic stress: achievements and limitations. Curr Opin Biotechnol 16:123–132

von Wettberg EJ, Chang PL, Başdemir F, Carrasquila-Garcia N, Korbu LB et al (2018) Ecology and genomics of an important crop wild relative as a prelude to agricultural innovation. Nat Commun 9:649

Waltz E (2018) With a free pass, CRISPR-edited plants reach market in record time. Nat Biotechnol 36:6–7

Wang Y, Cheng X, Shan Q et al (2014) Simultaneous editing of three homoeoalleles in hexaploid bread wheat confers heritable resistance to powdery mildew. Nat Biotechnol 32:947–951

Wang M, Lu Y, Botella J, Mao Y, Hua K, Zhu JK (2017) Gene targeting by homology- directed repair in Rice using a Geminivirus-based CRISPR/Cas9 system. Mol Plant 10:1007–1010
Warschefsky E, Varma Penmetsa R, Cook DR, Von Wettberg EJB (2014) Back to the wilds: tapping evolutionary adaptations for resilient crops through systematic hybridization with crop wild relatives. Am J Bot 101:1791–1800
Wendt T, Holm PB, Starker CG et al (2013) TAL effector nucleases induce mutations at a pre-selected location in the genome of primary barley transformants. Plant Mol Biol 83:279–285
Wittkopp PJ, Kalay G (2012) Cis-regulatory elements: molecular mechanisms and evolutionary processes underlying divergence. Nat Rev Genet 13:59–69
Wolt JD, Wang K, Sashital D, Lawrence-Dill CJ (2016) Achieving plant CRISPR targeting that limits off-target effects. Plant Genome 9(3):1–8. https://doi.org/10.3835/plantgenome2016.05.0047
Wright DA, Townsend JA, Winfrey RJ Jr, Irwin PA, Rajagopal J et al (2005) High-frequency homologous recombination in plants mediated by zinc-finger nucleases. The Plant J 44:693–705
Wyman C, Kanaar R (2006) DNA double-strand break repair: all's well that ends well. Annu Rev Genet 40:363–383
Xiong J, Ding J, Li Y (2015) Genome-editing technologies and their potential application in horticultural crop breeding. Horticu Res 2:15019
Zaidi SS, Mansoor S, Ali Z, Tashkandi M, Mahfouz MM (2016) Engineering plants for geminivirus resistance with CRISPR/Cas9 system. Trends Plant Sci 21:279–281
Zamir D (2001) Improving plant breeding with exotic genetic libraries. Nat Rev Genet 2:983–989
Zanga D, Capell T, Zhu C et al (2016) Freedom-to-operate analysis of a transgenic multivitamin corn variety. Plant Biotechnol J 14:1225–1240
Zentella R, Zhang Z-L, Park M et al (2007) Global analysis of della direct targets in early gibberellin signaling in Arabidopsis. Plant Cell 19:3037–3057
Zetsche B, Gootenberg JS, Abudayyeh OO et al (2015) Cpf1 is a single RNA-guided endonuclease of a class 2 CRISPR-Cas system. Cell 163:759–771
Zhang F, Maeder ML, Unger-Wallace E et al (2010) High frequency targeted mutagenesis in Arabidopsis thaliana using zinc finger nucleases. Proc Natl Acad Sci U S A 107:12028–12033
Zhang Y, Shan Q, Wang Y et al (2013) Rapid and efficient gene modification in rice and brachypodium using TALENs. Mol Plant 6:1365–1368
Zhang H, Zhang J, Wei P et al (2014) The CRISPR/Cas9 system produces specific and homozygous targeted gene editing in rice in one generation. Plant Biotechnol J 12:797–807
Zhou H, Liu B, Weeks DP, Spalding MH, Yang B (2014). Large chromosomal deletions and heritable small genetic changes induced by CRISPR/Cas9 in rice. Nucleic Acid Res 42:10903–10914. https://doi.org/10.1093/nar/gku806
Zong Y, Wang YP, Li C, Zhang R, Chen KL, Ran YD, Qiu JL, Wang DW, Gao CX (2017) Precise base editing in rice, wheat and maize with a Cas9-cytidine deaminase fusion. Nat Biotechnol 35:438–440
Zu Y, Tong X, Wang Z et al (2013) TALEN-mediated precise genome modification by homologous recombination in zebrafish. Nat Methods 10:329–331

# Metabolomics-Assisted Breeding for Crop Improvement: An Emerging Approach

# 11

Mohammed Jamaloddin, A. Maliha, C. G. Gokulan, Namami Gaur, and Hitendra Kumar Patel

**Abstract**

The "omics" technologies, namely, genomics, transcriptomics, proteomics, and metabolomics are pillars of modern molecular biology and biotechnology. Metabolomics is a rapidly developing branch of "omics" and it involves the detection and quantification of metabolites of cellular pathways in different biological species. Metabolomics plays a vital role in discovering gene–environment interactions, characterization of mutants, phenotyping, and identification of metabolic markers. Metabolomics is a favorable approach to decode various metabolic networks that are linked with biotic and abiotic stress tolerance in plants. In this context, metabolomics-assisted breeding allows efficient screening for stress tolerance and yield of crops at the metabolic level. Advanced metabolomics analytical tools, like mass spectroscopy (MS), nuclear magnetic resonance spectroscopy (NMR), direct infusion mass spectroscopy (DIMS), direct flow injection (DFI) mass spectrometry, high-resolution mass spectroscopy (HRMS), Fourier transform ion cyclotron resonance mass spectroscopy (FI-ICR-MS), high-performance thin-layer chromatography (HPTLC), ultra-performance liquid chromatography (UPLC), capillary electrophoresis mass spectroscopy (CE-MS), gas chromatography-mass spectroscopy (GC-MS), and liquid chromatography-mass spectroscopy (LC-MS) have been developed for speeding up metabolic profiling. Presently, with the development of genome sequencing and metabolic profiling technologies, multi-omics integrative analysis of genomes, transcriptomes, and metabolomes has enabled efficient dissection of the genetic basis of metabolic pathways, diversity of nutrient metabolites, and phenotypic association in crop plants. Here, we describe advanced metabolomics

M. Jamaloddin · A. Maliha · C. G. Gokulan · N. Gaur · H. K. Patel (✉)
CSIR-Centre for Cellular and Molecular Biology (CCMB), Hyderabad, India
e-mail: hkpatel@ccmb.res.in





A. Kumar et al. (eds.), *Omics Technologies for Sustainable Agriculture and Global Food Security Volume 1*, https://doi.org/10.1007/978-981-16-0831-5_11

tools, methodologies, and application of multi-omics, metabolic engineering in crop improvement.

**Keywords**

Biotic stress · Abiotic stress · Crop improvement · Analytical tools · Metabolic profiling · Metabolomics · Metabolomics-assisted breeding

## 11.1 Introduction

The ability of crops to produce economically significant yields depends on multiple internal and external factors. Plant architecture and cellular metabolism constitute the internal factors while external factors include biotic and abiotic stresses as well as climate change. The scientific community has majorly concentrated on optimization of plant architecture (Jiang et al. 2013; Cai et al. 2016) and introgression of one or more gene(s)/QTL(s) that confer resistance/tolerance against various biotic and abiotic stresses (Limbalkar et al. 2018; van Zonneveld et al. 2019; Dixit et al. 2020) in order to improve yield. These approaches have certainly contributed significantly to crop improvement but have not proved to be sufficient enough to attain the desired yield quantum. One major reason to be borne in mind is that the crops are capable of producing higher yields and combat various challenges as a direct consequence of their metabolic activities that involve organic compounds generally known as metabolites. Metabolites in general refer to organic substances or compounds formed during and/or necessary to carry out biochemical reactions in a living cell. They are broadly classified into primary and secondary metabolites. Primary metabolites are compounds directly involved in plant growth, development, and reproduction and are conserved across the plant kingdom (carbohydrates, proteins, lipids, enzymes, vitamins, etc.). They conciliate Kreb's cycle and glycolysis during photosynthesis and thus any changes in the synthesis of primary metabolites may be detrimental to the plants. Secondary metabolites are synthesized by primary metabolites and may not be directly involved in growth and development, but are important to plants for protection, competition, and species interaction (Tissier and Ziegler 2015). Secondary metabolites are species-specific and are often produced in smaller quantities during the time of need (essential oils, phenolics, alkaloids, pigments, etc.). Some of the discovered secondary metabolites are so unique that they are being used as biomarkers to assess plant performance under stress in many crop improvement programs (Che-Othman et al. 2020). All metabolites (both primary and secondary) present within an organelle or a cell or a tissue or an organ or a whole organism constitute the metabolome and the scientific study of a metabolome is called metabolomics.

Metabolomics found its application in plant biology during the mid-90s and was extensively applied to model plant *Arabidopsis thaliana* due to the availability of copious genome information and was later introduced to agriculture with an aim to improve crops. Though gradually, metabolomics instituted itself as a crucial tool in understanding plant metabolic pathways. Metabolomics research majorly deals with the recognition and quantum assessment of small molecules (<1500 Da), their chemical structure, and interactions within an organism (Deborde et al. 2017). Metabolomics is pivotal as it enables a comprehensive understanding of cellular metabolites that ultimately are representative of the physiological state of a cell. It not only allows the identification of genes involved in the metabolic network of a cell but also helps understand their function and effect on linked pathways (Wen et al. 2015; Hong et al. 2016). Metabolomics also allows a better understanding of environment–plant interactions as plant growth and development under varying environmental conditions highly depends upon its metabolism (Han and Micallef 2016). Integration of metabolomics in crop improvement programs is salient as this science in real essence fetches proper understanding of genotype–phenotype–environment interactions at metabolite level because the plant metabolites are the ultimate key players in these interactions (Carreno-Quintero et al. 2013).

Crops, like all other living organisms, are hosts to several pests and diseases (biotic stress) and due to their sessile nature are affected adversely by various abiotic factors, all of which result in yield losses. Rapid advances in genomics, high-throughput methodologies, and next-generation sequencing have contributed significantly to the identification and introgression of the gene(s) and QTL(s) controlling resistance/tolerance to major biotic and abiotic stresses in crops through marker-assisted breeding (Perez-de-Castro et al. 2012). Though molecular markers have become an integral part of breeding programs, they experience difficulty in crossing hurdles faced due to polygenic traits, epistatic loci, and traits highly influenced by the environment (Steinfath et al. 2010). Under such situations, integration of metabolomics paves way for multifarious opportunities that enable decoding such complexities (Carreno-Quintero et al. 2013). Moreover, metabolite based biomarkers allow phenotype prediction irrespective of the availability of genomic information (Che-Othman et al. 2020). An amalgamated approach considering deductions from genomics, transcriptomics, proteomics, and metabolomics will allow researchers to efficiently categorize and prioritize gene(s)/QTL(s) that will consistently tackle biotic and abiotic stresses significantly (Piasecka et al. 2019).

Crop yields can be improved by combating various biotic and abiotic stresses and by developing climate-resilient crop varieties. The objective of breeding programs so far has been to improve yields but the importance of enhancing the nutrient status of crops is being felt and acted upon only for the past few years. The present global food scenario demands not just an increase in quantity but also an improvement in the quality of food to address human nutritional requirements. The efficiency of crops to accumulate nutrients solely depends upon the availability of nutrients and the source-sink metabolism, complete knowledge of which is lacking (Rossi et al. 2015; Sonnewald and Fernie 2018). In other words, metabolomics plays a major role in understanding metabolic pathways contributing to the quality parameters of crops in terms of nutrient status (Alseekh and Fernie 2018). Thus, an integrated –omics approach will help achieve food security not only in terms of improved yield but also a better quality with enhanced nutrient index. This book chapter attempts to shed light on the importance of metabolomics-assisted breeding in crop improvement and the applicability of various –omics technologies in an integrated manner. The chapter also highlights the scope for the application of metabolite engineering and the success of biofortification in crops.

## 11.2 Metabolomics: Tools and Methodologies

Past research experiences have led to the conception of various approaches to detect and identify specific metabolites (Wishart 2011). Whole metabolome profiling using a single metabolomics tool becomes difficult due to diverse chemical composition, complex structure, and abundance of metabolites in cellular organelles. Apart from being used for metabolite profiling, metabolomics is also used to examine the prototype of a crucial gene to attain a genotype-metabolite-phenotype level of knowledge (Yang et al. 2018). Plants, unlike most other living organisms, are capable of synthesizing sundry metabolites varying in size, polarity, solubility, volatility, adaptability, and stability (Hein et al. 2016; Ma et al. 2018; Kang et al. 2019). The metabolomics experiment may be targeted, semi-targeted, or untargeted (Shulaev et al. 2008), but irrespective of the approach, factors like metabolite detection and quantification, a protocol of sample preparation and assessment of the desired metabolite decide the efficiency of metabolite profiling (Kumar et al. 2017; Sharma et al. 2018).

Over the last few years, plant metabolomics has made significant leaps forward in terms of technical progress. These advances have increased the throughput of metabolite profiling by allowing analysis of multiple (>200) known and unknown metabolites in a single experiment and imaging of metabolites in a whole organ with a spatial resolution (Bjarnholt et al. 2014). Despite these advances, no single approach is sufficient to analyze all metabolites present in a metabolome, rather, an approach combining multiple technologies is necessary to achieve a near-complete metabolite coverage. The basic instrumental setup generally used to carry out plant metabolite profiling includes spectroscopy and chromatography, or high-performance chromatography combined with spectroscopy. Various metabolomics techniques like mass spectroscopy (MS), nuclear magnetic resonance spectroscopy (NMR), direct infusion mass spectroscopy (DIMS), high-resolution mass spectroscopy (HRMS), Fourier transform ion cyclotron resonance mass spectroscopy (FI-ICR-MS), high-performance thin-layer chromatography (HPTLC), ultra-performance liquid chromatography (UPLC), capillary electrophoresis mass spectroscopy (CE-MS), gas chromatography-mass spectroscopy (GC-MS), and liquid chromatography-mass spectroscopy (GC-MS) have been developed for various experimental objectives and sample types (Razzaq et al. 2019).

The NMR-based approach is non-destructive and depends upon the magnetic properties of nuclei of atoms under an imposed magnetic field. This approach is effective in identifying metabolites having a lower molecular weight (< 50 kDa) and can be used for deducing the atomic structure of compounds, measuring metabolic flux, metabolite profiling, and metabolite fingerprinting (Winning et al. 2009). But, poor sensitivity of this technique makes detection of low-abundance yet informative metabolites difficult, thus restricting its extensive use. This drawback of NMR is efficiently addressed by MS, thus providing researchers with a broader metabolite coverage resulting in the identification of novel and highly informative metabolites crucial in the reconstruction of metabolic pathways. Combining chromatography techniques with MS increases the throughput of analysis. Among all the available techniques, CE-MS, GC-MS, LC-MS, and NMR-based approaches have been amply used for analyses (Razzaq et al. 2019).

The analytical tool to be used to carry out metabolomics experiment is decided by its sensitivity, selectivity, speed, accuracy, and precision. The NMR approach is preferred to attain a comprehensive understanding of metabolites in many organisms including plants. The NMR-based metabolite profiling is quick, convenient, and competent and is often chosen because it is non-destructive, specific, highly reproducible, and very efficient at delineating metabolite pathways (Boiteau et al. 2018). NMR is a highly quantitative and impartial approach that does not require chromatography separation. Recent technical advances have yielded more efficient tools like multi-dimensional NMR, multi-coil NMR, and isotope-labeled NMR, which address major limitations of this approach.

The integration of mass spectrometry with chromatography significantly increases the throughput. The GC-MS approach is used for non-targeted analysis (Dutta et al. 2012) and has a high sensitivity for metabolite profiling. It is counted among high-throughput techniques owing to its efficiency in detection, separation, and identification. However, the application of this technique is limited to the detection of thermally unstable and volatile compounds (Jorge et al. 2016). This technique was applied in Soybean along with DIMS to understand the regulation of metabolism during *Rhizoctonia solani* infection (Aliferis et al. 2014), in Sorghum along with FT-IR to assess the biochemical profile of few varieties under drought stress (Ogbaga et al. 2016) and in Maize to understand the relationship between metabolism and grain yield through metabolite profiling under drought, heat and combined stress (Obata et al. 2015). Unlike GC-MS, LC-MS is used to analyze polar and thermo-labile high molecular weight molecules (Turner et al. 2016). Though it can be applied to both primary and secondary metabolites for targeted as well as untargeted analysis, it has been largely used to analyze secondary metabolites. This approach was used in Rice along with NMR to identify and assess the structure of 36 specialized metabolites (Yang et al. 2014). The UPLC-MS approach was applied in Soybean to understand metabolite profile and tolerance mechanism under molybdenum toxicity (Xu et al. 2018) and in Wheat to carryout comparative metabolomics of temperature-sensitive resistance to wheat streak mosaic virus in resistant and susceptible cultivars (Farahbakhsh et al. 2019). In addition to these approaches, CE-MS gives high-resolution separation of different groups of compounds (neutral, charged, polar, and hydrophobic) through both targeted and untargeted approaches (Ramautar and De Jong 2014). This technique was applied in Rice to study the metabolome of two rice lines infected by *Rhizoctonia solani* (Suharti et al. 2016).

A metabolomics experiment is imposed with several challenges, overcoming which decides the credibility of data. A separate set of hurdles are experienced in the case of known and unknown metabolites. An untargeted approach yields

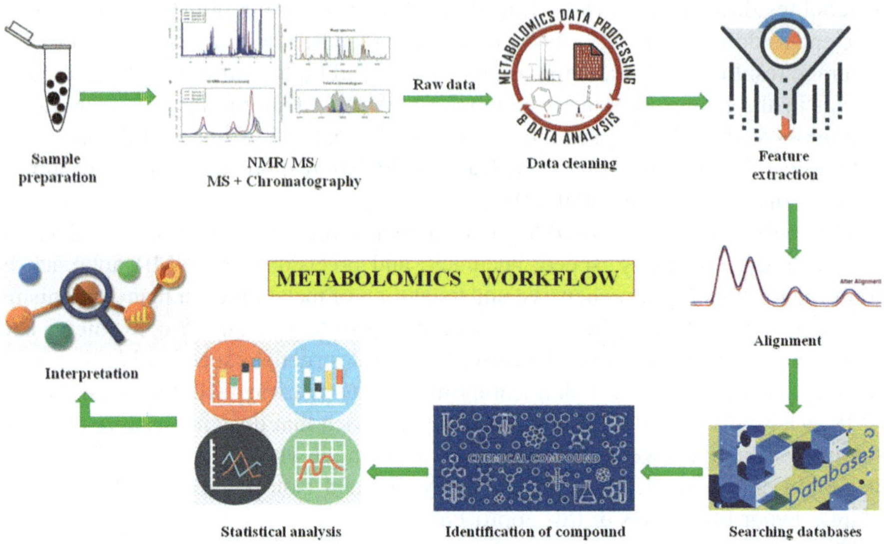


**Fig. 11.1** Schematic representation of general metabolomics workflow. The prepared sample is injected into the instrumental set up to procure raw data which is subjected to a quality control process to remove false positives. Feature extraction aids in distinguishing co-eluting peaks. Further, chemical compounds can be identified based on their structural and spectral data available in databases. Statistical analysis tools assist in the identification of reliable and stable biomarkers through correlation tests

metabolites most of which are either putatively identified or un-identified (Weckwerth 2011), which is a significant hindrance in metabolomics studies (Matsuda et al. 2009; Dunn et al. 2013). But, gradual congenial technical advances in metabolomics tools have contributed to the ease and efficiency of this omics study thus addressing most of the challenges experienced during the execution of experiments. The general workflow of a metabolomics experiment from sample preparation to data interpretation has been depicted in Fig. 11.1 (the steps may slightly vary between targeted and untargeted approach).

Metabolomics is considered a data-rich technique as it generates humongous raw data that needs to be processed (Kuhn et al. 2008). The handling of data of such quantum requires highly accommodating and very efficient data processing platforms. Few commonly used metabolomics data processing tools are MET-COFEA (Zhang et al. 2014), XCMS (Chang et al. 2016), Met-Align (Lommen and Kools 2012), MAVEN (Clasquin et al. 2012), ChromaTOF (Pegasus 2007), etc. After data processing, identification of metabolites is possible using databases like METLIN (Smith et al. 2005), CFM-ID (Allen et al. 2014), MMCD (Cui et al. 2008), GOLM (Johnson and Lange 2015), etc. For further ascertainment, the identified compounds are subjected to statistical analysis. This can be done using statistics packages like MetaboAnalyst (Xia et al. 2009), COVAIN (Sun and Weckwerth 2012), Babelomics (Alonso et al. 2015), etc.

## 11.3 Metabolomics in Crop Improvement

### 11.3.1 Stress

The metabolome, the biochemical phenotype of a plant, is a quantitative and qualitative measure to understand the genetic correlation of an organism with its environment. Metabolomics-assisted breeding is a systematic screening of crops to know metabolic and chemical footprints of plant regulatory processes to expedite stress tolerance and high yield. Metabolomics can be integrated with other omics tools such as proteomics, transcriptomics, and genomics, to study plant resistance against the pathogen, tolerance against abiotic stresses, the richness of ecotypes and to improvise breeding crops. It plays a significant role in linking the gene to its function.

Biotic and abiotic stresses are the main cause of global annual crop yield loss worldwide. Though the biochemical response against these stresses is different, the way plants react to biotic and abiotic stress are metabolically relatable. In response to the invading pathogen, plants activate innate immune response comprising pattern triggered immunity (PTI) and effector-triggered immunity (ETI) whereas plants synthesize various phytohormones, metabolites, and oxidative stress response against abiotic stress. Advancements in the field of metabolomics can decipher various metabolic networks that are linked to biotic and abiotic stresses in plants.

#### 11.3.1.1 Biotic Stress

The crops are ceaselessly jeopardized by biotic stresses induced by pathogens such as bacteria, fungi, viruses, and various pests which try to manipulate the host metabolism and manifest diseases in plants. This reduces the productivity of crops, leading to severe yield losses and ingenerate economic problems. There is always an evolutionary arms race between the host and the pathogen. Pathogens exploit host metabolisms to induce favorable conditions for their survival while plants recognize these biochemical changes and activate their defense response. In addition to the existing plant defense barriers such as the cell wall, plants also own highly efficient biochemical and metabolic defense mechanisms such as primary and secondary metabolites. Hence, metabolic profiling during host–pathogen and/or pest interactions can serve as a base to understand the plant's defense and physiology for sustainable crop improvement.

Metabolic analysis in barley spotlighted 496 metabolites including kaempferol and highlighted their pertinent role in enhanced plant defense in the resistant cultivar of barley (BOLLINA et al. 2010). In wheat, metabolic pathways of phenylpropanoid, terpenoid, and fatty acids were found to have a contributing role in imparting resistance against F. graminearum (Gunnaiah et al. 2012). Benzoxazinones such as 2,4-dihydroxy-7-methoxy-2H-1,4-benzoxazin-3(4H) (DIMBOA), one of the effective biocidal metabolites has been established in the resistance against *setosphaeria turcica* and aphids in maize (Ahmad et al. 2011). Erb et al. (2009) also evaluated the metabolomic role of 2,4-dihydroxy-7-methoxy-2H-1,4-benzoxazin-3(4H) (DIMBOA) in maize roots against western corn rootworm

*Diabrotica virgifera virgifera* and insect pest, *Spodoptera littoralis* (Erb et al. 2009). MS and nuclear magnetic resonance (NMR) based metabolomics was used to analyze the plant response against *Magnaporthe grisea* at different time points. Jones et al. (2011) proposed that excessive production of Alanine in the plant might be responsible for induced cell death to inhibit fungal invasion (Balmer et al. 2013). Oats (*Avena sativa*) react upon nematodes invasion of *Pratylenchus* and *Heterodera* by inducing a metabolic pathway of O-methyl-apigenin-C-deoxyhexoside-O-hexoside and protect the plant efficiently (Soriano et al. 2004). Tissue-specific metabolomics analysis of maize infected with *Ustilago maydis* revealed induction of flavonoid and shikimate pathways in response to fungal attack (Doehlemann et al. 2008). Screening of elite barley crops resistant against *Gibberellazeae* to identifies potential biomarkers against the fungi that have been successfully done through the metabolomics approach (Kumaraswamy et al. 2011). Constructed a model using GC/MS-based metabolomics approach to differentiate non-infected cereals from naturally mycotoxic-contaminated cereals. The analysis was done on cereals such as barley, oats, and rye (Perkowski et al. 2012). Accumulation of momilactone A, major phytoalexin of rice provides an inhibitory effect against *Magnaporthe grisea* and *Xanthomonas oryzae* (Sawada et al. 2004). Hence, it has been shown that further advances in metabolomics tools combining other "omics" will impart a better understanding of the role of metabolites during biotic stresses.

#### 11.3.1.2 Abiotic Stress

Plants thriving under abiotic stress conditions have stunted growth and turn down metabolisms. Main abiotic stresses like waterlogging, drought, temperature extremes such as freezing or high temperature, heavy metal soil contamination, high salinity, mineral deficiency, or a combination of more than one stress negatively regulate a plant's development and physiology. Metabolomics is a budding tool which can help in deciphering abiotic stress tolerance in plants, where the main objective is a remodeling of plant homeostasis and normalization of metabolic pathways.

High salinity in barley has been shown to cause osmotic imbalance due to excessive influx of Na + ions into the cell which deranges various physiological conditions in plants (Wu et al. 2013). Time course metabolic profiling of *Arabidopsis thaliana* suggests that salt stress induces various metabolic pathways such as phenylpropanoid pathway for lignin production, methylation cycle, and glycine-betaine biosynthesis (Kim et al. 2007). Osmotic stress in the shoots of *Zea mays* shows the induction of various amino acids such as alanine, asparagine, glutamate, and glycine-betaine in response to high salinity (Gavaghan et al. 2011). Polyamines, salicylic acid, glycine-betaine, and brassinosteroids improve drought tolerance in *Oryza sativa* L (Farooq et al. 2010). Several mechanisms exist by which plants overcome stresses, for instance, accumulation of osmoprotectants such as glycerol, proline synthesis, production of quaternary compounds such as glycine-betaine and dimethylsulfoniopropionate (Rontein et al. 2002; Sharma et al. 2019). The Waterlogging of soil is another challenging abiotic stress that halts the crop's metabolic pathway and affects crop yield. It creates a hypoxic condition for roots and inhibits photosynthesis in shoots (Wei et al. 2013). Against this, plants activate a varied array of metabolic pathways to promote adaptation to the stress by altering cellular functions. Biochemical pathways such as antioxidation upsurges when the plant is under stress to provide tolerance against a chilling environment (Zhang et al. 2016). It has also been found that phytohormones such as abscisic acid (ABA) impart drought tolerance to plants (Yamaguchi-Shinozaki and Shinozaki 2006). Metabolomics study in *Cyamopsis tetragonoloba* (L.) Taub showed an upsurge of carbohydrates under metal-induced stress of Zn and Cu (Manivasagaperumal 2011). It was reported by (Kim et al. 2017) that a drought-responsive pathway is activated during drought conditions wherein the metabolic pathway of glycolysis is converted to acetate synthesis to stimulate the jasmonate signaling pathway to impart drought tolerance. Also, wild species of barley have been screened through metabolomics and proteomics approaches for tolerance against salt by (Shen et al. 2016). Hence, metabolomics is one of the most promising approaches for the detection and quantification of stress response in plants which in turn can help in better understanding plant physiology.

### 11.3.2 Metabolomics for Yield and Quality Improvement

Research is solemnly interested in selecting desirable phenotypes and genotypes from a large crop population. Initial screening processes were based on phenotypic appearances of the plant which was time-consuming. Furtherance in crop improvement came with the emergence of marker-assisted selection and screening of crops that shortened the time course of previous methods. Recent advancement in breeding has brought about the concept of Genome-wide association studies (GWASs) assisted by metabolomics techniques (mGWAS) and metabolic quantitative trait loci (mQTLs) (Fernie and Schauer 2009). Metabolomics has the potential to expedite the selection and improvement of superior traits and breeding materials. Researchers are utilizing an effective combination of metabolomics with other omics technologies to pinpoint the functional candidate genes and metabolites by offering trait-specific markers to enhance the yield and growth of commercially important traits of crops. Interestingly, it has been found that the nutritional status of a crop depends on the metabolic composition of the plant which directly affects human health. Identification of metabolomics markers can be an efficient tool for the discovery of agronomic traits, for the detection of genetic variations linked to metabolic traits, and for the investigation of biochemical pathways responsible for various phenotypes.

(Chen et al. 2014) has analyzed about 840 metabolites in 524 rice cultivars and pinpointed the potential of these metabolites in crop breeding for rice improvement. (Rao et al. 2014) has generated a metabolic map by amalgamating transcriptomics, metabolomics, and proteomic data for enhanced kernel quality and yield in maize. In the study, compounds such as inositol, sorbitol, and hexaphosphate were reported in high quantities that could be utilized for improved kernel quality and yield in the future. Metabolomic techniques are applicable for studying mechanisms associated with enhanced photosynthesis and yield leading to biomass accumulation in many agricultural important plants (Hu et al. 2014). Increased levels of oxylipins that are responsible for rhizobial node factor in Medicago and that this pathway regulates nod factor signaling during early states of legume-rhizobia symbiosis in a plant (Zhang et al. 2012). Metabolomic analysis of 1181 metabolites in *Sorghum bicolor* leaf tissue showed that shikimate might be associated with enhanced plant growth and final biomass accumulation (Turner et al. 2016). A study by (Cañas et al. 2017) identifies the role of chlorogenate content for high grain to yield quality of maize by combining biochemical, fluxomic, and metabolic approaches. Hence, metabolomics has been acknowledged as one of the breakthroughs in plant sciences, which pave the way for better crop growth, improvement, and enhanced yield.

## 11.4 Application of Multi-Omics in Crop Improvement

The "omics" technologies, namely, genomics, transcriptomics, proteomics, and metabolomics are pillars of modern molecular biology and biotechnology. Genomics can further be divided into structural and functional genomics. Structural genomics deals with initial genome analysis leading to the deduction of the complete genome sequence of an organism, while functional genomics utilizes this sequence information to assess the functions of various genes on a large scale (Leister 2005). Transcriptomics, proteomics, and metabolomics together contribute to the functional study of the genome. The advent of platforms like next-generation sequencing and integrated mass spectroscopy and chromatography have led to the production of voluminous high-throughput data (Fondi and Liò 2015). Though the individual—omics technologies are producing humongous data, a comprehensive understanding can only be achieved by correlating data of different—omics studies. In other words, an integrated multi-omics approach is necessary to understand a biological system as a whole and this can only be accomplished using a meticulously designed and well-defined scheme as the biological systems in nature are very complex and so are the omics datasets (Hughes 2015; Wang et al. 2018).

While the multi-omics approach has been extensively used in humans (Cho et al. 2019), animals (García-Sevillano et al. 2014), microbes (Gutleben et al. 2018) and their combinations (Cavill et al. 2016; Pinu et al. 2019), its application to plants has been limited due to their complex biology and interactions, diverse metabolomes, and large genomes with incomplete annotation (in few crops). However, handling large biological datasets can be overwhelming and requires highly experienced researchers to obtain accurate results and interpretations. The researcher must be experienced enough to be able to select the right combination of software from all the available platforms (Pinu et al. 2019) in a way that ensures efficient and accurate analysis of large-scale data. Owing to the limitations and drawbacks of earlier "conceptual," "statistical," and "model-based" integration approaches (de Oliveira Dal'Molin and Nielsen 2018; Seaver et al. 2018), the integration methodology has been re-defined into three levels, namely "element-based approach," "pathway-based approach," and "mathematical-based approach" (Jamil et al. 2020).

### 11.4.1 Element-Based Approach

The unbiased element-based approach is very simple and intuitive and has three subclasses, namely correlation, clustering, and multivariate analyses. Correlation analysis allows the assessment of the correlation between two or more different omics datasets. This analysis can be done using Pearson's (Benesty et al. 2009) and Spearman's (Myers and Sirois 2006) correlation coefficients to assess linear and ranked relationships, respectively. There have also been reports where Fisher's transformation has been applied to skewed datasets to make them normally distributed and then assess the correlation (Mata et al. 2018). Correlation analysis has been mostly preferred to assess the correlation between transcripts and their cognate proteins on an assumption that a change in transcript level would correspond with an increase in respective protein concentration, but this is not always the case. For example, an insignificant correlation ($r = 0.03$) was reported between transcripts and corresponding proteins under salt stress between salt-tolerant Earlistaple 7 and salt-sensitive Nan Dan Ba Di Da Hua cultivars of cotton irrespective of the genetic background (Peng et al. 2018).

Clustering analysis groups omics datasets based on a common characteristic such as expression level or concentration to assess the underlying associations. Clustering can further be hierarchical or non-hierarchical and the latter integrates data using machine learning algorithms like *k*-means clustering and random forest (Silva et al. 2019). The *k*-means clustering categorizes data points into distinct groups based on the common attribute (such as differential expression), while random forest sorts a group of genes/proteins/metabolites based on the previous training datasets (Ma et al. 2014).

Multivariate analysis is capable of handling more convoluted omics datasets while permitting significant pliability in experimental design and metadata analysis (Rai et al. 2017). This approach allows the researcher to predict various facets of the datasets along with the assessment of variance or covariance associations as well as to explore the relationships and dynamic conformational networks between transcript-protein-metabolite components (Weckwerth 2019). The most commonly used multivariate analysis techniques include principal component analysis (PCA), partial least squares (PLS), and orthogonal projection to latent structures discriminant analysis (OPLS-DA). The success of this analysis is dictated by the selection of different multivariate techniques, optimal parameters, and model validation and this demands thorough knowledge and expertise. A different form of multivariate analysis technique called MCIA (multiple co-inertia analysis) was used in a maize near-isogenic line and its transgenic version (glyphosate-tolerant maize, NK603) to integrate their proteome and metabolome. This study succeeded in identifying metabolic differences in sugar metabolism and polyamine biosynthesis pathways between these two lines (Mesnage et al. 2016). Yet another variation of multivariate analysis called GFLASSO (graph-guided fused least absolute shrinkage and selection operator) was applied in maize to decode its lipid biosynthesis pathway by integrating the transcriptome and metabolome (de Abreu e Lima et al. 2018).

### 11.4.2 Pathway-Based Approach

This approach can be applied either through pathway mapping or co-expression analysis. The objective of the former approach is to map the omics dataset to the available metabolic pathway database. The database Kyoto Encyclopedia of Genes and Genomes (KEGG; https://www.genome.jp/kegg/) is one of the most commonly used databases, while other organism-specific databases like AraCyc for *Arabidopsis* (https://www.arabidopsis.org/biocyc/), SolCyc for *Solanaceae* species (https://solgenomics.net/tools/solcyc/index.pl), and CitrusCyc for citrus (https://www.citrusgenomedb.org/node/1136703) are also available. However, it is also possible to manually rebuild biochemical pathways without seeking help from complex software tools. KEGG is a repository of established pathways that can be rebuilt specifically for any desired species considering its annotated enzymes and/or metabolites. KEGG database was used to analyze the transcriptome and metabolome of soybean infected with fungus *Phytophthora sojae* (Zhu et al. 2018) and cyst nematode (Kang et al. 2018) and the studies observed transcriptional and metabolic modulation towards isoflavonoid and phenylpropanoid biosynthetic pathways, respectively.

Co-expression analysis greatly depends on statistical correlations between different omics datasets as discussed in the element-based approach, to assess the extent of relatedness between the compounds. Such relationships are later transformed into a weighted network and can be projected using a few tools including Weighted Gene Coexpression Network Analysis (WGCNA) in R program or Cytoscape. This approach has unraveled crucial clusters, modules, and hubs that aid in a better understanding of specific pathways or regulatory molecules in various plant studies. A study in maize development employed WGCNA to integrate data from transcriptomics, proteomics, and phosphoproteomics. An expression atlas was developed using 23 different maize tissues from vegetative to reproductive stage and their relatedness was further assessed using weighted networks. This study also utilized MapMan functional annotation to decode enriched pathways for their highly connected co-expressed hubs (Walley et al. 2016). The same dataset was further analyzed by Jiang et al. 2019, where a consensus network assisted by corroborations from multiple omics studies was generated by amalgamating different weighted networks from respective omics into a fused network. This exercise further highlighted the constitutive role of various transcription factors in the molecular regulation of maize development.

### 11.4.3 Mathematical-Based Approach

The mathematical-based approach further has two subclasses, namely differential analysis, and genome-scale analysis. This is the most complex approach of all and demands substantial omics data coverage and well-characterized plants. This analysis starts with the development of a well-defined differential equation and modeling for systems-level understanding. Differential analysis can be carried out in four steps: identification of systems components, determination of systems regulation and topology, development of appropriate mathematical equations, and finally, parameter selection and optimization (Voit 2017). Differential analysis can further be classified into targeted and non-targeted pathway approaches. Differential analysis by the targeted pathway approach can be employed to model a specific pathway for its metabolic flux and dynamics. The success of integration using a targeted approach is decided by the extent or completeness of annotation of a metabolic pathway in the case of both model and non-model plant species. However, differential analysis studies aid in further omics integration using genome-scale analysis (Cavill et al. 2016).

Differential analysis is a top-down approach where a stoichiometric equation is developed with a specific objective like quantification of translation rate or metabolic flux and the model is built based on the experimental results (Voit 2017). On the contrary, genome-scale analysis is a bottom-up approach as the model is built after exhaustive curation before validation (Goh 2018). This approach targets the deduction of a metabolic pathway at the organism and cellular level to an extent that every reaction can be considered for a comprehensive mathematical evaluation (de Oliveira Dal'Molin and Nielsen 2018). This analysis can be completed in four major steps: draft reconstruction using annotated genome, pathway refinement using experimental results, network modeling in mathematical format, and validation and iteration for model accuracy (Thiele and Palsson 2010). A genome-scale modeling (GSM) database called PlantSEED is a repository of 10 well-annotated plant genomes (Seaver et al. 2018) and this information can be used to annotate metabolic pathways in new plants. However, this database can be used for genome-scale reconstruction of primary metabolism, while the rebuilding of secondary metabolism requires manual curation owing to species specificity and diversity (Pinu et al. 2019). Since the plants need genome-wide metabolic reactions are mostly based on $C_3$ plants such as Arabidopsis, a $C_4$ GSM was developed to integrate omics data of $C_4$ plants (de Oliveira Dal'Molin et al. 2016). GSM was also been employed in soybean (*Glycine max*) using tools such as Plant/Eukaryotic and Microbial Metabolomics Systems Resource (PMR, http://metnetweb.gdcb.iastate.edu/PMR/) and MetNetDB (https://omictools.com/metnetdb-tool) to integrate transcriptomics and metabolomics data sets to operationalize seed filling metabolic model including starch utilization and fatty acid build-up (Li et al. 2015). GSM was developed in rapeseed (*Brassica napus*) using flux variability analysis (FVA) and it was also metabolically reconstructed through its Bna572+ database. Transcriptomics and $^{13}C$ metabolic flux experiments were used to construct and validate the model and higher flux for fatty acid biosynthesis was observed in high oil plant genotype (Hay et al. 2014). Similarly in maize, GSM was updated and validated through transcriptomics and biochemical assays for its leaf development (Bogart and Myers 2016).

## 11.5 Metabolic Engineering and Applications

Metabolic engineering can be defined as the intentional modification of cellular metabolism and assets through the manipulation of metabolic pathways to produce desired compounds by using recombinant DNA technology, RNA, gene editing, and other upcoming modern technologies (Fu et al. 2018). Plants themselves in their native form constitute a remarkable feat of metabolic engineering and is composed of over more than 200,000 metabolites (Fiehn 2002). The characterization of all these molecules has been very crucial to study for plant growth and development along with their response to external stimuli. Metabolomics is very crucial to studying biotic resistance, abiotic tolerance, vigorous ecotype, and metabolomics-assisted crop breeding (Razzaq et al. 2019). Over the last two decades, great research strides have been made in plant metabolic engineering. A great deal comes together with modern metabolic engineering and conventional breeding are exploited to explain complex biological pathways and explore secretly hidden regulatory discover novel regulatory networks and pathways controlling crop growth, development, productivity and enhance the fitness of plants (Dangl et al. 2013). Here we are providing insight into recent metabolomics studies and application on biotic, abiotic stress tolerance, and recent progress on GM crops (Table 11.1 and 11.2).

## 11.6 Biofortification

The nutrients required by a human body for proper growth and development can be classified into two categories; macronutrients and micronutrients. Macronutrients, as the name suggests are required in larger amounts as they are a major energy source for the human body, while the micronutrients are often misunderstood by their name. Though required in smaller or trace quantities, micronutrients are vital for proper metabolism, growth, and development. Essential micronutrients include all vitamins (A, B complex, C, D, E, and K) and minerals (Fe, Cu, Zn, Mn, I, Se, Co, Mo, and Ni). Food is the major source of these nutrients but the global staple food crops fail to satisfy the micronutrient requirement due to low concentration (Waters and Grusak 2008), while the natural micronutrient-rich foods are either unavailable or not affordable. This leads to micronutrient deficiency in humans called "hidden hunger" and affects every third person globally (FAO 2013) causing harmful and sometimes lethal consequences (Tulchinsky 2010; Bailey et al. 2015).

The key objective of most of the crop improvement programs has been to increase yield and to impart resistance/tolerance against biotic/abiotic stresses. The current global malnutrition scenario has brought about changes in the breeders' perspective to produce nutrient-rich food crops sustainably than just producing more food. This process of producing nutritionally enhanced food crops developed using agronomic practices, conventional breeding, and biotechnology tools are called "Biofortification." This approach can efficiently be used to win the battle against micronutrient malnutrition in a world where the staple food crops are micronutrient-poor (Khush et al. 2012). From an economic perspective,

**Table 11.1** Applications of metabolomics platforms to decipher abiotic and biotic stress tolerance in major crops

| S. No | Crop | Phenotype | Specific tissue | Key metabolites produced/genetic modification | Analytical platform | Data analysis | References |
|---|---|---|---|---|---|---|---|
| 1 | Rice | Drought stress | Leaves | 4-hydroxycinnamic acid, ferulic acid, stearic acid, and xylitol | GC-MS | PCA, PLS-DA | Ma et al. (2016) |
| 2 | Rice | Drought stress | Leaf | Glutamate, proline, GABA, arginine, and spermidine serine | GC/EI-TOF-MS | TagFinder and NIST | Do et al. (2013) |
| 3 | Rice | Drought stress | Leaf-blades | Serine, threonine, and asparagine | GC/MS | PCA | Degenkolbe et al. (2013) |
| 4 | Rice | Salt stress | Leaf | Mannitol and sucrose | GC/MS | ANOVA and mass hunter MS | Chang (2019) |
| 5 | Rice | Salt stress | Seedlings | Leucine, isoleucine, valine, proline, and GABA | GC-MS | ANOVA and DMRT | Gayen et al. (2019) |
| 6 | Rice | Salt stress | Leaf and root | Acetic acid, GABA, sucrose and non-polar metabolites | NMR | PLS-DA | Ma et al. (2018) |
| 7 | Rice | Salt stress | Leaf | Vanillic acid, 4-hydroxybenzoic acid, palmitic acid, stearic acid, raffinose, L-tryptophan, and pyruvic acid | GC-MS | PCA, PLS-DA, and Metabo-analyst 3.0 | Gupta and De (2017) |
| 8 | Rice | Nitrogen stress | Root | Amino acid metabolism, carbon and nitrogen metabolism, phenylpropanoid metabolism, and phytohormones' signal transduction | (LC-ESI-MS/MS) | ANOVA | Xin et al. (2019) |
| 9 | Rice | Waterlogging | Leaf | Glycine, alanine, and GABA | GC/MS | PCA and MarkerLynx XS | Barding et al. (2013) |
| 10 | Rice | Waterlogging | Leaf | 6-phosphogluconate, phenylalanine, and lactate | NMR and GC/MS | ANOVA and PCA | Locke et al. (2018) |
| 11 | Rice | Gall midge (*Orseoliaoryzae*) | Leaf | Heneicosanoic acid, threonic acid, palmitoleic acid, palmitic acid, | GC/MS | ANOVA | Agarrwal (2014) |

(continued)

**Table 11.1** (continued)

| S. No | Crop | Phenotype | Specific tissue | Key metabolites produced/genetic modification | Analytical platform | Data analysis | References |
|---|---|---|---|---|---|---|---|
| | | | | nonadecanoic acid, and linoleic acid | | | |
| 12 | Rice | Bacterial blight *(Xanthomonas oryzae pv. Oryzae)* | Leaf | Phenylalanine and tyrosine | GC/TOF and LC/TOF | KEGG, MassHunter, GeneSpring-MS 1.2, and METLIN | Sana et al. (2010) |
| 13 | Rice | Blast *(Magnaporthe grisea)* | Leaf | Cinnamate, proline, Rice, glutamine, and malate | NMR, GC/MS, and LC/MS | PCA and MATLAB | Jones et al. (2011) |
| 14 | Rice | Blast *(Magnaporthe grisea)* | Leaf | Glucosylceramide (GlcCer) | GC/MS and LC/MS | PCA | Liu et al. (2019) |
| 15 | Rice | Sheath blight (*Rhizoctonia solani*) | Leaf | Jasmonic acid, mucic acid, and glyceric acid | CE/TOF-MS | MPP software | Suharti et al. (2016) |
| 16 | Rice | Sheath blight (*Rhizoctonia solani*) | Leaf | Mitogen-activated protein kinase 6, probable protein phosphatase 2C1, probable trehalose-phosphate phosphatase 2 and heat shock protein,14–3–3GF14f protein | 2-DE/MS | PD quest software version 8.0 | Karmakar et al. (2019) |
| 17 | Rice | Brown planthopper (BPH) (*Nilaparvata lugens*) | Leaf-sheath | GABA and glyoxylate | GC/MS | PCA and PLS-DA | Peng et al. (2016) |
| 18 | Rice | Striped stemborer (*Chilo suppressalis*) | Leaf | Terpenoids and phenylpropanoids | HPLC-MS and GC-MS | KEGG | Liu et al. (2016) |
| 19 | Maize | Heat stress | Leaf | Sucrose, fructose, GABA, aspartate, asparagine, valine, inositol, analine, and proline | NMR | PCA and SIMCA | Sun et al. (2016) |

| | | | | | | | |
|---|---|---|---|---|---|---|---|
| 20 | Maize | Gibberella ear rot (GER) or stalk rot *(Fusarium graminearum)* | Roots | Metabolites smiglaside and smilaside A | LC/MS | ANOVA and SAS software | Zhou et al. (2019) |
| 21 | Maize | Southern *corn* leaf blight (SCLB) *(Bipolaris maydis)* | Leaf | Lignin, flavonoids, and polyphenols | FT-IR and NMR | PCA | Vasmatkar et al. (2019) |
| 22 | Maize | Asian *corn* borer *(Ostrinia furnacalis)* | Leaf | Phytohormones and benzoxazinoids | HPLC-MS/MS | KEGG, PLS-DA | Guo et al. (2019) |
| 23 | Wheat | Heat stress | Leaf | Melibiose, serine, lysine, glycine, malic acid, mannitol, xylitol, inositol, fructose, proline, glutamic acid, and alanine | GC-MS | LSD | Qi et al. (2017) |
| 24 | Wheat | Nitrogen stress | Leaf | Tyrosine, lysine, Allo-inositol, and L-ascorbic acid | GC-MS and LC-MS | MS-excel package | Khan (2019) |
| 25 | Wheat | Nitrogen stress | Leaf | Fucose, ribulose, lyxose, galactinol, and erythritol | GC-TOF-MS | PCA | Heyneke et al. (2017) |
| 26 | Wheat | Nitrogen stress | Flag leaf | Methylisoorientin-2”-O-rhamnoside, isoorientin, and isovitexin | UPLC-QTOF | PCA, OPLS-DA, Markerlynx XS™, SIMCA-P | Zhang et al. (2017) |
| 27 | Wheat | Agronomic traits | Spikes | Auxin accumulation | LC-MS/MS | ANOVA | Shi et al. (2020) |
| 28 | Wheat | Septoria leaf blotch *(Zymoseptoria Tritici)* | Leaf | Flavonoids, hydroxycinnamic acid amides, and cinnamyl alcohols | FT-ICR-MS | MetaboScape 4.0, DataAnalysis5.0 and KEGG | Seybold et al. (2019) |
| 29 | Wheat | *Fusarium* head blight *(Fusarium graminearum)* | Leaf | Trehalose, asparagine, phenylalanine, myoinositol, 3-hydroxybutarateandL-alanine | NMR | PCA, MestReNova 9.1.0 and Matlab | Cuperlovic-Culf et al. (2019) |

(continued)

**Table 11.1** (continued)

| S. No | Crop | Phenotype | Specific tissue | Key metabolites produced/genetic modification | Analytical platform | Data analysis | References |
|---|---|---|---|---|---|---|---|
| 30 | Wheat | *Fusarium* head blight (*Fusarium graminearum)* | Spikelet | Spermine, putrescine, GABA, inositols, galactose, and lactic acid | NMR | PCA, MestReNova 9.1.0 and Matlab | Cuperlovic-Culf et al. (2016) |
| 31 | Wheat | Wheat streak mosaic virus (WSMV) | Leaf | Reduction in some amino acids such as L-tyrosine, tryptophan, isoleucine, and phenylalanine | UPLC-QTOF/MS | PCA, KEGG, METLIN, MetFrag, and Metabo analyst | Farahbakhsh et al. (2019) |
| 32 | Wheat | *Fusarium* head blight (*Fusarium graminearum)* | Rachis and spikelet | Fatty acids, terpenoid, phenolic glycosides, flavonoid, and phenylpropanoids | LC-LTQ-Orbitrap | MetaXCMS | Gunnaiah et al. (2012) |
| 33 | Wheat | *Triticum turgidum* | Leaf | Benzoxazinoids | LC/MS | PCA, XCMS, and CAMERA | Shavit et al. (2018) |
| 34 | Wheat | Weeds | Root and shoot extract | Benzoxazinoids | LC-MS/MS QTrap | Analyst software | Mwendwa (2016b) |
| 35 | Wheat | *Lolium rigidum* *Urochloa panicoides* | Root and shoot extract | Hydroxamic acids and Benzoxazinoids | LC-MS/MS QTrap | Analyst software | Mwendwa (2016a) |
| 36 | Wheat | Pathogen resistance | Soil rhizosphere | Glutarimide, consabatine, methylpyrrole, arachidonic acid, gibberellic acid, and diacetyllycopsamine | Py-FIMS | PCA | Monreal and Schnitzer (2015) |
| 37 | Soybean | Drought stress | Leaf | Mannitol and sucrose | GC/MS | ANOVA and MassHunter MS | Chang et al. (2019) |
| 38 | Soybean | Waterlogging | Leaf | Phosphoenolpyruvate, NADH2, glycine, and gamma aminobutyric acid | CE/MS | ANOVA | Komatsu et al. (2014) |

| | | | | | | | |
|---|---|---|---|---|---|---|---|
| 39 | Soybean | Waterlogging | Roots and leaves | Isoflavones and kaempfero | NMR | ANOVA, PCA, and MATLAB | Coutinho et al. (2018) |
| 40 | Soybean | Heat stress | Pollens | Flavonoids | LC-QTOF-MS | MetAlign, METLIN, PCA, and ANOVA | Chebrolu et al. (2016) |
| 41 | Soybean | Metal stress | Roots and leaves | Citric acid, D-glucarate, gluconic, L-nicotine, and flavonoids/isoflavone | UPLC | PCA, KEGG, Metlin | Xu et al. (2018) |
| 42 | Sorghum | Drought stress | Leaf | Sugars and sugar alcohol | FT-IR and GC/MS | PC-DFA | Ogbaga et al. (2016) |
| 43 | Sorghum | *Burkholderia andropogonis* (*Sorghum bacterial stripe*) | Sorghum pathogen | Lipopolysaccharides | NMR/MALDI-TOF MS | MassLynx™ XS software | Mareya et al. (2020) |
| 44 | Barley | Drought stress | Fifth leaf and Palea | Aromatic amino acid, proline, glutamine, threonine, aspartate, glycine, and serine | MS-EI | PROC UNIVARIATE, SAS v. 9.4 | Hein et al. (2016) |
| 45 | Barley | Salt stress | Roots | Proline, sucrose, xylose, and maltose | GC/MS | Metabo analyst | Shelden et al. (2016) |
| 46 | Barley | Sulfur stress | Roots and leaves | Sulfur metabolites, organic acids, and amino acids | UPLC | ANOVA, PCA, MassLynx, and Progenesis QI | Ghosson et al. (2018) |
| 47 | Canola | Metal stress | Roots and leaves | Hydroxycinnamic acids and glucosinolates | NMR | ANOVA, PCA, and multi-experiment viewers | Jahangir et al. (2008) |
| 48 | Legumes | Weeds | Root and shoot extracts | Flavonoids | UHPLC QTOF-MS | METLIN | Latif et al. (2020) |

Updated from A. Razzaq et al. 2019

**Table 11.2** Metabolomics studies on genetically modified organisms (GMO's)

| S. No | GM Crop | Tissue | Donor Specie | Genetic Modification | Phenotype | Analytical Technique | Data Analysis | References |
|---|---|---|---|---|---|---|---|---|
| 1 | Rice | Seed | *B. thuringiensis* | *Cry1Ab* | Insect resistance | FTIR MS, NMR | LDA, PCA, NCSS and MESTREC | Keymanesh et al. (2009) |
| 2 | Rice | Seed | *B. Thuringiensis* | *Cry1Ac, sck* | Insect resistance | GC-FID, GC-EI-Q MS | PLS-DA and PCA | Zhou et al. (2009) |
| 3 | Rice | Leaf | *Z. mays* | *C1, R-S* | Flavonoid production | LC-ESI-Q MS, LC-DAD | – | Shin et al. (2006) |
| 4 | Rice | Leaf, seed, root | *O. sativa* | *YK1* | Stress tolerance | CE-ESI-Q MS | – | Takahashi et al. (2006) |
| 5 | Rice | Seed | *O. sativa* | *RCH10, RAC22, β-Glu, B-RIP* | Antifungal activity | NIRS, GC-EI-Q MS, LC-DAD, ICP-AES | PLS-DA and PCA | Jiao et al. (2010) |
| 6 | Rice | Seed | *O. sativa* | *Mod. (Xa23, Xa21* genes) | Insect resistance | GC-EI-Q MS | PLS-DA and PCA | Wu et al. (2012) |
| 7 | Rice | Seed | *B. Thuringiensis* | *Cry1Ac, sck* | Insect resistance | LC-ESI-Q/TOF MS | PCA and PLS-DA | Chang et al. (2012) |
| 8 | Rice | Seed | *E. coli* | *GlgC-TM* | Nutritionally enhanced | LC-ESI-Q MS | LRA | Nagai et al. (2009) |
| 9 | Rice | Seed | *N. tabacum* | *ASA2* | Nutritionally enhanced | LC-ESI-Q MS | ICA | Matsuda et al. (2010) |
| 10 | Rice | Seed | *A. tumefaciens* | *Bar* | Herbicide tolerance | GC-EI-TOF MS | PCA, ANOVA | Kim et al. (2013) |
| 11 | Rice | Leaf, seed | *E. coli/O. sativa* | *LysC, dapA/LKR/SDH* | Nutritionally enhanced | LC-FTIR MS, GC-EI-Q MS | ANOVA | Long et al. (2013) |
| 12 | Maize | Grain | *B. Thuringiensis* | *Cry1Ab* | Insect resistance | NMR | PCA and PLS-DA | Manetti et al. (2004) |

| | | | | | | | | |
|---|---|---|---|---|---|---|---|---|
| 13 | Maize | Grain | *Z. mays* | *Mod.* (*Rpd3* gene) | Seed development | NMR | PCA and N-PLS | Castro and Manetti (2007) |
| 14 | Maize | Grain | *B. Thuringiensis* | *Cry1Ab* | Insect resistance | NMR | PCA, ANOVA | Piccioni et al. (2009) |
| 15 | Maize | Grain | *B. Thuringiensis* | *Cry1Ab* | Insect resistance | CE-ESI-TOF MS | PCA | Levandi et al. (2008) |
| 16 | Maize | Grain | *B. Thuringiensis* | *Cry1Ab* | Insect resistance | FT-ICR MS | – | Leon et al. (2009) |
| 17 | Maize | Grain | *B. Thuringiensis* | *Bt toxin* | Insect resistance | GC-EI-Q MS | ANOVA | Bernal et al. (2008) |
| 18 | Maize | Grain | *B. Thuringiensis* | *Cry1Ab* | Insect resistance | GC-EI-Q MS | – | Jiménez et al. (2009) |
| 19 | Maize | Grain | *B. Thuringiensis* *A. Thuringiensis* | *Cry1Ab* *CP4 EPSPS* | Insect resistance, Herbicide tolerance | GC-EI-Q MS | ANOVA | Frank et al. (2012) |
| 20 | Maize | Grain | *Z. mays* | *Mod.* (*Zmpsy1*, *Pacrtl*, *Gllycb*, *Glbch*, *ParacrtW* genes) | Nutritionally enhanced | UHPLC-PDA-MS/MS, LC-DA, LC-ESI-APCI MS | – | Rivera et al. (2013) |
| 21 | Maize | Grain | *B. Thuringiensis* | *Bt toxin* | Herbicide tolerance insect resistance | NMR, GC-EI-Q-MS | PCA | Barros et al. (2010) |
| 22 | Wheat | Seed | *T. aestivum* | *Glu-A1*, *Glu-D1* | Nutritionally enhanced | NMR | PCA | Baker et al. (2006) |
| 23 | Wheat | Leaf | *U. maydis* | *Chit/Gluc*, *RIP*, *Mod.* (*KP4* gene) | Fungal resistance | LC-DAD, LC-ESI-Q *MS* | PCA | Ioset et al. (2007) |
| 24 | Wheat | *Seed* | *T. aestivum* | *Glu-A1*, *Glu-D1* | *Nutritionally enhanced* | NMR | PCA | Baker et al. (2006) |

(continued)

**Table 11.2** (continued)

| S. No | GM Crop | Tissue | Donor Specie | Genetic Modification | Phenotype | Analytical Technique | Data Analysis | References |
|---|---|---|---|---|---|---|---|---|
| 25 | Wheat | *Tuber* | *A. pullulans, S. tuberosum* | *W2, FK, Mal1, SamDC* | *Starch biosynthesis, leaf morphology, ethylene production* | *GC-EI-Q MS* | PCA and ANOVA | Shepherd et al. (2006) |
| 26 | Soybean | Leaf, EC, seed | *N. tabacum* | *ASA2* | Nutritionally enhanced | GC-EI-Q MS | ANOVA | Inaba et al. (2007) |
| 27 | Soybean | Seed | *A. tumefaciens* | *CP4 EPSPS* | Herbicide tolerance | CE-ESI-TOF MS | – | Giuffrida et al. (2009) |
| 28 | Soybean | Seed | *Avenaspp* | *Mod.* (*HPPD* gene) | Herbicide tolerance | LC-ESI-Q MS, GC-EI-Q MS | PCA | Clarke et al. (2013) |
| 29 | Soybean | Seed | *A. tumefaciens* | *CP4 EPSPS* | Herbicide tolerance | CE-ESI-TOF MS, GC-EI-TOF MS, LC-ESI-Q/TOF MS, ICP MS | OPLS-DA | Kusano et al. (2015) |
| 30 | Soybean | Seed | *A. tumefaciens* | *CP4 EPSPS* | Herbicide tolerance | GC-EI-Q MS | OPLS-DA | Bernal et al. (2008) |
| 31 | Soybean | Seed | *Agrobacterium spp* | *837ASDIS* | Herbicide tolerance | GC-EI-Q MS | – | Jiménez et al. (2009) |
| 32 | Soybean | Seed | *A. tumefaciens* | *CP4 EPSPS* | Herbicide tolerance | CE-ESI-TOF MS | – | García-Villalba et al. (2008) |
| 33 | Barley | Seed | *B. amyloliquefaciens* | *GluB, ChGP* | Antifungal activity | LC-ESI-IT MS | PCA | Kogel et al. (2010) |

Updated from Carolina Simó et al. (2014)

biofortification is a single-time investment, and thus is a lucrative, durable, and unceasing approach to tackle hidden hunger by cutting expenses on costly external nutrient supplements and inputs (Meenakshi et al. 2010; Hefferon 2016). Biofortification of crops can be achieved through three strategies, namely, agronomic, conventional breeding, and biotechnology/transgenics.

### 11.6.1 Agronomic Approach

The agronomic approach involves the application of mineral fertilizers (containing essential minerals) to soil or foliage and inoculation of beneficial microbes in soil. The success of fortifying crops using mineral fertilizers is dictated by solubility and mobility (in soil) and phytoavailability of the concerned minerals (White and Broadley 2009). However, factors like environmental pollution, soil composition, varying mineral mobility, presence of anti-nutrient compounds, and antagonistic effect on other minerals are significant hindrances in this approach (Frossard et al. 2000; White and Broadley 2009). Alternatively, foliar application of mineral fertilizers is resorted to when the mineral elements are not easily available through soil or cannot easily perfuse to edible plant parts (White and Broadley 2009). For certain minerals, a foliar application has been more effective than soil application (Shivay et al. 2015). On the other hand, plant rhizosphere symbiotic microbes like rhizobia, mycorrhizal fungi, actinomycetes, and diazotrophic bacteria contribute to plant growth and development by aiding in mineralization and availability of nutrients and also by producing various plant growth hormones (FAO 2019). In spite of their natural presence in soil, these beneficial microorganisms are often used as seed inoculants to produce growth hormones, plant protection substances and for nutrient mineralization (Mahaffee and Kloepper 1994). Owing to the natural presence and beneficial nature of these microbes, this approach is innocuous both to the plant and soil and thus the safest agronomic strategy for the fortification of crops. The agronomic approach of biofortification has been tested and proved to be effective in cereals (rice, wheat, maize, and barley), millets (sorghum), legumes (soybean, chickpea, peas, and common bean), and oilseeds (canola and mustard) (Garg et al. 2018).

### 11.6.2 Conventional and Molecular Breeding Approach

Biofortification through conventional breeding is another cost-efficient and sustainable approach that has been used to address the deficiency of micronutrients like carotenoids, Fe, and Zn (White and Broadley 2005). Products achieved through this approach are a one-time investment, thus can be cultivated by farmers throughout the year at a comparatively lower production cost and also fetch a better price. The presence of significant genetic diversity in the gene pool is a vital pre-requisite for this approach (White and Broadley 2009). Screening of a wide range of germplasm yields competent genotypes that can be used in crossing programs, genetic studies and to map genetic loci that can be used in marker-assisted breeding. Outperformers can be shortlisted for multi-location trials over seasons to assess the environmental effect on the trait and after thorough testing can be released for cultivation by farmers (Bouis and Saltzman 2017). Many products have been delivered by this approach by exploiting genetic diversity present in the primary secondary and tertiary gene pools of major staple food crops and are being commercially cultivated by farmers worldwide (Table 11.3). The accessibility and easy exchange of international germplasm of most agriculture crops have fetched significantly to product achievement (Nestel et al. 2006; Bouis et al. 2011). However, conventional breeding approaches fail when diversity for the desired trait is not naturally found in the gene pool or when the desired nutrient does not exist naturally in the concerned crop species.

### 11.6.3 Genetic Engineering and Gene-Editing Approach

Biofortification through genetic engineering/transgenic approach is often resorted to when the desired outcome cannot be achieved using conventional breeding (Pérez-Massot et al. 2013). The availability and accessibility to whole-genome sequences of crops have contributed significantly to this approach. This approach enables the fortification of crops not only in terms of enhanced nutrient status but also can be used to eliminate or suppress anti-nutritional factors in some crops (Garg et al. 2018). The transgenic approach for biofortification has exploited genes from across genera and also across kingdoms (Newell-McGloughlin 2008). To add to the success of this approach, a recent discovery of site-specific gene-editing tools like zinc finger nucleases (ZFNs), transcription-activator like effector nucleases (TALENs), clustered regularly interspaced short palindromic repeat (CRISPR) – CRISPR associated protein 9 (CRISPR-Cas9) have enabled precise modification of desired gene (s) (Jaganathan et al. 2018). Though voluminous research has been done, this approach is commercially under-utilized as only a few products are being commercially cultivated (Table 11.3) and most of them remain in the pipeline due to political and societal non-acceptance in many countries (Watanabe et al. 2005).

## 11.7 Conclusion and Future Prospects

Metabolomics has successfully carved its path into agricultural sciences by contributing significantly to research towards understanding responses to various stresses, pivotal metabolic pathways, deciphering gene functions, improving crop quality, and a better understanding of the genotype-metabolome-phenotype relationship. The gradual advent of new and high-throughput metabolomics tools has enabled more efficient metabolome profiling in plants that have a complex metabolome. The pace of development in this field has brought about a major shift in the approach from analysis of a single metabolite to generating footprints of multiple metabolites in a single assay. It has also allowed the researchers to precisely target a specific metabolite through metabolic engineering. The metabolomics tools

**Table 11.3** List of biofortified crop varieties released through breeding and transgenic approaches

| S. No. | Crop | Type of biofortification | Importance | Approach | Examples of varieties | References |
|---|---|---|---|---|---|---|
| Cereals | | | | | | |
| 1 | Rice | Zinc and iron | Zinc plays a role in cell division, cell growth, wound healing and carbohydrate breakdown. Iron helps in blood production and transfers oxygen from lungs to tissues through red blood cells | Breeding | BRRIdhan 62, 72, 64 | CIAT, HarvestPlus |
| 2 | | Zinc | – | Breeding (traditional variety) | Jalmagna | Gregorio et al. (2000) |
| 3 | | Iron | – | Breeding (traditional variety) | Jalmagna | Gregorio et al. (2000) |
| 4 | Wheat | Zinc and Iron | – | Breeding | WB2 | Indian Institute of Wheat and Barley Research, India |
| 5 | | Zinc | – | Breeding | BHU 1, 3, 5, 6, 17, 18, NR 419, 42, 421, Zincol | CIAT, CIMMYT, HarvestPlus |
| 6 | | | – | Breeding | PBW1Zn | Punjab agricultural university, India |
| 7 | | Carotene | Carotene is a precursor for vitamin A and has powerful antioxidant properties | Breeding | HI 8627 | IARI |
| 8 | | Anthocyanins | They are antioxidants, eliminate free radicals, anti-inflammatory, anti-viral, and anti-cancer | Breeding | Black-grained wheat, indigo | Havrlentová et al. (2014) |
| 9 | Maize | Vitamin A (Orange maize) | Vit. A is also called retinol as it produces pigments in retina of a | Breeding | | CIMMYT, International Institute of Tropical |

(continued)

**Table 11.3** (continued)

| S. No. | Crop | Type of biofortification | Importance | Approach | Examples of varieties | References |
|---|---|---|---|---|---|---|
| | | | human eye. Its deficiency causes xerophthalmia and night blindness | | GV662A, 664A, 665A, Ife maizehyb-3, 4, Sammaz 38, 39, CSIR-CRI Honampa | Agriculture (IITA), HarvestPlus |
| 10 | | Lysine and tryptophan (quality protein maize) | Lysine promotes calcium uptake, tissue growth, function and healing and improves immune system. Tryptophan helps in synthesis of melatonin and serotonin, liver uses tryptophan to produce vit. B3 | Breeding | HQPM-1, 4, 5, 7, VivekQPM-9, FQH-4567, Obatampa, Susuma | Surinder Vasal and Evangelina Villegas, CIMMYT |
| 11 | | Phytate degradation | Phytate is an anti-nutrient compound that reduces availability of metals | Transgenics | BVLA4 30,101 | Origin Agritech |
| 12 | | Lysine | – | Transgenics | MavreaTM YieldGard maize, MaveraTM Mazie (LY038) | Monsanto, Renessen LLC |
| Millets | | | | | | |
| 13 | Sorghum | Iron | – | Breeding | ICSR 14001, ICSH 14002, 12KNICSV (Deko)-188 12KNICSV-22 (Zabuwa) | ICRISAT, HarvestPlus |
| 14 | Pearl millet | Iron and zinc | – | Breeding | Dhanashakti, Hybrid ICMH 1201 (Shakti-1201) | ICRISAT, HarvestPlus |
| Legumes/pulses | | | | | | |
| 15 | Lentils | Iron and zinc | – | Breeding | Barimasur-4, 5, 6, 7, 8, Khajurah-1, 2, Shital, Sisir, Shekhar, Simal, Pusa Vaibhav, Alemaya, | ICARDA, HarvestPlus |
| 16 | Cowpea | Iron | – | Breeding | Pant Lobia-1, 2, 3, 4 | G.B. pant agriculture university, HarvestPlus |

| | | | | | | |
|---|---|---|---|---|---|---|
| 17 | Common beans | Iron and zinc | – | Breeding | RWR 2245; 2154; MAC 42; 44; CAB 2; RWV 1129; 3006; 3316; 3317; 2887 | HarvestPlus |
| 18 | Soybean | Oleic acid | Oleic acid is a mono unsaturated fatty acid that improves heart health by reducing blood cholesterol and also reduces inflammation | Transgenics | G94–1, 19, G16, Treus TM, Plenish TM, Vistuve gold TM | Dupont, Monsanto |
| 19 | | Stearidonic acid (STA) | STA is an omega-3 fatty acid that reduces plasma triacylglycerols, improves platelet function, reduces blood viscosity, inflammation, and blood pressure | Transgenics | MON87769 X MON89788 | Monsanto |
| Oilseeds | | | | | | |
| 20 | Linseed/ flax | Essential amino acids | Essential amino acids cannot be synthesized in the body and have to be obtained through food. They contribute to protein synthesis, tissue repair, and nutrient absorption | Transgenics | CDC Triffid flax (FP967) | University of Saskatchewan, Canada |
| 21 | Canola | Phytate degradation | Phytate is principal storage form of phosphorus and thus have to be degraded to increase P availability | Transgenics | PhytaseedTM canola (MPS 961, 962, 963, 964, 965) | BASF |

have yielded many mQTLs and metabolite markers that have been utilized in breeding programs. A significant amount of credit for this success goes to the cost-effective NGS techniques that contribute largely to understanding the metabolomes. Further, the integration of metabolomics with other omics tools has added to the pace of progress by providing a holistic understanding of plants as a living system.

In spite of all these developments, there is still a lot of scope for exploiting more areas of metabolomics. Identification and development of more and more metabolite markers and mQTLs would aid in the success of metabolomics-assisted breeding with a perspective of imparting tolerance to stresses, improving yields, and crop quality. In the future, new technologies like CRISPR – Cas9 can also be utilized for metabolite profiling of genome-edited plants or to edit the genes involved in metabolic pathways. The application of metabolomics to agriculture has more scope of extension and expansion in the future to address challenges posed to agriculture.

## References

Agarrwal R, Bentur JS, Nair S (2014) Gas chromatography mass spectrometry based metabolic profiling reveals biomarkers involved in rice-gall midge interactions. J Integr Plant Biol 56:837–848. https://doi.org/10.1111/jipb.12244

Ahmad S, Veyrat N, Gordon-Weeks R et al (2011) Benzoxazinoid metabolites regulate innate immunity against aphids and fungi in maize. Plant Physiol 157:317–327. https://doi.org/10.1104/pp.111.180224

Aliferis KA, Faubert D, Jabaji S (2014) A metabolic profiling strategy for the dissection of plant defense against fungal pathogens. PLoS One 9(11):e111930. https://doi.org/10.1371/journal.pone.0111930

Allen F, Pon A, Wilson M et al (2014) CFM-ID: a web server for annotation, spectrum prediction and metabolite identification from tandem mass spectra. Nucleic Acids Res 42:W94. https://doi.org/10.1093/nar/gku436

Alonso R, Salavert F, Garcia-Garcia F et al (2015) Babelomics 5.0: functional interpretation for new generations of genomic data. Nucleic Acids Res 43:W117–W121. https://doi.org/10.1093/nar/gkv384

Alseekh S, Fernie AR (2018) Metabolomics 20 years on: what have we learned and what hurdles remain? Plant J 94:933–942

Bailey RL, West KP, Black RE (2015) The epidemiology of global micronutrient deficiencies. Ann Nutr Metab 66:22–33. https://doi.org/10.1159/000371618

Baker JM, Hawkins ND, Ward JL et al (2006) A metabolomic study of substantial equivalence of field-grown genetically modified wheat. Plant Biotechnol J 4:381–392. https://doi.org/10.1111/j.1467-7652.2006.00197.x

Balmer D, Flors V, Glauser G, Mauch-Mani B (2013) Metabolomics of cereals under biotic stress: current knowledge and techniques. Front Plant Sci 4:82

Barding GA, Béni S, Fukao T et al (2013) Comparison of GC-MS and NMR for metabolite profiling of rice subjected to submergence stress. J Proteome Res 12:898–909. https://doi.org/10.1021/pr300953k

Barros E, Lezar S, Anttonen MJ et al (2010) Comparison of two GM maize varieties with a near-isogenic non-GM variety using transcriptomics, proteomics and metabolomics. Plant Biotechnol J 8:436–451. https://doi.org/10.1111/j.1467-7652.2009.00487.x

Benesty J, Chen J, Huang Y, Cohen I (2009) Pearson correlation coefficient. Springer, Berlin, pp 1–4

Bernal JL, Nozal MJ, Toribio L et al (2008) Use of supercritical fluid extraction and gas chromatography-mass spectrometry to obtain amino acid profiles from several genetically modified varieties of maize and soybean. J Chromatogr A 1192:266–272. https://doi.org/10.1016/j.chroma.2008.03.047

Bjarnholt N, Li B, D'Alvise J, Janfelt C (2014) Mass spectrometry imaging of plant metabolites-principles and possibilities. Nat Prod Rep 31:818–837

Bogart E, Myers CR (2016) Multiscale metabolic modeling of C4 plants: connecting nonlinear genome-scale models to leaf-scale metabolism in developing maize leaves. PLoS One 11: e0151722. https://doi.org/10.1371/journal.pone.0151722

Boiteau RM, Hoyt DW, Nicora CD et al (2018) Structure elucidation of unknown metabolites in metabolomics by combined NMR and MS/MS prediction. Metabolites 8(1):8. https://doi.org/10.3390/metabo8010008

Bollina V, Kumaraswamy GK, Kushalappa AC et al (2010) Mass spectrometry-based metabolomics application to identify quantitative resistance-related metabolites in barley against Fusarium head blight. Mol Plant Pathol 11(6):769–782. https://doi.org/10.1111/j.1364-3703.2010.00643.x

Bouis HE, Hotz C, McClafferty B et al (2011) Biofortification: a new tool to reduce micronutrient malnutrition. Food Nutr Bull 32:S31–S40. https://doi.org/10.1177/15648265110321s105

Bouis HE, Saltzman A (2017) Improving nutrition through biofortification: a review of evidence from HarvestPlus, 2003 through 2016. Glob Food Sec 12:49–58

Cai G, Yang Q, Chen H et al (2016) Genetic dissection of plant architecture and yield-related traits in Brassica napus. Sci Rep 6:1–16. https://doi.org/10.1038/srep21625

Cañas RA, Yesbergenova-Cuny Z, Simons M et al (2017) Exploiting the genetic diversity of maize using a combined metabolomic, enzyme activity profiling, and metabolic modeling approach to link leaf physiology to kernel yield. Plant Cell 29:919–943. https://doi.org/10.1105/tpc.16.00613

Carreno-Quintero N, Bouwmeester HJ, Keurentjes JJB (2013) Genetic analysis of metabolome-phenotype interactions: from model to crop species. Trends Genet 29:41–50

Carolina S, Ibáñez C, Valdés A, Cifuentes A, Garcia-Cañas V (2014) Metabolomics of Genetically Modified Crops. Int J Mol Sci 15:18941–18966. https://doi.org/10.3390/ijms151018941

Castro C, Manetti C (2007) A multiway approach to analyze metabonomic data: a study of maize seeds development. Anal Biochem 371:194–200. https://doi.org/10.1016/j.ab.2007.08.028

Cavill R, Jennen D, Kleinjans J, Briedé JJ (2016) Transcriptomic and metabolomic data integration. Brief Bioinform 17:891–901. https://doi.org/10.1093/bib/bbv090

Chang H-Y, Chen C-T, Lih TM et al (2016) iMet-Q: a user-friendly tool for label-free metabolomics quantitation using dynamic peak-width determination. PLoS One 11:e0146112. https://doi.org/10.1371/journal.pone.0146112

Chang J, Cheong BE, Natera S, Roessner U (2019) Morphological and metabolic responses to salt stress of rice (Oryza sativa L.) cultivars which differ in salinity tolerance. Plant Physiol Biochem 144:427–435. https://doi.org/10.1016/j.plaphy.2019.10.017

Chang Y, Zhao C, Zhu Z et al (2012) Metabolic profiling based on LC/MS to evaluate unintended effects of transgenic rice with cry1Ac and sck genes. Plant Mol Biol 78:477–487. https://doi.org/10.1007/s11103-012-9876-3

Chebrolu KK, Fritschi FB, Ye S et al (2016) Impact of heat stress during seed development on soybean seed metabolome. Metabolomics 12:1–14. https://doi.org/10.1007/s11306-015-0941-1

Chen W, Gao Y, Xie W et al (2014) Genome-wide association analyses provide genetic and biochemical insights into natural variation in rice metabolism. Nat Genet 46:714–721. https://doi.org/10.1038/ng.3007

Che-Othman MH, Jacoby RP, Millar AH, Taylor NL (2020) Wheat mitochondrial respiration shifts from the tricarboxylic acid cycle to the GABA shunt under salt stress. New Phytol 225:1166–1180. https://doi.org/10.1111/nph.15713

Cho JS, Gu C, Han TH et al (2019) Reconstruction of context-specific genome-scale metabolic models using multiomics data to study metabolic rewiring. Curr Opin Syst Biol 15:1–11

Clarke JD, Alexander DC, Ward DP et al (2013) Assessment of genetically modified soybean in relation to natural variation in the soybean seed metabolome. Sci Rep 3:3082. https://doi.org/10.1038/srep03082

Clasquin MF, Melamud E, Rabinowitz JD (2012) LC-MS data processing with MAVEN: a metabolomic analysis and visualization engine. Curr Protoc Bioinforma Chapter 37 (1):14.11.1–14.11.23. https://doi.org/10.1002/0471250953.bi1411s37

Coutinho ID, Henning LMM, Döpp SA et al (2018) Flooded soybean metabolomic analysis reveals important primary and secondary metabolites involved in the hypoxia stress response and tolerance. Environ Exp Bot 153:176–187. https://doi.org/10.1016/j.envexpbot.2018.05.018

Cui Q, Lewis IA, Hegeman AD et al (2008) Metabolite identification via the Madison metabolomics consortium database [3]. Nat Biotechnol 26:162–164

Cuperlovic-Culf M, Vaughan MM, Vermillion K et al (2019) Effects of atmospheric CO2 level on the metabolic response of resistant and susceptible wheat to Fusarium graminearum infection. Mol Plant-Microbe Interact 32:379–391. https://doi.org/10.1094/MPMI-06-18-0161-R

Cuperlovic-Culf M, Wang L, Forseille L et al (2016) Metabolic biomarker panels of response to Fusarium head blight infection in different wheat varieties. PLoS One 11:e0153642. https://doi.org/10.1371/journal.pone.0153642

Dangl JL, Horvath DM, Staskawicz BJ (2013) Pivoting the plant immune system from dissection to deployment. Science 341:746–751

de Abreu Lima F, Li K, Wen W et al (2018) Unraveling lipid metabolism in maize with time-resolved multi-omics data. Plant J 93:1102–1115. https://doi.org/10.1111/tpj.13833

de Oliveira Dal'Molin CG, Orellana C, Gebbie L et al (2016) Metabolic reconstruction of Setaria italica: a systems biology approach for integrating tissue-specific Omics and pathway analysis of bioenergy grasses. Front Plant Sci 7:1138. https://doi.org/10.3389/fpls.2016.01138

de Oliveira Dal'Molin G, Nielsen LK (2018) Plant genome-scale reconstruction: from single cell to multi-tissue modelling and omics analyses. Curr Opin Biotechnol 49:42–48

Deborde C, Moing A, Roch L et al (2017) Plant metabolism as studied by NMR spectroscopy. Prog Nucl Magn Reson Spectrosc 102–103:61–97

Degenkolbe T, Do PT, Kopka J et al (2013) Identification of drought tolerance markers in a diverse population of Rice cultivars by expression and metabolite profiling. PLoS One 8(5):e63637. https://doi.org/10.1371/journal.pone.0063637

Dixit S, Singh UM, Singh AK et al (2020) Marker assisted forward breeding to combine multiple biotic-abiotic stress resistance/tolerance in Rice. Rice 13:29. https://doi.org/10.1186/s12284-020-00391-7

Do PT, Degenkolbe T, Erban A et al (2013) Dissecting Rice polyamine metabolism under controlled Long-term drought stress. PLoS One 8:e60325. https://doi.org/10.1371/journal.pone.0060325

Doehlemann G, Wahl R, Horst RJ et al (2008) Reprogramming a maize plant: transcriptional and metabolic changes induced by the fungal biotroph Ustilago maydis. Plant J 56:181–195. https://doi.org/10.1111/j.1365-313X.2008.03590.x

Dunn WB, Erban A, Weber RJM et al (2013) Mass appeal: metabolite identification in mass spectrometry-focused untargeted metabolomics. Metabolomics 9:44–66

Dutta A, Shetty P, Bhat S et al (2012) A mass spectrometric study for comparative analysis and evaluation of metabolite recovery from plasma by various solvent systems. J Biomol Tech 23:128–135. https://doi.org/10.7171/jbt.12-2304-001

Erb M, Flors V, Karlen D et al (2009) Signal signature of aboveground-induced resistance upon belowground herbivory in maize. Plant J 59:292–302. https://doi.org/10.1111/j.1365-313X.2009.03868.x

FAO (2013) The state of food and agriculture. Food and Agriculture Organization, Rome, Italy

FAO (2019) The plant production and protection division (AGP)—soil biological management with beneficial microorganisms. Food and Agriculture Organization, Rome, Italy

Farahbakhsh F, Hamzehzarghani H, Massah A et al (2019) Comparative metabolomics of temperature sensitive resistance to wheat streak mosaic virus (WSMV) in resistant and susceptible wheat cultivars. J Plant Physiol 237:30–42. https://doi.org/10.1016/j.jplph.2019.03.011

Farooq M, Wahid A, Lee D-J et al (2010) DROUGHT STRESS: comparative time course action of the foliar applied Glycinebetaine, salicylic acid, nitrous oxide, Brassinosteroids and Spermine in improving drought resistance of Rice. J Agron Crop Sci 196:336–345. https://doi.org/10.1111/j.1439-037X.2010.00422.x

Fernie AR, Schauer N (2009) Metabolomics-assisted breeding: a viable option for crop improvement? Trends Genet 25:39–48

Fiehn O (2002) Metabolomics - the link between genotypes and phenotypes. Plant Mol Biol 48:155–171. https://doi.org/10.1023/A:1013713905833

Fondi M, Liò P (2015) Multi -omics and metabolic modelling pipelines: challenges and tools for systems microbiology. Microbiol Res 171:52–64

Frank T, Röhlig RM, Davies HV et al (2012) Metabolite profiling of maize kernels-genetic modification versus environmental influence. J Agric Food Chem 60:3005–3012. https://doi.org/10.1021/jf204167t

Frossard E, Bucher M, Mächler F et al (2000) Potential for increasing the content and bioavailability of Fe, Zn and Ca in plants for human nutrition. J Sci Food Agric 80:861–879

Fu R, Martin C, Zhang Y (2018) Next-generation plant metabolic engineering, inspired by an ancient chinese irrigation system. Mol Plant 11:47–57

García-Sevillano MÁ, García-Barrera T, Abril N et al (2014) Omics technologies and their applications to evaluate metal toxicity in mice M. spretus as a bioindicator. J. Proteomics 104:4–23

García-Villalba R, León C, Dinelli G et al (2008) Comparative metabolomic study of transgenic versus conventional soybean using capillary electrophoresis-time-of-flight mass spectrometry. J Chromatogr A 1195:164–173. https://doi.org/10.1016/j.chroma.2008.05.018

Garg M, Sharma N, Sharma S et al (2018) Biofortified crops generated by breeding, agronomy, and transgenic approaches are improving lives of millions of people around the world. Front Nutr 5:1

Gavaghan CL, Li JV, Hadfield ST et al (2011) Application of NMR-based metabolomics to the investigation of salt stress in maize (Zea mays). Phytochem Anal 22:214–224. https://doi.org/10.1002/pca.1268

Gayen D, Barua P, Lande NV et al (2019) Dehydration-responsive alterations in the chloroplast proteome and cell metabolomic profile of rice reveals key stress adaptation responses. Environ Exp Bot 160:12–24. https://doi.org/10.1016/j.envexpbot.2019.01.003

Ghosson H, Schwarzenberg A, Jamois F, Yvin JC (2018) Simultaneous untargeted and targeted metabolomics profiling of underivatized primary metabolites in sulfur-deficient barley by ultra-high performance liquid chromatography-quadrupole/time-of-flight mass spectrometry. Plant Methods 14:62. https://doi.org/10.1186/s13007-018-0329-0

Giuffrida A, León C, García-Cañas V et al (2009) Modified cyclodextrins for fast and sensitive chiral-capillary electrophoresis-mass spectrometry. Electrophoresis 30:1734–1742. https://doi.org/10.1002/elps.200800333

Goh HH (2018) Integrative multi-omics through bioinformatics. In: Advances in experimental medicine and biology. Springer, New York, pp 69–80

Gregorio GB, Senadhira D, Htut H, Graham RD (2000) Breeding for trace mineral density in Rice. Food Nutr Bull 21:382–386. https://doi.org/10.1177/156482650002100407

Gunnaiah R, Kushalappa AC, Duggavathi R et al (2012) Integrated Metabolo-proteomic approach to decipher the mechanisms by which wheat QTL (Fhb1) contributes to resistance against Fusarium graminearum. PLoS One 7:e40695. https://doi.org/10.1371/journal.pone.0040695

Guo J, Qi J, He K et al (2019) The Asian corn borer Ostrinia furnacalis feeding increases the direct and indirect defence of mid-whorl stage commercial maize in the field. Plant Biotechnol J 17:88–102. https://doi.org/10.1111/pbi.12949

Gupta P, De B (2017) Metabolomics analysis of rice responses to salinity stress revealed elevation of serotonin, and gentisic acid levels in leaves of tolerant varieties. Plant Signal Behav 12(7): e1335845. https://doi.org/10.1080/15592324.2017.1335845

Gutleben J, Chaib De Mares M, van Elsas JD et al (2018) The multi-omics promise in context: from sequence to microbial isolate. Crit Rev Microbiol 44:212–229

Han S, Micallef SA (2016) Environmental metabolomics of the tomato plant surface provides insights on Salmonella enterica colonization. Appl Environ Microbiol 82:3131–3142. https://doi.org/10.1128/AEM.00435-16

Havrlentová M, Pšenáková I, Žofajová A et al (2014) Anthocyanins in wheat seed - a mini review. Nov Biotechnol Chim 13:1–12

Hay JO, Shi H, Heinzel N et al (2014) Integration of a constraint-based metabolic model of Brassica napus developing seeds with 13C-metabolic flux analysis. Front Plant Sci 5:724. https://doi.org/10.3389/fpls.2014.00724

Hefferon KL (2016) Can biofortified crops help attain food security? Curr Mol Biol Rep 2:180–185. https://doi.org/10.1007/s40610-016-0048-0

Hein JA, Sherrard ME, Manfredi KP, Abebe T (2016) The fifth leaf and spike organs of barley (Hordeum vulgare L.) display different physiological and metabolic responses to drought stress. BMC Plant Biol 16:248. https://doi.org/10.1186/s12870-016-0922-1

Heyneke E, Watanabe M, Erban A et al (2017) Characterization of the wheat leaf Metabolome during grain filling and under varied N-supply. Front Plant Sci 8:2048. https://doi.org/10.3389/fpls.2017.02048

Hong J, Yang L, Zhang D, Shi J (2016) Plant metabolomics: an indispensable system biology tool for plant science. Int J Mol Sci 17(6):767

Hu C, Shi J, Quan S et al (2014) Metabolic variation between japonica and indica rice cultivars as revealed by non-targeted metabolomics. Sci Rep 4:1–10. https://doi.org/10.1038/srep05067

Hughes M (2015) Systems biology tools for integrated Omics analysis. Genet Eng Biotechnol News 35:18–19. https://doi.org/10.1089/gen.35.03.11

Inaba Y, Brotherton JE, Ulanov A, Widholm JM (2007) Expression of a feedback insensitive anthranilate synthase gene from tobacco increases free tryptophan in soybean plants. Plant Cell Rep 26:1763–1771. https://doi.org/10.1007/s00299-007-0381-0

Ioset JR, Urbaniak B, Ndjoko-Ioset K et al (2007) Flavonoid profiling among wild type and related GM wheat varieties. Plant Mol Biol 65:645–654. https://doi.org/10.1007/s11103-007-9229-9

Jaganathan D, Ramasamy K, Sellamuthu G et al (2018) CRISPR for crop improvement: an update review. Front Plant Sci 9:985

Jahangir M, Abdel-Farid IB, Choi YH, Verpoorte R (2008) Metal ion-inducing metabolite accumulation in Brassica rapa. J Plant Physiol 165:1429–1437. https://doi.org/10.1016/j.jplph.2008.04.011

Jamil IN, Remali J, Azizan KA et al (2020) Systematic multi-Omics integration (MOI) approach in plant systems biology. Front Plant Sci 11:944

Jiang J, Xing F, Wang C et al (2019) Investigation and development of maize fused network analysis with multi-omics. Plant Physiol Biochem 141:380–387. https://doi.org/10.1016/j.plaphy.2019.06.016

Jiang K, Liberatore KL, Park SJ et al (2013) Tomato yield Heterosis is triggered by a dosage sensitivity of the Florigen pathway that fine-tunes shoot architecture. PLoS Genet 9:1004043. https://doi.org/10.1371/journal.pgen.1004043

Jiao Z, Si XXI, Li GKE et al (2010) Unintended compositional changes in transgenic rice seeds (Oryza sativa L.) studied by spectral and chromatographic analysis coupled with chemometrics methods. J Agric Food Chem 58:1746–1754. https://doi.org/10.1021/jf902676y

Jiménez JJ, Bernal JL, Nozal MJ et al (2009) Profile and relative concentrations of fatty acids in corn and soybean seeds from transgenic and isogenic crops. J Chromatogr A 1216:7288–7295. https://doi.org/10.1016/j.chroma.2009.08.015

Johnson SR, Lange BM (2015) Open-access metabolomics databases for natural product research: present capabilities and future potential. Front Bioeng Biotechnol 3:22

Jones OAH, Maguire ML, Griffin JL et al (2011) Using metabolic profiling to assess plant-pathogen interactions: an example using rice (Oryza sativa) and the blast pathogen Magnaporthe grisea. Eur J Plant Pathol 129:539–554. https://doi.org/10.1007/s10658-010-9718-6

Jorge TF, Rodrigues JA, Caldana C et al (2016) Mass spectrometry-based plant metabolomics: metabolite responses to abiotic stress. Mass Spectrom Rev 35:620–649

Kang W, Zhu X, Wang Y et al (2018) Transcriptomic and metabolomic analyses reveal that bacteria promote plant defense during infection of soybean cyst nematode in soybean. BMC Plant Biol 18:86. https://doi.org/10.1186/s12870-018-1302-9

Kang Z, Babar MA, Khan N et al (2019) Comparative metabolomic profiling in the roots and leaves in contrasting genotypes reveals complex mechanisms involved in post-anthesis drought tolerance in wheat. PLoS One 14:e0213502. https://doi.org/10.1371/journal.pone.0213502

Karmakar S, Datta K, Molla KA et al (2019) Proteo-metabolomic investigation of transgenic rice unravels metabolic alterations and accumulation of novel proteins potentially involved in defence against Rhizoctonia solani. Sci Rep 9:1–16. https://doi.org/10.1038/s41598-019-46885-3

Keymanesh K, Darvishi MH, Sardari S (2009) Metabolome comparison of transgenic and non-transgenic Rice by statistical analysis of FTIR and NMR spectra. Rice Sci 16:119–123. https://doi.org/10.1016/S1672-6308(08)60067-0

Khan FU (2019) Comparative metabolite profiling of two wheat genotypes as affected by nitrogen stress at seedling stage. J Anim Plant Sci 29:260–268

Khush GS, Lee S, Il CJ, Jeon JS (2012) Biofortification of crops for reducing malnutrition. Plant Biotechnol Rep 6:195–202

Kim JK, Bamba T, Harada K et al (2007) Time-course metabolic profiling in Arabidopsis thaliana cell cultures after salt stress treatment. J Exp Bot 58:415–424. https://doi.org/10.1093/jxb/erl216

Kim JK, Park SY, Lee SM et al (2013) Unintended polar metabolite profiling of carotenoid-biofortified transgenic rice reveals substantial equivalence to its non-transgenic counterpart. Plant Biotechnol Rep 7:121–128. https://doi.org/10.1007/s11816-012-0231-6

Kim JM, To TK, Matsui A et al (2017) Acetate-mediated novel survival strategy against drought in plants. Nat Plants 3:1–7. https://doi.org/10.1038/nplants.2017.97

Kogel KH, Voll LM, Schäfer P et al (2010) Transcriptome and metabolome profiling of field-grown transgenic barley lack induced differences but show cultivar-specific variances. Proc Natl Acad Sci U S A 107:6198–6203. https://doi.org/10.1073/pnas.1001945107

Komatsu S, Nakamura T, Sugimoto Y, Sakamoto K (2014) Proteomic and Metabolomic analyses of soybean root tips under flooding stress. Protein Pept Lett 21:865–884

Kuhn S, Egert B, Neumann S, Steinbeck C (2008) Building blocks for automated elucidation of metabolites: machine learning methods for NMR prediction. BMC Bioinform 9:400. https://doi.org/10.1186/1471-2105-9-400

Kumar R, Bohra A, Pandey AK, Pandey MK, Kumar A (2017) Metabolomics for plant improvement: status and prospects. Front Plant Sci 8:1302

Kumaraswamy KG, Kushalappa AC, Choo TM et al (2011) Mass spectrometry based metabolomics to identify potential biomarkers for resistance in barley against Fusarium head blight (Fusarium graminearum). J Chem Ecol 37:846–856. https://doi.org/10.1007/s10886-011-9989-1

Kusano M, Baxter I, Fukushima A et al (2015) Assessing metabolomic and chemical diversity of a soybean lineage representing 35 years of breeding. Metabolomics 11:261–270. https://doi.org/10.1007/s11306-014-0702-6

Latif S, Gurusinghe S, Weston PA et al (2020) Metabolomic approaches for the identification of flavonoids associated with weed suppression in selected Hardseeded annual pasture legumes. Plant Soil 447:199–218. https://doi.org/10.1007/s11104-019-04225-4

Leister D (2005) Plant functional genomics. Food Products Press, New York

Leon C, Rodriguez-Meizoso I, Lucio M et al (2009) Metabolomics of transgenic maize combining Fourier transform-ion cyclotron resonance-mass spectrometry, capillary electrophoresis-mass spectrometry and pressurized liquid extraction. J Chromatogr A 1216:7314–7323. https://doi.org/10.1016/j.chroma.2009.04.092

Levandi T, Leon C, Kaljurand M et al (2008) Capillary electrophoresis time-of-flight mass spectrometry for comparative metabolomics of transgenic versus conventional maize. Anal Chem 80:6329–6335. https://doi.org/10.1021/ac8006329

Li L, Hur M, Lee JY et al (2015) A systems biology approach toward understanding seed composition in soybean. BMC Genomics 16:S3–S9. https://doi.org/10.1186/1471-2164-16-S3-S9

Limbalkar OM, Meena VK, Singh M, Sunilkumar VP (2018) Genetic improvement of wheat for biotic and abiotic stress tolerance. Int J Curr Microbiol App Sci 7:1962–1971. https://doi.org/10.20546/ijcmas.2018.712.226

Liu Q, Wang X, Tzin V et al (2016) Combined transcriptome and metabolome analyses to understand the dynamic responses of rice plants to attack by the rice stem borer Chilo suppressalis (Lepidoptera: Crambidae). BMC Plant Biol 16:259. https://doi.org/10.1186/s12870-016-0946-6

Liu XH, Liang S, Wei YY et al (2019) Metabolomics analysis identifies sphingolipids as key signaling moieties in appressorium morphogenesis and function in Magnaporthe oryzae. MBio 10:e01467-19. https://doi.org/10.1128/mBio.01467-19

Locke AM, Barding GA, Sathnur S et al (2018) Rice SUB1A constrains remodelling of the transcriptome and metabolome during submergence to facilitate post-submergence recovery. Plant Cell Environ 41:721–736. https://doi.org/10.1111/pce.13094

Lommen A, Kools HJ (2012) MetAlign 3.0: performance enhancement by efficient use of advances in computer hardware. Metabolomics 8:719–726. https://doi.org/10.1007/s11306-011-0369-1

Long X, Liu Q, Chan M et al (2013) Metabolic engineering and profiling of rice with increased lysine. Plant Biotechnol J 11:490–501. https://doi.org/10.1111/pbi.12037

Ma C, Zhang HH, Wang X (2014) Machine learning for big data analytics in plants. Trends Plant Sci 19:798–808

Ma NL, Che Lah WA, Kadir NA et al (2018) Susceptibility and tolerance of rice crop to salt threat: physiological and metabolic inspections. PLoS One 13(2):e0192732. https://doi.org/10.1371/journal.pone.0192732

Ma X, Xia H, Liu Y et al (2016) Transcriptomic and metabolomic studies disclose key metabolism pathways contributing to well-maintained photosynthesis under the drought and the consequent drought-tolerance in rice. Front Plant Sci 7:1886. https://doi.org/10.3389/fpls.2016.01886

Mahaffee WF, Kloepper J (1994) Applications of plant growth-promoting rhizobacteria in sustainable agriculture. In: Pankhurst CE, Doube BM, Gupta VVSR, Grace PR (eds) Soil biota: management in sustainable farming systems. CSIRO, Melbourne, Australia, pp 23–31

Manetti C, Bianchetti C, Bizzarri M et al (2004) NMR-based metabonomic study of transgenic maize. Phytochemistry 65:3187–3198. https://doi.org/10.1016/j.phytochem.2004.10.015

Manivasagaperumal R (2011) Effect of zinc on germination, seedling growth and biochemical content of cluster bean (Cyamopsis tetragonoloba (L.) Taub). Curr Bot 2(5):11–15

Mareya CR, Tugizimana F, Di Lorenzo F et al (2020) Adaptive defence-related changes in the metabolome of Sorghum bicolor cells in response to lipopolysaccharides of the pathogen Burkholderia andropogonis. Sci Rep 10:1–17. https://doi.org/10.1038/s41598-020-64186-y

Mata CI, Fabre B, Parsons HT et al (2018) Ethylene receptors, CTRs and EIN2 target protein identification and quantification through parallel reaction monitoring during tomato fruit ripening. Front Plant Sci 9:1626. https://doi.org/10.3389/fpls.2018.01626

Matsuda F, Ishihara A, Takanashi K et al (2010) Metabolic profiling analysis of genetically modified rice seedlings that overproduce tryptophan reveals the occurrence of its inter-tissue translocation. Plant Biotechnol 27:17–27. https://doi.org/10.5511/plantbiotechnology.27.17

Matsuda F, Yonekura-Sakakibara K, Niida R et al (2009) MS/MS spectral tag-based annotation of non-targeted profile of plant secondary metabolites. Plant J 57:555–577. https://doi.org/10.1111/j.1365-313X.2008.03705.x

Meenakshi JV, Johnson NL, Manyong VM et al (2010) How cost-effective is biofortification in combating micronutrient malnutrition? An ex ante assessment. World Dev 38:64–75. https://doi.org/10.1016/j.worlddev.2009.03.014

Mesnage R, Agapito-Tenfen SZ, Vilperte V et al (2016) An integrated multi-omics analysis of the NK603 roundup-tolerant GM maize reveals metabolism disturbances caused by the transformation process. Sci Rep 6:1–14. https://doi.org/10.1038/srep37855

Monreal CM, Schnitzer MI (2015) Labile organic matter in soil solution: II. Separation and identification of metabolites from plant-microbial communication in soil solutions of wheat rhizospheres. Wiley, New York, pp 173–193

Mwendwa JM (2016a) Metabolic profiling for benzoxazinoids in weed-suppressive and early vigour wheat genotypes. In: Australasian Weeds Conference at Perth

Mwendwa J (2016b) Mechanisms of weed suppression by wheat genotypes. GRDC Grains Research Update Grain Research and Development Cooporation, Canberra, Australia, pp 1–17

Myers L, Sirois MJ (2006) Spearman correlation coefficients, differences between. In: Encyclopedia of statistical sciences. Wiley, Hoboken, NJ

Nagai YS, Sakulsingharoj C, Edwards GE et al (2009) Control of starch synthesis in cereals: metabolite analysis of transgenic rice expressing an up-regulated cytoplasmic ADP-glucose pyrophosphorylase in developing seeds. Plant Cell Physiol 50:635–643. https://doi.org/10.1093/pcp/pcp021

Nestel P, Bouis HE, Meenakshi JV, Pfeiffer W (2006) Biofortification of staple food crops. J Nutr. 136(4):1064–1067

Newell-McGloughlin M (2008) Nutritionally improved agricultural crops. Plant Physiol 147:939–953

Obata T, Witt S, Lisec J et al (2015) Metabolite profiles of maize leaves in drought, heat, and combined stress field trials reveal the relationship between metabolism and grain yield. Plant Physiol 169:2665–2683. https://doi.org/10.1104/pp.15.01164

Ogbaga CC, Stepien P, Dyson BC et al (2016) Biochemical analyses of Sorghum varieties reveal differential responses to drought. PLoS One 11(5):e0154423. https://doi.org/10.1371/journal.pone.0154423

Pegasus HT (2007) ChromaTOF Software Instruction Manual, Version 3.3. http://www.scielo.br/scielo.php?script=sci_nlinks&ref=000133&

Peng L, Zhao Y, Wang H et al (2016) Comparative metabolomics of the interaction between rice and the brown planthopper. Metabolomics 12:1–15. https://doi.org/10.1007/s11306-016-1077-7

Peng Z, He S, Gong W et al (2018) Integration of proteomic and transcriptomic profiles reveals multiple levels of genetic regulation of salt tolerance in cotton. BMC Plant Biol 18:128. https://doi.org/10.1186/s12870-018-1350-1

Perez-de-Castro MA, Vilanova S, Canizares J et al (2012) Application of genomic tools in plant breeding. Curr Genomics 13:179–195. https://doi.org/10.2174/138920212800543084

Pérez-Massot E, Banakar R, Gómez-Galera S et al (2013) The contribution of transgenic plants to better health through improved nutrition: opportunities and constraints. Genes Nutr 8:29–41

Perkowski J, Stuper K, Buśko M et al (2012) Differences in metabolomic profiles of the naturally contaminated grain of barley, oats and rye. J Cereal Sci 56:544–551. https://doi.org/10.1016/j.jcs.2012.07.012

Piasecka A, Kachlicki P, Stobiecki M (2019) Analytical methods for detection of plant metabolomes changes in response to biotic and abiotic stresses. Int J Mol Sci 20(2):379

Piccioni F, Capitam D, Zolla L, Mannina L (2009) NMR metabolic profiling of transgenic maize with the Cry1A(b) gene. J Agric Food Chem 57:6041–6049. https://doi.org/10.1021/jf900811u

Pinu FR, Beale DJ, Paten AM et al (2019) Systems biology and multi-omics integration: viewpoints from the metabolomics research community. Metabolites 9(4):76. https://doi.org/10.3390/metabo9040076

Qi X, Xu W, Zhang J et al (2017) Physiological characteristics and metabolomics of transgenic wheat containing the maize C4 phosphoenolpyruvate carboxylase (PEPC) gene under high temperature stress. Protoplasma 254:1017–1030. https://doi.org/10.1007/s00709-016-1010-y

Rai A, Saito K, Yamazaki M (2017) Integrated omics analysis of specialized metabolism in medicinal plants. Plant J 90:764–787. https://doi.org/10.1111/tpj.13485

Ramautar R, De Jong GJ (2014) Recent developments in liquid-phase separation techniques for metabolomics. Bioanalysis 6:1011–1026

Rao J, Cheng F, Hu C et al (2014) Metabolic map of mature maize kernels. Metabolomics 10:775–787. https://doi.org/10.1007/s11306-014-0624-3

Razzaq A, Sadia B, Raza A et al (2019) Metabolomics: a way forward for crop improvement. Metabolites 9(12):303. https://doi.org/10.3390/metabo9120303

Rivera SM, Vilaró F, Zhu C et al (2013) Fast quantitative method for the analysis of carotenoids in transgenic maize. J Agric Food Chem 61:5279–5285. https://doi.org/10.1021/jf400694z

Rontein D, Basset G, Hanson AD (2002) Metabolic engineering of osmoprotectant accumulation in plants. Metab Eng 4:49–56. https://doi.org/10.1006/mben.2001.0208

Rossi M, Bermudez L, Carrari F (2015) Crop yield: challenges from a metabolic perspective. Curr Opin Plant Biol 25:79–89

Sana TR, Fischer S, Wohlgemuth G et al (2010) Metabolomic and transcriptomic analysis of the rice response to the bacterial blight pathogen Xanthomonas oryzae pv. Oryzae. Metabolomics 6:451–465. https://doi.org/10.1007/s11306-010-0218-7

Sawada K, Hasegawa M, Tokuda L et al (2004) Enhanced resistance to blast fungus and bacterial blight in transgenic rice constitutively expressing OsSBP, a rice homologue of mammalian selenium-binding proteins. Biosci Biotechnol Biochem 68:873–880. https://doi.org/10.1271/bbb.68.873

Seaver SMD, Lerma-Ortiz C, Conrad N et al (2018) Plant SEED enables automated annotation and reconstruction of plant primary metabolism with improved compartmentalization and comparative consistency. Plant J 95:1102–1113. https://doi.org/10.1111/tpj.14003

Seybold H, Demetrowitsch T, Hassani MA, et al (2019) Hemibiotrophic fungal pathogen induces systemic susceptibility and systemic 1 shifts in wheat metabolome and microbiome composition bioRxiv:702373. doi:https://doi.org/10.1101/702373

Sharma A, Shahzad B, Kumar V et al (2019) Phytohormones regulate accumulation of osmolytes under abiotic stress. Biomol Ther 9:285. https://doi.org/10.3390/biom9070285

Sharma K, Sarma S, Bohra A, Mitra A, Sharma NK, Kumar A (2018) Plant metabolomics: an emerging technology for crop improvement. In: Çelik O (ed) New visions in plant science, 1st edn. IntechOpen, London, pp 65–79

Shavit R, Batyrshina ZS, Dotan N, Tzin V (2018) Cereal aphids differently affect benzoxazinoid levels in durum wheat. PLoS One 13:e0208103. https://doi.org/10.1371/journal.pone.0208103

Shelden MC, Dias DA, Jayasinghe NS et al (2016) Root spatial metabolite profiling of two genotypes of barley (Hordeum vulgare L.) reveals differences in response to short-term salt stress. J Exp Bot 67:3731–3745. https://doi.org/10.1093/jxb/erw059

Shen Q, Fu L, Dai F et al (2016) Multi-omics analysis reveals molecular mechanisms of shoot adaption to salt stress in Tibetan wild barley. BMC Genomics 17:889. https://doi.org/10.1186/s12864-016-3242-9

Shepherd LVT, McNicol JW, Razzo R et al (2006) Assessing the potential for unintended effects in genetically modified potatoes perturbed in metabolic and developmental processes. Targeted analysis of key nutrients and anti-nutrients. Transgenic Res 15:409–425. https://doi.org/10.1007/s11248-006-0012-5

Shi T, Zhu A, Jia J et al (2020) Metabolomics analysis and metabolite-agronomic trait associations using kernels of wheat (Triticum aestivum) recombinant inbred lines. Plant J 103:279–292. https://doi.org/10.1111/tpj.14727

Shin YM, Park HJ, Yim SD et al (2006) Transgenic rice lines expressing maize C1 and R-S regulatory genes produce various flavonoids in the endosperm. Plant Biotechnol J 4:303–315. https://doi.org/10.1111/j.1467-7652.2006.00182.x

Shivay YS, Prasad R, Pal M (2015) Effects of source and method of zinc application on yield, zinc biofortification of grain, and Zn uptake and use efficiency in chickpea (Cicer arietinum L.). Commun Soil Sci Plant Anal 46:2191–2200. https://doi.org/10.1080/00103624.2015.1069320

Shulaev V, Cortes D, Miller G, Mittler R (2008) Metabolomics for plant stress response. Physiol Plant 132:199–208

Silva JCF, Teixeira RM, Silva FF et al (2019) Machine learning approaches and their current application in plant molecular biology: a systematic review. Plant Sci 284:37–47

Smith CA, O'Maille G, Want EJ et al (2005) METLIN: a metabolite mass spectral database. Ther Drug Monit 27(6):747–751

Sonnewald U, Fernie AR (2018) Next-generation strategies for understanding and influencing source–sink relations in crop plants. Curr Opin Plant Biol 43:63–70

Soriano IR, Asenstorfer RE, Schmidt O, Riley IT (2004) Inducible flavone in oats (Avena sativa) is a novel defense against plant-parasitic nematodes. Phytopathology 94:1207–1214. https://doi.org/10.1094/PHYTO.2004.94.11.1207

Steinfath M, Strehmel N, Peters R et al (2010) Discovering plant metabolic biomarkers for phenotype prediction using an untargeted approach. Plant Biotechnol J 8:900–911. https://doi.org/10.1111/j.1467-7652.2010.00516.x

Suharti WS, Nose A, Zheng SH (2016) Metabolomic study of two rice lines infected by Rhizoctonia solani in negative ion mode by CE/TOF-MS. J Plant Physiol 206:13–24. https://doi.org/10.1016/j.jplph.2016.09.004

Sun CX, Gao XX, Li MQ et al (2016) Plastic responses in the metabolome and functional traits of maize plants to temperature variations. Plant Biol 18:249–261. https://doi.org/10.1111/plb.12378

Sun X, Weckwerth W (2012) COVAIN: a toolbox for uni- and multivariate statistics, time-series and correlation network analysis and inverse estimation of the differential Jacobian from metabolomics covariance data. Metabolomics 8:81–93. https://doi.org/10.1007/s11306-012-0399-3

Takahashi H, Hayashi M, Goto F et al (2006) Evaluation of metabolic alteration in transgenic rice overexpressing dihydroflavonol-4-reductase. Ann Bot 98:819–825. https://doi.org/10.1093/aob/mcl162

Thiele I, Palsson B (2010) A protocol for generating a high-quality genome-scale metabolic reconstruction. Nat Protoc 5:93–121. https://doi.org/10.1038/nprot.2009.203

Tissier A, Ziegler JVT (2015) Specialized plant metabolites: diversity and biosynthesis. In: Krauss GJ Nies DH (ed) Ecological biochemistry: environmental and interspecies interactions. Wiley, Weinheim, Germany, pp 15–37

Tulchinsky TH (2010) Micronutrient deficiency conditions: global health issues. Public Health Rev 32:243

Turner MF, Heuberger AL, Kirkwood JS et al (2016) Non-targeted metabolomics in diverse Sorghum breeding lines indicates primary and secondary metabolite profiles are associated with plant biomass accumulation and photosynthesis. Front Plant Sci 7:953. https://doi.org/10.3389/fpls.2016.00953

van Zonneveld M, Rakha M, Tan S, et al (2019) Understanding patterns of abiotic and biotic stress resilience to unleash the potential of crop wild relatives for climate-smart legume breeding. bioRxiv 596072.

Vasmatkar P, Kaur K, Pannu PPS et al (2019) Unraveling the metabolite signatures of maize genotypes showing differential response towards southern corn leaf blight by 1H-NMR and FTIR spectroscopy. Physiol Mol Plant Pathol 108:101441. https://doi.org/10.1016/j.pmpp.2019.101441

Voit EO (2017) The best models of metabolism. Wiley Interdiscip Rev Syst Biol Med 9(6). https://doi.org/10.1002/wsbm.1391

Walley JW, Sartor RC, Shen Z et al (2016) Integration of omic networks in a developmental atlas of maize. Science 353:814–818. https://doi.org/10.1126/science.aag1125

Wang JP, Matthews ML, Williams CM et al (2018) Improving wood properties for wood utilization through multi-omics integration in lignin biosynthesis. Nat Commun 9:1–16. https://doi.org/10.1038/s41467-018-03863-z

Watanabe KN, Sassa Y, Suda E et al (2005) Global political, economic, social and technological issues on transgenic crops. Plant Biotechnol 22:515–522

Waters BM, Grusak MA (2008) Quantitative trait locus mapping for seed mineral concentrations in two Arabidopsis thaliana recombinant inbred populations. New Phytol 179:1033–1047. https://doi.org/10.1111/j.1469-8137.2008.02544.x

Weckwerth W (2011) Unpredictability of metabolism-the key role of metabolomics science in combination with next-generation genome sequencing. Anal Bioanal Chem 400:1967–1978

Weckwerth W (2019) Toward a unification of system-theoretical principles in biology and ecology—the stochastic Lyapunov matrix equation and its inverse application. Front Appl Math Stat 5:29. https://doi.org/10.3389/fams.2019.00029

Wei W, Li D, Wang L et al (2013) Morpho-anatomical and physiological responses to waterlogging of sesame (Sesamum indicum L.). Plant Sci 208:102–111. https://doi.org/10.1016/j.plantsci.2013.03.014

Wen W, Li K, Alseekh S et al (2015) Genetic determinants of the network of primary metabolism and their relationships to plant performance in a maize recombinant inbred line population. Plant Cell 27:1839–1856. https://doi.org/10.1105/tpc.15.00208

White PJ, Broadley MR (2005) Biofortifying crops with essential mineral elements. Trends Plant Sci 10:586–593

White PJ, Broadley MR (2009) Biofortification of crops with seven mineral elements often lacking in human diets - Iron, zinc, copper, calcium, magnesium, selenium and iodine. New Phytol 182:49–84

Winning H, Roldán-Marín E, Dragsted LO et al (2009) An exploratory NMR nutri-metabonomic investigation reveals dimethyl sulfone as a dietary biomarker for onion intake. Analyst 134:2344–2351. https://doi.org/10.1039/b918259d

Wishart DS (2011) Advances in metabolite identification. Bioanalysis 3:1769–1782

Wu D, Cai S, Chen M et al (2013) Tissue metabolic responses to salt stress in wild and cultivated barley. PLoS One 8:e55431. https://doi.org/10.1371/journal.pone.0055431

Wu J, Yu H, Dai H et al (2012) Metabolite profiles of rice cultivars containing bacterial blight-resistant genes are distinctive from susceptible rice. Acta Biochim Biophys Sin Shanghai 44:650–659. https://doi.org/10.1093/abbs/gms043

Xia J, Psychogios N, Young N, Wishart DS (2009) MetaboAnalyst: a web server for metabolomic data analysis and interpretation. Nucleic Acids Res 37:W652. https://doi.org/10.1093/nar/gkp356

Xin W, Zhang L, Zhang W et al (2019) An integrated analysis of the rice transcriptome and metabolome reveals root growth regulation mechanisms in response to nitrogen availability. Int J Mol Sci 20(23):5893. https://doi.org/10.3390/ijms20235893

Xu S, Hu C, Hussain S et al (2018) Metabolomics analysis reveals potential mechanisms of tolerance to excess molybdenum in soybean seedlings. Ecotoxicol Environ Saf 164:589–596. https://doi.org/10.1016/j.ecoenv.2018.08.062

Yamaguchi-Shinozaki K, Shinozaki K (2006) Transcriptional regulatory networks in cellular responses and tolerance to DEHYDRATION and cold stresses. Annu Rev Plant Biol 57:781–803. https://doi.org/10.1146/annurev.arplant.57.032905.105444

Yang L, Fountain JC, Ji P et al (2018) Deciphering drought-induced metabolic responses and regulation in developing maize kernels. Plant Biotechnol J 16:1616–1628. https://doi.org/10.1111/pbi.12899

Yang Z, Nakabayashi R, Okazaki Y et al (2014) Toward better annotation in plant metabolomics: isolation and structure elucidation of 36 specialized metabolites from Oryza sativa (rice) by using MS/MS and NMR analyses. Metabolomics 10:543–555. https://doi.org/10.1007/s11306-013-0619-5

Zhang J, Luo W, Zhao Y et al (2016) Comparative metabolomic analysis reveals a reactive oxygen species-dominated dynamic model underlying chilling environment adaptation and tolerance in rice. New Phytol 211:1295–1310. https://doi.org/10.1111/nph.14011

Zhang N, Venkateshwaran M, Boersma M et al (2012) Metabolomic profiling reveals suppression of oxylipin biosynthesis during the early stages of legume-rhizobia symbiosis. FEBS Lett 586:3150–3158. https://doi.org/10.1016/j.febslet.2012.06.046

Zhang W, Chang J, Lei Z et al (2014) MET-COFEA: a liquid chromatography/mass spectrometry data processing platform for metabolite compound feature extraction and annotation. Anal Chem 86:6245–6253. https://doi.org/10.1021/ac501162k

Zhang Y, ming MX, chun WX et al (2017) UPLC-QTOF analysis reveals metabolomic changes in the flag leaf of wheat (Triticum aestivum L.) under low-nitrogen stress. Plant Physiol Biochem 111:30–38. https://doi.org/10.1016/j.plaphy.2016.11.009

Zhou J, Ma C, Xu H et al (2009) Metabolic profiling of transgenic rice with cryIAc and sck genes: an evaluation of unintended effects at metabolic level by using GC-FID and GC-MS. J Chromatogr B Anal Technol Biomed Life Sci 877:725–732. https://doi.org/10.1016/j.jchromb.2009.01.040

Zhou S, Zhang YK, Kremling KA et al (2019) Ethylene signaling regulates natural variation in the abundance of antifungal acetylated diferuloylsucroses and Fusarium graminearum resistance in maize seedling roots. New Phytol 221:2096–2111. https://doi.org/10.1111/nph.15520

Zhu L, Zhou Y, Li X et al (2018) Metabolomics analysis of soybean hypocotyls in response to Phytophthora sojae infection. Front Plant Sci 9:1530. https://doi.org/10.3389/fpls.2018.01530

# Safety and Ethics in Omics Biology

# 12

Praveen Kumar Sharma, Ashwani K. Rai, and Naveen K. Sharma

**Abstract**

The recent advances in "omics" technologies have impacted biological science research in many ways. This includes the realm diversity studies, physiological, biochemical, and molecular level researches to the development of breakthrough products and approaches. They are strengthening our attempts to fight with emerging diseases, global food crisis, environmental degradation, production of quality value-added products values, and in better understanding of the underlying mechanism(s) of host–pathogen interactions and stress (both biotic and abiotic) tolerance in plants and microbes. However, many of these techniques are still under refinement and resulting data need careful integration for attending meaningful conclusions. Although not a big concern, but we must be attentive to some of the biosafety measures and ethical concerns associated with these techniques, which we have tried to discuss in this chapter.

**Keywords**

Bioinformatics · Ethics · Omics · Technologies · Productivity · Safety · Stress tolerance · Sustainability

P. K. Sharma
Department of Life Sciences, Central University of Jharkhand, Ranchi, Jharkhand, India

A. K. Rai
Centre for Advance Study in Botany, Banaras Hindu University, Varanasi, Uttar Pradesh, India

N. K. Sharma (✉)
Department of Botany, Indira Gandhi National Tribal University, Amarkantak, Madhya Pradesh, India
e-mail: naveen.sharma@igntu.ac.in





A. Kumar et al. (eds.), *Omics Technologies for Sustainable Agriculture and Global Food Security Volume 1*, https://doi.org/10.1007/978-981-16-0831-5_12

## 12.1 Introduction: The Omics and Emergence of Omics Technologies

The suffix "omics" is derived from Latin Suffix "ome" meaning mass or many and "omics" in biological sciences loosely implies a comprehensive, or global, assessment of a set of molecules (http://omics.org/). Table 12.1 summarizes various "omics" approaches and related sub-fields for each case. Genomics was the first member of the "omics" family. Subsequent technological advances have led to the rapid diversification of the "omics" field, which has grown both in type as well as in quality and application of data. The variety of data generated by multiple "omics" approaches (multi-omics) had led to the establishment of an inter-disciplinary field study known as Systems Biology (Breitling 2010) that creates holistic understanding of biological organization and its function.

Rapid advancements in genomics are based on techniques such as next-generation sequencing (NGS), also known as high-throughput sequencing (Mardis

**Table 12.1** Types and subtypes of omics

| Omics type | Study of sub-fields | |
|---|---|---|
| Genomics | Structure and function of organism's DNA sequence, including coding and non-coding part of gene and regulatory regions. Also include evolution, mapping, and editing of genomes | Structural genomics<br>Functional genomics<br>Comparative genomics<br>Mutation genomics |
| Transcriptomics | Complete set of RNA transcripts that are produced by the genome, snapshot of expression of genes in a biological system at a specific point in time | |
| Proteomics | All the proteins produced or modified in biological system at a specific point in time | Protein mining<br>Expression profiling<br>Functional proteomics<br>Structural proteomics<br>Protein networks<br>Post translational modifications |
| Metabolomics | Systematic identification and quantification of the metabolic products of a biological system at a specific point in time | Exometabolomics<br>Metabonomics |
| Epigenomics | Complete set of epigenetic modifications on the genetic material of a biological system at a specific point in time | DNA modifications<br>Histone modifications |
| Ionomics | Quantitative measurement of the total elemental composition (metals, metalloids, non-metals, etc.) in a living organism, and subsequent changes in their production under varied external stimuli using techniques such as Inductively Coupled Plasma-Mass Spectrometry (ICP-MS) and ICP-Atomic Emission Spectrometry (ICP-AES) (Salt et al. 2008) | Physiology and molecular mechanisms<br>Ecological functions |

2008). This encompasses various sequencing technologies allowing rapid and cost-effective sequencing of DNA and RNA over the previous techniques (e.g., Sanger sequencing). The NGS technologies include Illumina sequencing, Pyrosequencing, and Ion torrent Sequencing. The analyses of NGS data by bioinformatics tools allow discovering new genes and regulatory sequences. The RNA-Seq (abbreviation of RNA sequencing) (Wang et al. 2009) is another technique with multiple applications such as in gene expression profiling, spliced variants of transcripts, post-translational modifications, SNPs, and quantity of RNA.

Proteomics is a relatively new approach to that of the genomics, but historically protein sequencing is older than the DNA sequencing. These sequencing technologies are comparatively slow, labor intensive and require a lot of starting material. During the last 50 years, mass spectroscopy was the choice for proteome analysis, but had its own limitations and challenges. Subsequently, the proteome analysis was greatly supported by the development of MALDI (Matrix-assisted laser desorption/ionization), which made mass spectrometry suitable for high-throughput proteomic studies. The recent development in the field of proteomics analysis is represented by SWATH MS based proteomics, which is a data independent acquisition (DIA) method, complementing traditional mass spectrometry-based proteomics techniques. This allows a complete recording of all detectable peptides present in a biological sample (Ludwig et al. 2018).

The study of complete set of epigenetic modifications called epigenomics is largely based on the genomic and proteomic tools (Wang and Chang 2018). Other epigenomic tools are - Chromatin immuno-precipitation (ChIP), ChIP sequencing (ChIP-seq), DNase-seq, fluorescent in situ hybridization (FISH), chromosome conformation capture (3C), chromatin interaction analysis by paired-end tag sequencing (ChIA-PET) and single cell RNASequencing (scRNA-seq), etc. (McKeown and Spillane 2014; Sati and Cavalli 2017).

Metabolomics is for comprehensive analysis of metabolites in a biological specimen, is a technology of an enormous economic value. Traditional analysis of metabolites depended on the chromatographic and spectrometry techniques. The current high-throughput technologies associated with metabolomics include liquid chromatography–mass spectrometry (LC-MS), UPLC-MS, and GC-MS (Nassar et al. 2017; Beale et al. 2018; Pinu et al. 2019). These techniques can measure tens to hundreds of metabolites with excellent precision. Besides quantifying the amount of a compound in a sample, these technologies have the capacity of identifying unknown molecules and determining the structure of molecules.

The rapid development in the field of omics could not be possible without the incredible advancement in the computational capability that included both storage, processing and phenomenal growth in bioinformatics. Large-scale data acquisition, storage, integration, and analysis have been made possible by these developments. Although the enhanced availability and accessibility of large-scale omics datasets have revolutionized our understanding of biological systems and processes, but it has also generated newer opportunities and challenges for biologists to ponder upon.

## 12.2 Uses and Achievements of Plant Omics

Like any other system in biological sciences the *Omics* approach has enhanced our understanding of plant systems, even in non-model species (Zhang et al. 2018). The resulting advancements in these technologies have created new datasets for plant species, which would be helpful in supporting food, health, energy securities, while preserving and remediating the environment on the other hand. Hittalmani et al. (2017) suggested for the use of Omics in enhancing the nutritional values of food. Besides, these technologies could assist in learning more about the genetic and biochemical pathways to be used for production of beneficial metabolites (Deborde et al. 2017). Genetic analysis of stilbenoid profiles in grapevine stems revealed a major mQTL hotspot on chromosome 18 associated with disease-resistance motifs (Teh et al. 2019). Integration of omics data may contribute to the identification of genes and pathways responsible for important agronomic phenotypes (Zhu et al. 2018) and disease resistance (Bhadauria 2016). Omics technologies could also be harnessed for novel insight into the origin of well-known domesticate plants as well as potential targets for further improvement of crops by pin pointing underutilized genetic traits in highly domesticated plants (Giovannoni 2018). Application of high-throughput genotyping technologies to large germplasm collections may contribute immensely in identification of novel alleles from diverse sources for effective use to breeding purposes (Dwivedi et al. 2017). Furthermore, omics technologies may be applied for identifying abiotic stress tolerance, genetic loci characterization, mechanism elucidation in varieties, landraces, and wild relatives of crops for genetic modification (GM) of current, high yielding elite cultivars (Rabara et al. 2014; Chaudhary et al. 2019).

Applications of omics may facilitate the analysis of non-model organisms and rapidly generate a large amount of novel data. This makes them an attractive option for studying poorly characterized interactions (Windram et al. 2014) such as various host–pest and host–pathogen interactions (Barah and Bones 2015; Mishra et al. 2019). Single nucleotide polymorphisms (SNPs) markers were used in 341 tropical maize lines to test the genetic basis of resistance of maize against the multiple insect pests. Similarly, a multi-locus genome-wide association study (GWAS) revealed that multiple quantitative trait nucleotides (QTNs) are responsible for the resistance. Such studies may contribute to the development of combined insect resistance in maize (Badji et al. 2020 and references therein).

The whole genome analysis of endophytic microbes could provide a greater understanding of their ecology and evolution, identifying the factors influencing their growth and survival and other related metabolic mechanisms (Kaul et al. 2016). An understanding of the impacts of beneficial microbes and the associated mechanism on crop plants is much required to enhance food production. Likewise, the therapeutic potential of plants is well known, but initially there was an apparent lack of interest from pharmaceutical industry due to a lengthy and tedious process of discovering, isolating, and characterizing the lead molecules/products. But, since early 2000, use of techniques such as LC-MS, ultra-high-performance liquid chromatography (UHPLC), NMR, and associated bioinformatics has fueled up the discovery of such molecules and products. Apart from metabolic profiling, genome data may also be used to evaluate the biosynthetic potential of any organism. The combined use of genomic and metabolomic methods has led to the identification of natural products from cyanobacteria to angiosperms (e.g., *Arabidopsis thaliana, Oryza sativa, and Catharanthus roseus)* (Wolfender et al. 2019).

A Search of "Plant AND Omics" in PubMed® for the last 1 year (2019 – 2020, i.e., up to the writing of this chapter) revealed that majority of publications in the field are following the same pattern as discussed above. Few important developments in the field of database, web resource, and other bioinformatic tools are listed in Table 12.2. A search of literature database also suggested that most of the researches in the field of plant omics are directed towards understanding the metabolic pathways, elucidation of the effect of various stresses (both biotic and abiotic), nutritional state and hormonal signaling on plant proteome and metabolome, and transcriptional control mechanisms (refer Table 12.3). Other major applications of omics included analysis of food microbial ecology (Xie et al. 2019), understanding of evolutionary metabolic divergence (Xu et al. 2019; Levsh et al. 2019), accessing the effect and mode of temperature dependent virulence and pathogenicity (Félix et al. 2019), bioprospecting of plants for medicinal use (Gonulalan et al. 2020), and understanding the role of non-coding RNA in fruit ripening (Wang et al. 2020b).

## 12.3 Challenges and Issues

The "Arabidopsis Research and Training for the 21st Century" (ART-21) funded by the National Science Foundation (NSF), USA, conducted a series of workshops to explore the areas/issues that are important for research and training in plant biology. The outcome presented in the form of a white paper (Argueso et al. 2019) suggested the directions for research and training in plant omics also. The white paper lays out certain "Big question" that could coalesce future research in plant biology. Such as (a) understanding the mechanism of receptor and ligand interactions and their role in regulating plant transcriptome activity, (b) use of omics to understand the programmed development in plant system and (c) finding out the underlying cause of variation within the plant species. The white paper should be adopted as trendsetter for future plant omics research. Although the knowledge and insights gained from omics studies have facilitated many discoveries and conceptual advancements in plant biology, however, just like any other technology this field also offers some challenges too. A few of them need immediate attention such as reproducibility of omics data acquisition, data analysis and to define the best practices (Buesen et al. 2017).

### 12.3.1 Challenges in Data Integration

The appropriate integration of datasets obtained from high-throughput omics research is the first and major problem associated with these technologies

**Table 12.2** Recently developed plant "omics" web resources, databases, frameworks, and other tools

| Name | Web address | References |
|---|---|---|
| TPIA (tea plant information archive) | http://tpia.teaplant.org/ | Xia et al. (2019) |
| Knowledge Base commons (KBCommons) v1.1 (multi-level OMICS web resource for data retrieval, sharing, analysis and visualization) | http://kbcommons.org/ | Zeng et al. (2019) |
| Plant Regulomics (Interface for retrieving upstream regulators from plant multi-omics data) | http://bioinfo.sibs.ac.cn/plant-regulomics | Ran et al. (2020) |
| AppleMDO (A multi-dimensional Omics database for apple co-expression networks and chromatin states) | http://bioinformatics.cau.edu.cn/AppleMDO/ | Da et al. (2019) |
| Plant Reactome (knowledgebase and resource for comparative pathway analysis) | https://plantreactome.gramene.org | Naithani et al. (2020) |
| JCDB: (knowledge base for *Jatropha curcas*) | http://jcdb.xtbg.ac.cn. | Zhang et al. (2019a) |
| QTL.gCIMapping.GUI v2.0 (software packages to identify all kinds of omics QTLs.) | https://cran.r-project.org/web/packages/ or https://bigd.big.ac.cn/biocode/tools/7078/releases/27 | Zhang et al. (2019c) |
| MaGenDB (A functional genomics hub for Malvaceae plants) | http://magen.whu.edu.cn | Wang et al. (2020c) |
| MBKbase-rice (Rice sub-database of an integrated omics knowledgebase) | www.mbkbase.org/rice | Peng et al. (2020) |
| MAPPS (metabolic network analysis and pathway prediction server) | https://mapps.lums.edu.pk | Riaz et al. (2020) |
| GOMCL (lightweight python toolkit, identifies clusters within a list of GO terms using the Markov clustering (MCL) algorithm, based on the overlap of gene members between GO terms | https://github.com/Guannan-Wang/GOMCL and www.lsugenomics.org. | Wang et al. (2020a) |
| ZEAMAP (comprehensive database adapted to the maize multi-Omics era | http://www.zeamap.com/ | Gui et al. (2020) |
| RGPDB (database of root-associated genes and promoters in maize, soybean, and sorghum) | https://crri.unl.edu/databases and http://sysbio.unl.edu/RGPDB. | Moisseyev et al. (2020) |
| LeGOO (knowledge database for the model legume *Medicago truncatula*) | https://www.legoo.org | Carrï Re et al. (2020) |

**Table 12.3** Representative recent achievements in the field of plant "omics"

| Achievements | References |
|---|---|
| Assembly of a draft genome of *Polygonum cuspidatum* (Japanese knotweed) producer of Stilbenes and quinones using Illumina sequencing technology | Zhang et al. (2019b) |
| Assembled and annotated draft genome of *Pleurozium schreberi*: A Pleurocarpous feather Moss | Pederson et al. (2019) |
| eQTL analyses of genes affecting cadmium content in cultivated rice | Lee et al. (2019) |
| Study of regulatory mechanism of secondary metabolites production during the flowering stages of *Lonicera japonica* Thunb., a native of East Asia | Yang et al. (2019) |
| Host-targeted transcriptomics detected the taxonomic and functional diversity of root microbiota in tomato (*Solanum lycopersicum*) growing on different native soils | Chialva et al. (2019) |
| DNA methylation associates with the expression of the epialleles AT2G34100, AT4G09360, *LSU4,* and AT5G56910 regulate the resistance of *Arabidopsis thaliana* to local pests and diseases, and help in the adaptation to local environments | Mei et al. (2020) |
| Phosphoproteomic analysis of plant DNA damage signaling pathways finds a functional role of histone H2AX phosphorylation in plant growth under genotoxic stress | Waterworth et al. (2019) |
| Hexaploid monocot wheat responds to beneficial or pathogenic microorganisms and prolongs the onset of take-all disease through modulation of cell reprogramming and signaling events | Kang et al. (2019) |
| Sugar cane treated with GA3, a growth hormone revealed total 1516 differentially expressing transcripts in bottom internodes and 1589 in top internodes. KEGG (enrichment) analysis grouped theses transcripts into 153 plant-related functional categories. Starch and sucrose metabolizing genes showed maximum fold change of 5.0 and 3.0 among top and bottom internodal samples | Chandra et al. (2019) |
| Label-free quantitative proteomics and metabolomics data (GC-TOF-MS), using a network-based approach, tested leaf from two contrasting commercial varieties of sugarcane, CTC15 (tolerant) and SP90–3414 (susceptible), to observe changes metabolism in response to drought | Budzinski et al. (2019) |
| The metabolome and transcriptome of 11 tea cultivars were profiled and then illustrated by a weighted gene co-expression network analysis (WGCNA) to interpret metabolomic flux, prediction of gene functions, and mine key regulators involved in the flavonoid biosynthesis pathway | Zheng et al. (2019) |
| A tandem mass tag (TMT)-labeled mass spectrometry-based quantitative proteomic analysis of *Capsicum frutescens* leaves and apical meristems in healthy and broad mite infcted condition demonstrated differential regulation of 1677 proteins | Patavardhan et al. (2020) |
| Plant derived smoke promotes soybean growth. The proteomic and metabolomic analysis revealed importance of arginine metabolism and ubiquitin proteasome pathway | Zhong et al. (2020) |
| Cause of post-harvest disorders in blackberries was illustrated by using metabolomics | Kim et al. (2019) |
| Deciphering of sub-cellular plant metabolism (chloroplast, cytosol, vacuole, and mitochondria) in *A. thaliana* | Fürtauer et al. (2019) |

(Fukushima et al. 2009; Tardieu et al. 2017). Development of various omics domains has created a layered architecture in omics research. A single-omic technique is one layer, which captures data for a small subset of the components of any particular pathway. Evolution of the layered system has created a necessity for integrating different omics layers in order to have proper meaningfulness and seamless use of data. Since omics layers are inherently variable, distinguishing the signal from the noise in datasets requires reproducible results as well as advanced statistical/machine-learning approaches for analyses (Yuan et al. 2017).

The data integration problem can further be appreciated from the fact that changes in the transcriptome or proteome do not always reciprocated to the results obtained from metabolomic findings (Ryan and Robards 2006). The interlinked nature of physiological processes (Kohl et al. 2010), feedback regulations, epistatic and hypostatic nature of gene interactions, epigenetic processes modulating gene expression are some of the other sources of "uncertainty" operating at different levels of plant organization. This creates a blur to the predictability from single and lower level results (do Amaral and Souza 2017). In many cases, expression of nuclear genes can be altered without altering the total concentration of the signaling molecule in the cell as a whole (Tuteja and Mahajan 2007).

Apart from the above-mentioned concerns integration across the time is another challenge known as the fourth dimension. During the integration of resulting data, it is not necessary that the sampled system be/was in a steady state at a given moment, as each physiological process is a dynamic metabolic network (Toubiana et al. 2013). Use of data from such system and its/their integration based on static networks mathematical models that often bypass the network modulation over time is likely to be error prone. Thus, it is necessary to develop new methods to allow investigations of dynamic aspects of large-scale models (Medeiros et al. 2015).

### 12.3.2 Challenges and Issues with Individual Techniques

Our ability to characterize proteomes in a comprehensive and quantitative manner lags far behind to that of the genome and transcriptome. The principle bottleneck is isolation and low quantity of samples and lack of proper enrichment methods prior to the analysis. The proteome analysis further gets compromised due to the complexity created by post-translational modifications (PTM), including less understanding of the types and their spatio-temporal differences in expression and change in protein mass due to modifications. Developing robust methods to monitor the protein function at the sub-cellular level would provide important additional information particularly for building networks of activities (Alvarez and Naldrett 2016; Argueso et al. 2019).

Due to the uniqueness in the number and diversity of metabolites produced by plants, metabolomics is an exciting frontier in plant omics. Metabolites are not only the products useful for their commercial values, but also serve as regulatory molecules during metabolism and physiology. Functional characterization of even a single metabolite is a time-consuming task, undertaking this task at omics level makes it further difficult. This difficulty is not only due to the number of molecules but in many cases also because of lack of information concerning their regulation by signaling system, spatio-temporal distribution, and impact of various biotic and abiotic factors; all impacting the quantity and quality of sample to be analyzed (do Amaral and Souza 2017; Argueso et al. 2019).

The building blocks of metabolites are diverse, hence causing further complication in the metabolome analysis (a disadvantage). The identification and quantification of a given metabolite require a matching standard molecule with same chromatographic profile and mass spectrum. If a metabolite is novel or modified by a novel mechanism, standards for its identification and characterization are commercially unavailable. This requires custom syntheses, which is a major financial hurdle. To overcome/bypass this, metabolites are often matched with "fingerprint" obtained through LC-MS. The approach may computationally identify "test" metabolites, but lack explicit identification of their chemical/molecular structure, which can be elucidated by NMR. However, the low sensitivity and low throughput of NMR preclude its application in large-scale investigations (Argueso et al. 2019).

## 12.4 Ethics in Omics

The majority of new ideas often met with controversy and omics technologies and their applications are no exception. Though, they are simply a tool to understand the biological system at the molecular level, but ultimately this omics-generated knowledge has to be integrated into the broader understanding about the plants and their different level of assemblages. For too many people (super-skeptics) any advancement in science is alarming (no matter how much unfounded it may appear to a scientific mind), with the potential to do great harm to people and societies. They argue for and are much concerned about the wise and just use of new discoveries (including technologies). For such willful super-skeptics/denialists (make no confusion they are good in number fairly distributed in both vertical and horizontal layers of the societies) any form of science is a threat, and opposition of technology is merely part of a broader philosophy that science is bad. However, for majority people concerns are based on well-reasoned arguments, therefore, these views are important.

Ethics is a branch of philosophy dealing with values related to human conduct with respect to the rightness and wrongness of actions, the goodness and badness of motives and ends, and the moral principles of an individual (Reynnells 2004). The term bioethics used first in relation to ethical issues including policy and practice emerging from advances in biology and medicine. Different theories, principles, and approaches applied as a measure to bioethics are given in Table 12.4. For a researcher, aside major ethical concerns related to the use of organisms and the resulting socio-ecological impacts, other ethical issues such as adherence to the good research practice; a sense of social responsibility, observance of all regulatory frameworks, integrity, honesty, respect, fairness, etc., also deserve due to attention.

**Table 12.4** Theories in bioethics (https://iep.utm.edu/bioethic/#SH5a: after Jahn 2011)

| Theory/approach | Features |
|---|---|
| Deontological | Based on application of strict moral rules or norms, includes both religious and non-religious approach |
| Utilitarianism | Based on (1) the consequences of a given action are the measure of its moral quality, (2) the moral rightness and wrongness of actions are determined by the greatest possible utility for the greatest possible number of all sentient beings, (3) the consequences of a given action are evaluated with reference to a particular value. This particular prime value can be as follows: (a) promoting pleasure, or (b) avoiding pain, or (c) satisfaction of interests or considered preferences, or (iv) satisfaction of some objective criteria of Well-being, and so forth, and (4) maximize the total utility for all sentient beings affected |
| The four principle approach (Principlism) | Based on four ethical principles: (1) autonomy (acting independently without the influence or distortion of others), (2) non-maleficence (there is an obligation not to inflict harm on others), (3) beneficence (moral obligation to act for the benefit of others), and (4) justice (equitably distribute benefits, risks, costs, and resources) |
| Virtue ethics | The action is morally good if the person in question acts on the basis of the right motive as well as his or her action is based on a firm and good character or disposition |
| Casuistry | Based on depiction of the case, classification of the case and moral judgment |
| Feminist bioethics | Make a well-informed ethical decision that is not gender biased and appeal to important core values |

For one and all sections of society (both, general public and scientists), one major concern against the omics research is that the data obtained from the wide use of these technologies may be used to alter naturalness of the environment and the system studied (say, for example, plants). In general, be naturalists, environmentalist, and policy makers (hopefully so) their ethical considerations revolve largely around the impact of these technologies on biodiversity and ownership of resources.

Out of the three subdivisions of bioethics, i.e., medical, animal, and environmental ethics, the third one (i.e., environmental ethics) deals with moral dimension of relationship between human beings and non-human nature (i.e., animals, plants, natural resources, ecosystem, landscape as well as biosphere and cosmos) (*https://iep.utm.edu/bioethic/*). This suggests that an ethical concern regarding plants is very much a part of environmental ethics also. The concept of "plant ethics" first appeared in the Swiss constitution dealing about the dignity or integrity of plants. This still dominates the philosophical discussions centered on the subject (Pouteau 2014). But so far, we do not have any regulatory mechanism for ethics relating to plant based research as we see for animal research or medical research like animal ethics committees or human ethical committees, which are there to validate and approve the study protocols based on ethical considerations.

For example, historically the plant resources, particularly agricultural resources have been treated as a global public good. This notion continued up to the mid of twentieth century thereafter, use of breeding technologies has resulted in spectacular gains, which nevertheless came forth with their own problems. The selective breeding of plants aimed to develop resistance varieties against pests or diseases, also lead to the rapid development of resistance amongst pests and disease-causing organisms. To avoid such problems the breeders always have to create new genetic variations by introgressive hybridization. The main source of genetic variation is either different populations of the crop in question or wild plants. The omics has potential to find out such variations in large quantity and more quickly than the traditional genetic techniques. Greater emphasis is being given to the use of metabolomics to screen out the important metabolites as well as to gather information about the new sources. If the sources of such variations are wild or endemic species and intended to be used for commercial gain, this poses a problem of ownership. Easy access to bioresources is important for the socio-economic growth of a country. Therefore, implementation of existing international legal frameworks (and subsequent modifications) be given highest priority and be followed in letter and spirit in plant omics research also. These frameworks are necessary to ensure adequate recognition of the contributions of different stockholders.

The Food and Agriculture Organization (FAO) in the year 1983 promulgated the International Undertaking on Plant Genetic Resources. It was revised several times, resulting in the International Treaty on Plant Genetic Resources for Food and Agriculture (http://www.fao.org/plant-treaty/overview/texts-treaty/en/) and came into effect from 2006. In 1993, the Convention on Biological Diversity (CBD) also created a legal binding framework for the access and benefit-sharing (ABS) related to biological resources. It accepted the sovereign rights of nations over their biological resources and reaffirmed that the nations have the authority to determine access norms to their genetic resources. The Nagoya Protocol on ABS was adopted on 29th October 2010, and entered into force on 12th October 2014, is a supplementary agreement to the CBD. It provides legal framework for the effective implementation of the fair and equitable sharing of benefits arising out of the utilization of genetic resources (https://www.cbd.int/abs/about/default.shtml/#objective). The existing rules and regulations apply over national and international levels, but little legal and ethical guidance exists on how individual researchers ought to deal with local communities. Researchers are often under pressure to speedily produce results and thus might not exercise appropriate levels of ethical scrutiny; they need to be aware of the potentially far-reaching impact of their actions (Engels et al. 2011).

### 12.4.1 Associated Risk and Ethics Considerations Regarding Use of the Omics Technology

One of the theories in bioethics is virtue ethics, which is a part of moral philosophy. It states that an action is morally right if it is performed by adhering to the ethical virtues that promote human flourishing and well-being. The action is morally good if the person in question acts with right motive as well as his/her action is based on a firm and good character or disposition (Internet Encyclopedia of Philosophy https://iep.utm.edu/bioethic/).

The ultimate goal of omics research is gathering knowledge and applies the same for human well-being. One of the applications of genomics is to alter the gene pool of a target species in favor of the good traits. The consequence of this alteration may be predicted, but its real effect on that species in natural habitat remains unknown for a very long time. Therefore, it asks for a moral obligation on a researcher to tread cautiously for such activities. The risk associated with the release of thus modified species in nature also includes large-scale changes in biotic component of ecosystem (s).

Transgenic plants showing resistance against a pest though appear to be a promising, cost-effective, and safe alternative over the application of chemical dependent processes. However, there is insufficient information on the unintended consequences of such modifications. This raises public concerns on the safety of genetically modified organisms. For example, in the case of genetically modified (GM) foods the fear includes chances of allergenicity, antibiotic resistance, and toxicity. Often assessment of these side effects is very complicated (Kiran et al. 2017). This fear regarding GM plants is not limited to edible plants only. Rather, what if the pest resistant characteristics of GM plants escaping to their weedy relatives and causing resistance in weeds thereby increasing their (weeds) population. Consequent death of pests caused by pest resistance plants and dominance of new pest(s), resulting due to the decreased competition may have unforeseen adverse consequences for the crop as well as for the entire ecosystem. Further, the expression of the transgene and its downstream products may lead to secondary and pleotropic effects of gene expression; and insertional mutagenesis could also result from gene integration. Many transgenes encode an enzyme that could alter biochemical pathways leading to change in metabolic profile and nutritional value of the plant (Conner and Jacobs 1999). Use of information obtained from the transcriptome analysis and RNA interference have been under trial for improving the crop output. The gene expression regulation through Exogenous RNA interference (exo-RNAi) could trigger unintended alterations, resulting in epigenetic modification in plants and cross kingdom nucleic acid trafficking (Wang et al. 2016; Dalakouras and Papadopoulou 2020).

An important part of bioethics is risk assessment and risk management. Risk assessment is an analysis and prediction of risks using scientific data and estimation of the effects of exposure to hazardous materials or conditions. Whereas risk management is the process of weighing alternatives and selection of most suitable strategy to reduce the risk (Macer 1997). Based on the above-cited examples, it may be concluded here that the application of these technologies without proper risk assessment may cause unintentional changes in genome and general structure and function, as well as risk to the organism, biodiversity, and ecosystem. However, information and knowledge gained through these approaches observing strict regulatory framework will be perfectly ethical. It will strengthen the global food security, reduce the need of agrochemicals, enhance nutritional status, broaden plant tolerance to stress, and may also help in the plants' growth in previously inhospitable environments.

## 12.5 Concluding Remarks

There cannot be two opinions that advancement in "omics" technologies is a welcome step. As a set of techniques, they have wide use in diverse areas of research, notably in agriculture, health sciences to environmental protection. Many of these techniques are still under refinement therefore, in order to draw meaningful conclusions; the resulting data need careful integration. We do not subscribe to the exaggeration that the techniques per se possess any kind of threat. But we must be attentive to some of the biosafety measures and ethical concerns associated with these techniques. Therefore, formulation of a necessary regulatory framework shall be a welcome step in the right direction.

**Acknowledgements** AKR gratefully acknowledges the National Academy of Sciences, India, for awarding NASI-Senior Scientist Fellowship.

## References

Alvarez S, Naldrett MJ (2016) Plant structure and specificity-challenges and sample preparation considerations for proteomics. Adv Exp Med Biol 919:63–81

Argueso CT, Assmann SM, Birnbaum KD, Chen S, Dinneny JR et al (2019) Directions for research and training in plant omics: big questions and big data. Plant Direct 3(4):e00133

Badji A, Kwemoi DB, Machida L, Okii D, Mwila N et al (2020) Genetic basis of maize resistance to multiple insect pests: integrated genome-wide comparative mapping and candidate gene prioritization. Genes 11(6):689

Barah P, Bones AM (2015) Multidimensional approaches for studying plant defense against insects: from ecology to omics and synthetic biology. J Exp Bot 66:479–493

Beale DJ, Pinu FR, Kouremenos KA, Poojary MM, Narayana VK et al (2018) Review of recent developments in GC-MS approaches to metabolomics-based research. Metabolomics 14 (11):152

Bhadauria V (2016) OMICS in plant disease resistance. Mol Biol 19:1–2

Breitling R (2010) What is systems biology? Front Physiol 1:9

Budzinski IGF, de Moraes FE, Cataldi TR, Franceschini LM, Labate CA (2019) Network analyses and data integration of proteomics and metabolomics from leaves of two contrasting varieties of sugarcane in response to drought. Front Plant Sci 10:1524

Buesen R, Chorley BN, da Silva Lima B, Daston G, Deferme L et al (2017) Applying omics technologies in chemicals risk assessment: report of an ECETOC workshop. Regul Toxicol Pharmacol 91(Suppl 1):S3–S13

Carrï Re SB, Verdenaud M, Gough C, Gouzy JRM, Gamas P (2020) LeGOO: An expertized knowledge database for the model legume *Medicago truncatula*. Plant Cell Physiol 61 (1):203–211

Chandra A, Roopendra K, Verma I (2019) Transcriptome analysis of the effect of GA3 in sugarcane culm. 3 Biotech 9(10):376

Chaudhary J, Khatri P, Singla P, Kumawat S, Kumari A et al (2019) Advances in omics approaches for abiotic stress tolerance in tomato. Biology 8(4):90

Chialva M, Ghignone S, Novero M, Hozzein WN, Lanfranco L et al (2019) Tomato RNA-seq data mining reveals the taxonomic and functional diversity of root-associated microbiota. Microorganisms 8(1):38

Conner AJ, Jacobs JM (1999) Genetic engineering of crops as potential source of genetic hazard in the human diet. Mutat Res 443(1–2):223–234

Da L, Liu Y, Yang J, Tian T, She J et al (2019) AppleMDO: a multi-dimensional omics database for apple co-expression networks and chromatin states. Front Plant Sci 10:1333

Dalakouras A, Papadopoulou KK (2020) Epigenetic modifications: An unexplored facet of exogenous RNA application in plants. Plan Theory 9:673

Deborde C, Moing A, Roch L, Jacob D, Rolin D et al (2017) Plant metabolism as studied by NMR spectroscopy. Prog Nucl Magn Reson Spectrosc 102–103:61–97

do Amaral MN, Souza GM (2017) The challenge to translate OMICS data to whole plant physiology: the context matters. Front Plant Sci 8:2146

Dwivedi SL, Scheben A, Edwards D, Spillane C, Ortiz R (2017) Assessing and exploiting functional diversity in germ-plasm pools to enhance abiotic stress adaptation and yield in cereals and food legumes. Front Plant Sci 8:1461

Engels JMM, Dempewolf H, Henson-Apollonio V (2011) Ethical considerations in agro-biodiversity research, collecting, and use. J Agric Environ Ethics 24:107–126

Félix C, Meneses R, Gonçalves M, TillemanL DAS et al (2019) A multi-omics analysis of the grapevine pathogen Lasiodiplodia theobromae reveals that temperature affects the expression of virulence- and pathogenicity-related genes. Sci Rep 9(1):13144

Fukushima A, Kusano M, Redestig H, Arita M, Saito K (2009) Integrated omics approaches in plant systems biology. Curr Opin Chem Biol 13(5–6):532–538

Fürtauer L, Küstner L, Weckwerth W, Heyer AG, Nägele T (2019) Resolving subcellular plant metabolism. Plant J 100(3):438–455

Giovannoni J (2018) Tomato multiomics reveals consequences of crop domestication and improvement. Cell 172(1–2):6–8

Gonulalan EM, Nemutlu E, Bayazeid O, Koçak E, Yalçın FN et al (2020) Metabolomics and proteomics profiles of some medicinal plants and correlation with BDNF activity. Phytomedicine 74:152920

Gui S, Yang L, Li J, Luo J, Xu X et al (2020) ZEAMAP, a comprehensive database adapted to the maize multi-Omics era. iScience 23(6):101241

Hittalmani S, Mahesh HB, Shirke MD, Biradar H, Uday G et al (2017) Genome and transcriptome sequence of finger millet (*Eleusine coracana* (L.) Gaertn.) provides insights into drought tolerance and nutraceutical properties. BMC Genomics 18(1):465

Jahn WT (2011) The 4 basic ethical principles that apply to forensic activities are respect for autonomy, beneficence, nonmaleficence, and justice. J Chiropr Med 10(3):225–226

Kang X, Wang L, Guo Y, Ul Arifeen MZ, Cai X et al (2019) A comparative transcriptomic and proteomic analysis of hexaploid wheat's responses to colonization by *Bacillus velezensis* and *Gaeumannomyces graminis*, both separately and combined. Mol Plant Microbe Interact 32 (10):1336–1347

Kaul S, Sharma TK, Dhar M (2016) "Omics" tools for better understanding the plant-Endophyte interactions. Front Plant Sci 7:955

Kim MJ, Lee MY, Shon JC, Kwon YS, Liu KH et al (2019) Untargeted and targeted metabolomics analyses of blackberries - understanding postharvest red drupelet disorder. Food Chem 300:125169

Kiran U, Abdin MZ, Pandey NK (2017) Biosafety, bioethics, and IPR issues in plant biotechnology. In: Abdin M, Kiran U, Kamaluddin AA (eds) Plant biotechnology: principles and applications. Springer, Singapore, pp 367–392

Kohl P, Crampin EJ, Quinn TA, Noble D (2010) Systems biology: an approach. Clin Pharmacol Ther 88(1):25–33

Lee SB, Kim GJ, Kim KW, Chu SH, Shin JD et al (2019) Functional haplotype and eQTL analyses of genes affecting cadmium content in cultivated rice. Rice 12(1):84

Levsh O, Pluskal T, Carballo V, Mitchell AJ, Weng JK (2019) Independent evolution of rosmarinic acid biosynthesis in two sister families under the Lamiids clade of flowering plants. J Biol Chem 294(42):15193–15205

Ludwig C, Gillet L, Rosenberger G, Amon S, Collins BC et al (2018) Data-independent acquisition-based SWATH-MS for quantitative proteomics: a tutorial. Mol Syst Biol 14(8):e8126

Macer DR (1997) Major concerns on plant biotechnology applications in plants: safety issues and bioethics. In: K. Watanabe, E. Pehu, R.G. Landes (eds) Plant biotechnology and plant genetic resources for sustainability and productivity. Elsevier, Amsterdam, pp. 87–99

Mardis ER (2008) Next-generation DNA sequencing methods. Annu Rev Genomics Hum Genet 9:387–402

McKeown PC, Spillane C (2014) Landscaping plant epigenetics. Methods Mol Biol 1112:1–24

Medeiros DB, Daloso DM, Fernie AR, Nikoloski Z, Araújo WL (2015) Utilizing systems biology to unravel stomatal function and the hierarchies underpinning its control. Plant Cell Environ 38 (8):1457–1470

Mei ZC, Wei ZJ, Yu JH, Ji FD, Xie LN (2020) Multi-omics association analysis revealed the role and mechanism of epialleles in environmental adaptive evolution of *Arabidopsis thaliana*. Yi Chuan 42(3):321–331

Mishra B, Kumar N, Mukhtar MS (2019) Systems biology and machine learning in plant pathogen interactions. Mol Plant-Microbe Interact 32(1):45–55

Moisseyev G, Park K, Cui A, Freitas D, Rajagopal D et al (2020) RGPDB: database of root-associated genes and promoters in maize, soybean, and sorghum. Database J Biol Databases Curation 2020:baaa038

Naithani S, Gupta P, Preece J, D'Eustachio P, Elser JL et al (2020) Plant Reactome: a knowledgebase and resource for comparative pathway analysis. Nucleic Acids Res 48(D1): D1093–D1103

Nassar AF, Wu T, Nassar SF, Wisnewski AV (2017) UPLC-MS for metabolomics: a giant step forward in support of pharmaceutical research. Drug Discov Today 22(2):463–470

Patavardhan SS, Subba P, Najar A, Awasthi K, D'Souza L et al (2020) Plant-pathogen interactions: broad mite (Polyphagotarsonemus latus)-induced proteomic changes in chili pepper plant (Capsicum frutescens). OMICS 24(12):714–725

Pederson ERA, Warshan D, Rasmussen U (2019) Genome sequencing of *Pleurozium schreberi*: the assembled and annotated draft genome of a *Pleurocarpous* feather Moss. G3 (Bethesda) 9 (9):2791–2797

Peng H, Wang K, Chen Z, Cao Y, Gao Q et al (2020) MBKbase for rice: an integrated omics knowledgebase for molecular breeding in rice. Nucleic Acids Res 48(D1):D1085–D1092

Pinu FR, Beale DJ, Paten AM, Kouremenos K, Swarup S et al (2019) Systems biology and multi-omics integration: viewpoints from the metabolomics research community. Meta 9(4):76

Pouteau S (2014) Beyond "second animals": making sense of plant ethics. J Agric Environ Ethics 27:1–25

Rabara RC, Tripathi P, Rushton PJ (2014) The potential of transcription factor-based genetic engineering in improving crop tolerance to drought. OMICS 18(10):601–614

Ran X, Zhao F, Wang Y, Liu J, Zhuang Y et al (2020) Plant Regulomics: a data-driven interface for retrieving upstream regulators from plant multi-omics data. Plant J 101(1):237–248

Riaz MR, Preston GM, Mithani A (2020) MAPPS: a web-based tool for metabolic pathway prediction and network analysis in the postgenomic era. ACS Synth Biol 9(5):1069–1082

Reynnells RD (2004) Bioethical considerations in animal production. Poult Sci 83(3):303–306

Ryan D, Robards K (2006) Metabolomics: the greatest omics of them all. Anal Chem 78 (23):7954–7958

Salt DE, Baxter I, Lahner B (2008) Ionomics and the study of the plant ionome. Annu Rev Plant Biol 59:709–733

Sati S, Cavalli G (2017) Chromosome conformation capture technologies and their impact in understanding genome function. Chromosoma 126(1):33–44

Tardieu F, Varshney RK, Tuberosa R (2017) Improving crop performance under drought - cross-fertilization of disciplines. J Exp Bot 68(7):1393–1398

Teh SL, Rostandy B, Awale M, Luby JJ, Fennell A et al (2019) Genetic analysis of stilbenoid profiles in grapevine stems reveals a major mQTL hotspot on chromosome 18 associated with disease-resistance motifs. Hortic Res 6:121

Toubiana D, Fernie AR, Nikoloski Z, Fait A (2013) Network analysis: tackling complex data to study plant metabolism. Trends Biotechnol 31(1):29–36

Tuteja N, Mahajan S (2007) Calcium signaling network in plants: an overview. Plant Signal Behav 2(2):79–85

Wang WQ, Wang J, Wu YY, Li DW, Allan AC et al (2020a) Genome-wide analysis of coding and non-coding RNA reveals a conserved miR164-NAC regulatory pathway for fruit ripening. New Phytol 225(4):1618–1634

Wang D, Fan W, Guo X, Wu K, Zhou S et al (2020b) MaGenDB: a functional genomics hub for Malvaceae plants. Nucleic Acids Res 48(D1):D1076–D1084

Wang G, Oh DH, Dassanayake M (2020c) GOMCL: a toolkit to cluster, evaluate, and extract non-redundant associations of gene ontology-based functions. BMC Bioinformatics 21(1):139

Wang KC, Chang HY (2018) Epigenomics: technologies and applications. Circ Res 122 (9):1191–1199

Wang M, Weiberg A, Lin FM, Thomma BP, Huang HD et al (2016) Bidirectional cross-kingdom RNAi and fungal uptake of external RNAs confer plant protection. Nat Plants 2:16151

Wang Z, Gerstein M, Snyder M (2009) RNA-Seq: a revolutionary tool for transcriptomics. Nat Rev Genet 10(1):57–63

Waterworth WM, Wilson M, Wang D, Nuhse T, Warward S et al (2019) Phosphoproteomic analysis reveals plant DNA damage signalling pathways with a functional role for histone H2AX phosphorylation in plant growth under genotoxic stress. The Plant J 100(5):1007–1021

Windram O, Penfold CA, Denby KJ (2014) Network modeling to understand plant immunity. Annu Rev Phytopathol 52:93–111

Wolfender J, Litaudon M, Touboul D, Queiroz EF (2019) Innovative omics-based approaches for prioritisation and targeted isolation of natural products - new strategies for drug discovery. Nat Prod Rep 36(6):855–868

Xia EH, Li FD, Tong W, Li PH, Wu Q et al (2019) Tea plant information archive: a comprehensive genomics and bioinformatics platform for tea plant. Plant Biotechnol J 17(10):1938–1953

Xie M, An F, Wu J, Liu Y, Shi H et al (2019) Meta-omics reveal microbial assortments and key enzymes in bean sauce mash, a traditional fermented soybean product. J Sci Food Agric 99 (14):6522–6534

Xu G, Cao J, Wang X, Chen Q, Jin W et al (2019) Evolutionary metabolomics identifies substantial metabolic divergence between maize and its wild ancestor, Teosinte. Plant Cell 31 (9):1990–2009

Yang B, Zhong Z, Wang T, Ou Y, Tian J et al (2019) Integrative omics of *Lonicera japonica* Thunb. Flower development unravels molecular changes regulating secondary metabolites. J Proteome 208:103470

Yuan GC, Cai L, Elowitz M, Enver T, Fan G et al (2017) Challenges and emerging directions in single-cell analysis. Genome Biol 18(1):84

Zeng S, Lyu Z, Narisetti SRK, Xu D, Joshi T (2019) Knowledge Base commons (KBCommons) v1.1: a universal framework for multi-omics data integration and biological discoveries. BMC Genomics 20(Suppl 11):947

Zhang S, Zhang L, Tai Y, Wang X, Ho CT et al (2018) Gene discovery of characteristic metabolic pathways in the tea plant (*Camellia sinensis*) using 'Omics'-based network approaches: a future perspective. Front Plant Sci 9:480

Zhang X, Pan BZ, Chen M, Chen W, Li J et al (2019a) JCDB: a comprehensive knowledge base for *Jatropha curcas*, an emerging model for woody energy plants. BMC Genomics 20(Suppl 9):958

Zhang YW, Wen YJ, Dunwell JM, Zhang YM (2019b) QTL.gCIMapping.GUI v2.0: an R software for detecting small-effect and linked QTLs for quantitative traits in bi-parental segregation populations. Comput Struct Biotechnol J 18:59–65

Zhang Y, Zheng L, Zheng Y, Zhou C, Huang P et al (2019c) Assembly and annotation of a draft genome of the medicinal plant Polygonum cuspidatum. Front Plant Sci 10:1274

Zheng C, Ma JQ, Chen JD, Ma CL, Chen W et al (2019) Gene coexpression networks reveal key drivers of flavonoid variation in eleven tea cultivars (*Camellia sinensis*). J Agric Food Chem 67 (35):9967–9978

Zhong Z, Kobayashi T, Zhu W, Imai H, Zhao R et al (2020) Plant-derived smoke enhances plant growth through ornithine-synthesis pathway and ubiquitin-proteasome pathway in soybean. J Proteome 221:103781

Zhu G, Wang S, Huang Z, Zhang S, Liao Q et al (2018) Rewiring of the fruit metabolome in tomato breeding. Cell 172(1–2):249–261

The manufacturer's authorised representative in the EU is Springer Nature Customer Service Centre GmbH, Europaplatz 3, 69115 Heidelberg, Germany. If you have any concerns regarding our products, please contact ProductSafety@springernature.com

Printed and bound by CPI Group (UK) Ltd, Croydon, CR0 4YY
30/09/2026
02225330-0009